AF800106

Plant Growth Regulators: Resilience for Sustainable Agriculture

Mohammad Faizan • Shamsul Hayat
Editors

Plant Growth Regulators: Resilience for Sustainable Agriculture

Springer

Editors
Mohammad Faizan
Department of Botany, School of Sciences
Maulana Azad National Urdu University
Hyderabad, Telangana, India

Shamsul Hayat
Department of Botany
Aligarh Muslim University
Aligarh, Uttar Pradesh, India

ISBN 978-981-97-2917-3 ISBN 978-981-97-2918-0 (eBook)
https://doi.org/10.1007/978-981-97-2918-0

© The Editor(s) (if applicable) and The Author(s), under exclusive license to Springer Nature Singapore Pte Ltd. 2024
This work is subject to copyright. All rights are solely and exclusively licensed by the Publisher, whether the whole or part of the material is concerned, specifically the rights of translation, reprinting, reuse of illustrations, recitation, broadcasting, reproduction on microfilms or in any other physical way, and transmission or information storage and retrieval, electronic adaptation, computer software, or by similar or dissimilar methodology now known or hereafter developed.
The use of general descriptive names, registered names, trademarks, service marks, etc. in this publication does not imply, even in the absence of a specific statement, that such names are exempt from the relevant protective laws and regulations and therefore free for general use.
The publisher, the authors and the editors are safe to assume that the advice and information in this book are believed to be true and accurate at the date of publication. Neither the publisher nor the authors or the editors give a warranty, expressed or implied, with respect to the material contained herein or for any errors or omissions that may have been made. The publisher remains neutral with regard to jurisdictional claims in published maps and institutional affiliations.

This Springer imprint is published by the registered company Springer Nature Singapore Pte Ltd.
The registered company address is: 152 Beach Road, #21-01/04 Gateway East, Singapore 189721, Singapore

If disposing of this product, please recycle the paper.

Preface

Plant growth regulators (PGRs) are chemical compounds, which are synthesized naturally (within the plants) or synthetically that regulate or modify various morphological and physiological processes. PGRs in low concentrations often lead to significant improvements and high yields in plants. These PGRs are engaged in the positive regulation of seed germination, seedling growth, root growth, ripening, flowering, nutrient uptake, redox homeostasis, antioxidant pool, photosynthetic attributes, and their response against abiotic stress in crop plants. In plants, PGRs is considered to be the part of important mechanisms participated in the plant metabolism. These regulators also play an important role in plant acclimatization to abiotic stress conditions and could rectify the plant growth and development. Apart from five PGRs (Auxin, Gibberellins, Cytokinin, Abscisic acid, and Ethylene), some molecules also act as a PGRs and help in abiotic stress tolerance. This book is dedicated to presenting the latest developments of the role of PGRs such as nitric oxide, strigolactones, karrikins, hemin, and melatonin.

This book is comprised of 14 chapters. Chapter 1 of the book discusses the latest update on brassinosteroids and future of crop plants under changing environments. Chapter 2 deals with the salicylic acid: food, functions, and future. Nitric oxide as a key bioactive regulator of plant tolerance mechanism under metal-induced oxidative stress is covered in Chap. 3. Chapter 4 describes the recent advances in the role of nitric oxide and cross-talk with signaling molecules under environmental stress in plants. Chapter 5 highlights the strigolactone action in plants. This chapter also includes strigolactone signaling, transport, and the control of plants growth and development. Chapter 6 addresses the genetic symphony of strigolactones in respect of development. Chapter 7 describes the role of karrikins in mitigating the nutrient deficiency and stress in plants. Chapter 8 briefly discusses the newly registered biostimulator in plants called "hemin." Chapter 9 describes the phyto-melatonin history, biosynthesis, and response. In Chap. 10, tools and techniques to quantify the PGRs content is discussed. Chapter 11 deals with the implications of phytohormones research: from lab to field. In Chap. 12, potential for PGRs to impact crop productivity in future agricultural system is discussed in brief. Chapter 13 demonstrates the integration of plant hormones in the biological system as an opportunity for sustainable crop production. Chapter 14 discusses the phytohormones for combat global challenges: an eco-friendly approach.

This book is not an encyclopedia of reviews but rather a compendium of newly composed, integrated, and illustrated contributions describing our knowledge of PGRs in plants. Apart from traditional PGRs, some new PGRs also come in light with very significant response in plants under normal as well as stress conditions. PGRs role in plant growth and development, physiology and metabolism, and abiotic stress resilience are explained briefly. In addition, mechanism, signaling, and cross-talk with other molecules/pathways are also included in the book.

Chapters incorporate both theoretical and practical aspects of PGRs impacts on nature, plants, and tolerance against it and may serve as baseline information for future research through which significant development can be made. The book will be useful to researchers, instructors, and students both in universities and research institutes, especially in relation to plant sciences.

With great pleasure, we extend our sincere thanks to all the contributors for their timely response, their excellent and up-to-date contributions, and their consistent support and cooperation. We are thankful to all who have helped us in any capacity during the preparation of this volume. We are extremely thankful to Springer Publishing for their expeditious acceptance of our proposal and completion of the review process. The subsequent cooperation and understanding by their staff are gratefully acknowledged. We express our sincere thanks to our family members for all the support they provided and the neglect and loss they suffered during the preparation of this book.

Finally, we are thankful to the Almighty who provided and guided all the channels to work in cohesion on the concept to the development of the final version of this treatise, *Plant Growth Regulators: Resilience for Sustainable Agriculture.*

Hyderabad, India Mohammad Faizan
Aligarh, India Shamsul Hayat

Contents

Editors and Contributors

About the Editors

Mohammad Faizan is working as an Assistant Professor, in the Department of Botany, Maulana Azad National Urdu University, Hyderabad, India. He completed his postdoc in 2021, from the Department of Silviculture, College of Forestry, Nanjing Forestry University, Nanjing, China. In 2018, he received his Ph.D. degree in Botany from Aligarh Muslim University, Aligarh, India, and his M.Sc. in 2014 from Chhatrapati Shahu Ji Maharaj University, Kanpur, India. His ongoing research is based on abiotic stress tolerance mediated by nanoparticles, and investigating the bioaccumulation, bio-transformations, uptake, translocation, and effects of bulk- and nano-forms of metals on plant physiology, morphology, anatomy, the ultrastructure of cellular and subcellular organelles, and modifications. With long experience and experimental work, he comprehensively detailed the state of research in environmental science regarding "how nanoparticles/heavy metals interact with plants, soil, microbial community, and the larger environment as well as possible remediation technology using nanoparticles. He has published (a total of 73 scientific publications) 65 peer-reviewed full-length articles, 6 books (Springer & MANUU DDE), 21 chapters (Scopus indexed), and achieved h-index: 23 by Google Scholar with >2691 citations, till July 2024. He is an internationally recognized reviewer and reviewed a lot of manuscripts.

Shamsul Hayat is a Professor in the Department of Botany, Aligarh Muslim University, Aligarh, India. He has also worked as an Associate Professor at King Saud University, Riyadh, Saudi Arabia, as a BOYSCAST Fellow at the National Institute of Agrobiological Sciences, Tsukuba, Japan, and as a visiting scientist through the INSA-Bilateral exchange program at the Department of Plant Biochemistry and Toxicology, University of Bialystok, Poland. The major area of research includes plant hormone, nanoscience, and abiotic stress in plants. Dr. Hayat has been awarded the Prof. Hira Lal Chakravorty Award by the Indian Science Congress Association, Kolkata, India, Associate of National Academy of Agricultural Sciences, New Delhi, India, BOYSCAST fellow by the Department of Science and Technology, Government of India, New Delhi and young scientist by the Association of the Advancement of Science, Aligarh, India. Dr. P.S. Khankhoje gold medal has also been awarded to him by Dr. Punjab Rao Deshmukh Krishi

Vidyapeeth, Akola. He has been the Principal Investigator of various projects sanctioned by different agencies and guided eight students to the award of Ph.D. degrees and two students to the award of M.Phil degrees besides many M.Sc. students. Dr. Hayat has published more than 195 research papers in leading journal of the world. Besides this, 35 book chapters have also been published. Dr. Hayat has presented his work at several national and international conferences in Japan, Brazil, Spain, China, Poland, and Saudi Arabia. He is a regular reviewer and is on the panel of editorial boards of National and International journals. Dr. Hayat is also a member of important National and International scientific societies. Recently, he has been appointed as an Indian representative in the Asia Association of Plant Scientists.

Contributors

Sadia Afzal Department of Botany, Lahore College for Women University, Lahore, Pakistan

Zahoor Ahmad Department of Botany, Constituent Punjab College, University of Central Punjab, Bahawalpur, Pakistan

S. Maqbool Ahmed Botany Section, School of Sciences, Maulana Azad National Urdu University, Hyderabad, India

Anita Plant Biotechnology and Molecular Biology Lab, Department of Botany (UGC-CAS), Jai Narain Vyas University, Jodhpur, India

Samia Muhammad Arif Department of Plant Sciences, Quaid-i-Azam University, Islamabad, Pakistan

Yamshi Arif Plant Physiology Section, Department of Botany, Faculty of Life Sciences, Aligarh Muslim University, Aligarh, India

Lalarukh Azhar Department of Botany, Lahore College for Women University, Lahore, Pakistan

Ritu Bala Department of Microbiology, School of Bioengineering and Biosciences, Lovely Professional University, Phagwara, Punjab, India

Kejun Cheng Chemical Biology Center, Lishui Institute of Agriculture and Forestry Sciences, Lishui, China

Ozair Choudhary Ontario Institute of Agrologist, Harriston, ON, Canada

Padmanabh Dwivedi Department of Plant Physiology, Institute of Agricultural Sciences, Banaras Hindu University, Varanasi, India

Mohammad Faizan Department of Botany, School of Sciences, Maulana Azad National Urdu University, Hyderabad, India

Ahmad Faraz Department of Biotechnology, School of Life Sciences, Glocal University, Saharanpur, India

Monika Garg Department of Agri-Biotechnology, National Agri-Food Biotechnology Institute (NABI), Mohali, Punjab, India

Shamsul Hayat Department of Botany, Aligarh Muslim University, Aligarh, India

Plant Physiology Section, Department of Botany, Faculty of Life Sciences, Aligarh Muslim University, Aligarh, India

Anjuman Hussain Department of Botany, Aligarh Muslim University, Aligarh, India

Havza Imtiaz Plant Physiology Section, Department of Botany, Faculty of Life Sciences, Aligarh Muslim University, Aligarh, India

Aina Inam Department of Botany, Lahore College for Women University, Lahore, Pakistan

Savita Jangde Department of Plant Physiology, Institute of Agricultural Sciences, Banaras Hindu University, Varanasi, India

Sumera Javad Department of Botany, Lahore College for Women University, Lahore, Pakistan

Fadime Karabulut Department of Biology, Faculty of Science, Firat University, Elazıg, Turkey

Manpreet Kaur Department of Biotechnology, School of Bioengineering and Biosciences, Lovely Professional University, Phagwara, Punjab, India

Samuel Kenyanya Department of Biomedical Sciences, Kisii University, Kisii, Kenya

Kiran Khandare Division of Chemical Engineering, Center of Innovative and Applied Bioprocessing (CIAB), Mohali, Punjab, India

Khushboo Department of Microbiology, School of Bioengineering and Biosciences, Lovely Professional University, Phagwara, Punjab, India

Prasann Kumar Department of Agronomy, School of Agriculture, Lovely Professional University, Phagwara, India

Katenahalli Rudrappa Maruthi Botany Section, School of Sciences, Maulana Azad National Urdu University, Hyderabad, India

Nihar Mathur Plant Biotechnology and Molecular Biology Lab, Department of Botany (UGC-CAS), Jai Narain Vyas University, Jodhpur, India

Maninder Meenu Department of Agri-Biotechnology, National Agri-Food Biotechnology Institute (NABI), Mohali, Punjab, India

Ghazala Mustafa Department of Plant Sciences, Quaid-i-Azam University, Islamabad, Pakistan

Chemical Biology Center, Lishui Institute of Agriculture and Forestry Sciences, Lishui, China

Mukesh Narwat Department of Microbiology, School of Bioengineering and Biosciences, Lovely Professional University, Phagwara, Punjab, India

Suman Parihar Plant Biotechnology and Molecular Biology Laboratory, Department of Botany (UGC-CAS), Jai Narain Vyas University, Jodhpur, Rajasthan, India

Urvasha Patyal Department of Biosciences and Technology, Maharishi Markandeshwar University, Ambala, Haryana, India

Department of Biosciences and Technology, Maharishi Markandeshwar University, Mullana, Ambala, Haryana, India

Danyal Rasheed Department of Botany, Bacha Khan University, Charsadda, Pakistan

Anis Ali Shah Department of Botany, University of Education, Lahore, Pakistan

Kiran Shahzadi Department of Botany, Lahore College for Women University Lahore, Lahore, Pakistan

Kanu Priya Sharma Department of Biochemistry, Punjab Agricultural University (PAU), Ludhiana, Punjab, India

Lata Sharma Plant Biotechnology and Molecular Biology Laboratory, Department of Botany (UGC-CAS), Jai Narain Vyas University, Jodhpur, Rajasthan, India

Saima Shaukat Department of Plant Sciences, Quaid-i-Azam University, Islamabad, Pakistan

Gyan Singh Shekhawat Plant Biotechnology and Molecular Biology Laboratory, Department of Botany (UGC-CAS), Jai Narain Vyas University, Jodhpur, Rajasthan, India

Saumya Shruti Department of Plant Physiology and Ecology, Anand Agricultural University, Anand, India

Ajit Singh School of Biosciences, University of Nottingham, Malaysia Campus, Semenyih, Selangor, Malaysia

Manali Singh Microbial Type Culture Collection and Gene Bank (MTCC), Institute of Microbial Technology (IMTECH), Chandigarh, India

Gafoor Unnisa Botany Section, School of Sciences, Maulana Azad National Urdu University, Hyderabad, India

Shafia Zauq Department of Botany, Lahore College for Women University Lahore, Lahore, Pakistan

1 Brassinosteroids and Future of Crop Plants Under Changing Environments

Mohammad Faizan, Anjuman Hussain, S. Maqbool Ahmed, Katenahalli Rudrappa Maruthi, and Shamsul Hayat

Abstract

A number of climate models indicate that crop productivity will drop dramatically as a result of global climate change, endangering the food security of an expanding population. It is anticipated that the arid and semi-arid regions would experience a rise in desertification as a result of climate change and shifting rainfall patterns, which will reduce the amount of land that is ploughed. Increased frequency and severity of abiotic stresses, including drought, salinity, heat, and heavy metals, are sign of climate change and are expected to severely reduce the yield of staple crops. In certain nations, the commercialization of plant growth regulators (PGRs) has occurred with the aim of increasing crop productivity. These PGRs are crucial for controlling a plant's growth and development because they cause morphological, physiological, and molecular changes in response to stress. Brassinosteroids (BRs) are important plant hormones, which play important role in plant growth and development. These BRs provides protection for the crops from abiotic stresses as well as increases the crop productivity in a sustainable way. BRs have potential to take forward the agriculture and food processing sector. Hence, the present chapter will focus on the future prospects of the crop plants in the presence of BRs.

Keywords

Abiotic stress · Brassinosteroids · Food productivity · Plant growth regulators

M. Faizan (✉) · S. M. Ahmed · K. R. Maruthi
Department of Botany, School of Sciences, Maulana Azad National Urdu University, Hyderabad, India
e-mail: mohammadfaizan@manuu.edu.in

A. Hussain · S. Hayat
Department of Botany, Aligarh Muslim University, Aligarh, India

© The Author(s), under exclusive license to Springer Nature Singapore Pte Ltd. 2024
M. Faizan, S. Hayat (eds.), *Plant Growth Regulators: Resilience for Sustainable Agriculture*, https://doi.org/10.1007/978-981-97-2918-0_1

1.1 Introduction

Abiotic stresses that negatively impact plant growth and development include temperature fluctuations, water scarcity or abundance, excessive salinity, heavy metals, UV radiation, and rising and dropping temperatures. These factors also have an impact on agricultural output, which lowers sustainability globally. Global agriculture yield and sustainability are severely constrained by these stresses. In order to mitigate the effects of environmental changes and support population growth, it is becoming more and more crucial to arm crops with multi-stress tolerance, as a variety of abiotic stresses frequently occur simultaneously in the field. Numerous reports exist regarding the various pressures that are currently impeding plant growth and productivity. Furthermore, not enough study has been done on sustainable methods to lessen the damaging effects of abiotic stress on plants.

A novel class of plant growth hormones known as brassinosteroids (BRs) has a potent growth-promoting influence. Since BRs influence several physiological processes in plants, including as growth, germination of seeds, rhizogenesis, senescence, and tolerance to abiotic stress, they are considered pleiotropic growth regulators. The leucine-rich repeat receptor kinases BRASSINOSTEROID INSENSTIVE 1 (BRI1) and BRI1 ASSOCIATED RECEPOTOR KINASE (BAK1), which are cell surface receptors bound by these plant-specific BR ligands, phosphorylate BSU1 protein and degrade BRASSINOSTEROID-INSENSTIVE 2 (BIN2) proteins through phosphorylation cascades in the cytoplasm. BRI1 EMS SUPPRESSOR1/Brassinazole-resistant 1 (BES1/BZR1) can enter the nucleus and affect target gene expression when BIN2 is inactivated. Moreover, BR interacts with other phytohormones in all of these signalling cascades (Mubarik et al. 2021; Kour et al. 2021). Plant developmental activities such xylem differentiation, photomorphogenesis, cell elongation, and seed germination are all regulated by BRs (Wei and Li 2016). But as a new method for overcoming stress-related physiological disruption, BR application has drawn a lot of interest lately as a prospective factor for raising agricultural output under stress and control conditions (Ahanger et al. 2020). While BRs have been shown to help reduce stress-related adverse conditions (Zeid et al. 2019), the physiological and molecular mechanisms by which this steroidal hormone works to counteract stress-related adverse events remain unclear (Raza et al. 2019; Mubarik et al. 2021). The reason for this unclear understanding is that BRs have close relationships with numerous other signalling systems that are associated with stress reduction (Divi et al. 2016). Some potential physiological and molecular pathways that the BRs may alter in response to abiotic stimuli are covered in the current review. Understanding abiotic stress remediation strategies, particularly for crops that are extremely essential to agriculture, can be done with the help of the literature. The purpose of this chapter is to shed light on the mechanisms behind BR-triggered abiotic stress tolerance as well as BR-mediated plant responses to a range of abiotic challenges.

1.2 Climate Change and Plant Yield

Climate variability has had a significant impact on plant physiology in a number of ways. The potential impact of various stresses on plants was increased by environmental extremes and climatic variability (Thornton et al. 2014). Crop production is impacted by climate change in a direct, indirect, and socioeconomic way. Additionally, according to the Food and Agriculture Organization (FAO), the frequency of climate-related disasters (drought, floods, high temperatures, storms, etc.) has grown considerably. Boyer (1982) reported that since 1982, crop yields have decreased by up to 70% due to climate change. A 2007 FAO study found that only 3.5% of farmed areas are free from constraints to the environment, with all cultivated areas worldwide being impacted by climatic changes (Van Velthuizen 2007). Although the effects of abiotic pressures on crop output are difficult to measure precisely, crop productivity is thought to be significantly impacted by abiotic stresses, depending on how much of the entire cultivable region is damaged. Future estimates of global warming, water scarcity, and other environmental effects indicate that many countries will see a decline in the yield of their key crops (Bonan and Doney 2018).

Significant variations in Europe's susceptibility to the current climate shifts were found based on national agricultural yields and questionnaire surveys. The Pannonian zone, which includes Hungary, Serbia, Bulgaria, and Romania, will experience the most detrimental effects of the continental climate. In Northern Europe, the short growing season and cool temperatures are the main concerns, while in Southern Europe, crop productivity is limited by temperature extremes and little precipitation (Olesen et al. 2011). Increased greenhouse gas emissions and sudden climate shifts were projected to result in higher crop yields in North-Western Europe and lower crop yields in the Mediterranean region (Olesen and Bindi 2002). In many countries, climate change has a significant impact on wheat production due to temperature extremes, which can lower crop yields by 6% for every degree Celsius that temperatures rise (Asseng et al. 2015). According to Barnabás et al. (2008), two major stressors that have a significant impact on cereal yields are drought and high temperatures. Rubisco, the primary enzyme in photosynthesis, is disrupted and stops the process if the temperature rises beyond 35 °C. According to Gong et al. (1997), heat stress has a detrimental effect on Zea mays antioxidant enzymes. Studies have examined at how heat and drought stress together affect crop output in sorghum, maize, and barley. Research showed that the combined effects of heat and drought stress were more harmful than either stress alone (Wang and Huang 2004). When Leymus Chinensis was treated to the combined stressors of heat and drought in 2006, Xu and Zhou discovered that PSII functioned less well (Xu and Zhou 2006). Temperature extremes and water scarcity are two factors that affect a plant's reproductive phase of growth as a result of climate change. Water stress in cereals has been shown to have a negative impact on flower commencement and inflorescence (Winkel et al. 1997). Similarly, sterility in cereals can result from temperature increases of roughly 30 °C during floret development (Saini and Aspinall 1982). Water deficiency caused a 35–75% drop in grain set in wheat and rice during the meiotic phase (Sheoran and Saini 1996). Drought stress significantly disrupts the anthesis and fertilization processes in rice. Grain set is lowered and the harvest index drops below 60% as a result of the water deficit (Garrity and O'Toole 1994). Climate change is predicted to cause agricultural production to drop to 25.7% by 2080, with maize predicted to be Mexico's most affected crop (Hellin et al. 2014). In an experiment to examine the effects of climate change on main crop yields, Zhao et al. (2017) found that maize, soybeans, rice, and wheat all had significant yield losses of 7.4%, 3.2%, 3.1%, and 6%, respectively. By creating climate-resistant crops, new genomics findings are enabling climate-smart agriculture as a response to climate change (Scheben et al. 2016).

All phases of wheat growth are impacted by drought stress, but the reproductive and grain-forming stages are the most critical (Pradhan et al. 2012). According to de Oliveira et al. (2012), wheat output dropped from 1 to 30% during a light drought stress at post-anthesis, and this loss escalated to 92% in the event of a protracted mild drought stress during flowering and grain formation. Important grain legume yields have been significantly lowered by drought stress. Drought stress has been shown to diminish the production of mash beans (*Vigna mungo* L.) from 31 to 57% during the flowering stage, and from 26 to 57% during the reproductive phase (Baroowa and Gogoi 2014). According to Maleki et al. (2013), drought stress has had a significant impact on soybean output, with a 42% decrease seen during the grain filling stage of the plant. According to Schlenker and Roberts (2009), maize yield increased at an ideal temperature of 29 °C, but rising temperatures beyond that point reduced the yield (Schlenker and Roberts 2009). It was discovered that the yield of maize was adversely affected by temperature increases of even 1 °C (Lobell et al. 2011). Similarly, with every 1 °C increase in temperature, rice yield dropped by 2.6% (Easterling et al. 2007). A 1 °C increase in temperature lowered the yield of sorghum by 7.8% (Kjellström et al. 2018). Water scarcity is another significant problem with sorghum, as noted by the majority of the world's leading producers (Otto et al. 2017).

This increased concern was evident in the increasing number of research publications devoted to abiotic challenges following Kitano's critical systems biology review (Kitano 2002). By using various techniques, the number of research studies on biotic and abiotic stressors in plants has significantly expanded. Food security is impacted by climate change in a complex way. It directly reduces agricultural production by upsetting the agro-ecological environment, and it indirectly reduces agricultural yield by increasing the demand for agricultural products through pressure on growth and income circulation. Numerous methods have been employed to calculate the effects of climate change on food security (Intergovernmental Panel on Climate Change (IPCC) 2007). Slight temperature increases may increase pasture productivity in humid grassland zones and temperate regions. These developments must be made in order to address the accelerated rate of climate change. For instance, the Mediterranean region experiences drought and temperature extremes, whereas temperate regions are more vulnerable to flooding during periods of heavy precipitation (Schmidhuber and Tubiello 2007). However, it can slow down livestock development and increase their mortality rates in dry and semiarid areas

(Intergovernmental Panel on Climate Change (IPCC) 2000). Different climate models anticipate that in drier places, there will be less soil moisture and an extended rate of evapo-transpiration (Intergovernmental Panel on Climate Change (IPCC) 2007). As a result, many agricultural land areas may no longer be suited for cultivation owing to climate change, and other tropical regions may produce more crops. Insect pests of crops will benefit from temperature instability by having more conducive environmental conditions, which will increase their ability to survive in cold climates and subsequently emerge in springtime outbreaks. It is important to note that, with regard to food accessibility, all recent estimates for food security and safety have primarily focused on the effects of climate change, without taking into account the likelihood of a significant change in the frequency of climate extremes on crop productivity. Additionally, they failed to take into account the possibility of abrupt changes in the climate and socioeconomic position, which has had a detrimental effect on global food security and safety (Schmidhuber and Tubiello 2007). Food security is extremely important to people all around the world. The main concerns for researchers continue to be food safety, supply, and quality in light of climate change. In order to estimate food security conditions completely and scientifically, future studies on food security will need to take population, crop production, water supply, and climate change into account (Kang et al. 2009).

1.3 Crop Adaptation to Overall Extreme Climatic Stresses

The Earth's temperature rises, causing drastic changes to the environment that lead to abiotic stress. Environmental shifts are extremely harmful and provide a number of risks to crop species that are naturally prevalent (Espeland and Kettenring 2018). Heat and drought are the two main stresses that plants face in the field and have a big impact on them (Pereira 2016). According to reports, plants need a certain temperature to grow and blossom normally. Temperature variations have a significant impact on plant physiology (Hatfield and Prueger 2015). Drought stress adversely affects the morpho-physiology of plants, heat stress reduces grain yield and productivity, and cold stress causes sterility (Salehi-Lisar and Bakhshayeshan-Agdam 2016). The adverse effects of these climate issues on plant growth and yield lead to a wide range of reactions, including physiological, molecular, biochemical, and morphological changes (Zandalinas et al. 2018). As stated in, both climate change and global warming have some beneficial and bad effects on human health and agricultural products. In light of this, it has become increasingly challenging for plant scientists to comprehend the mechanisms underlying plant resistance to stress in order to create stress-tolerant plants (Singh et al. 2018). The main cereal crops grown worldwide, including wheat, rice, and maize, is essential for meeting the world's daily food needs. Among these, wheat has been the most extensively grown staple crop (Tack et al. 2015). 38.8% of all agricultural land is used for wheat harvesting, which yields a notably higher concentration of proteins (15–23% per gram) than maize or rice (2–3%) (Faostat FAO 2017). Despite having a lot of growing land worldwide, its output has generally lagged behind that of rice and maize. With a

2 °C rise in temperature, a reasonable decrease in wheat productivity was predicted. A 6% decrease in wheat output was predicted by related study on environmental variability (Abhinandan et al. 2018). According to Challinor and his colleagues, the primary cause of crop yield decline in changing climatic conditions is a drop in the grain filling phase brought on by temperature increases (Challinor et al. 2007). Thus, producing crop plants that can withstand stress and maintaining crop output are crucial tasks in modern agriculture (Abhinandan et al. 2018).

1.4 Plant Hormone Responses in Abiotic Stresses

To withstand these harmful pressures, plants have evolved effective sensing, signalling, and response systems. Plant hormones (PHs) are a prominent example of the response mechanisms exhibited by plants. PHs are cellular signal molecules that function as chemical messengers in plants at low concentrations and play a crucial role in regulating the responses that plants display to abiotic stresses (Williams 2011). The intricate network of interconnected signalling channels regulates physiological processes, enabling prompt adaptation to environmental stressors.

Because of their diverse function in plant growth and development, BRs are a new class of phytohormones that are emerging (Wei and Li 2016). Since stresses cause morphological, physiological, biochemical, and molecular changes in plants, BRs play a critical role in promoting plant tolerance to biotic and abiotic stresses such as drought, salinity, temperature, and others (Table 1.1).

1.5 BRs and Plants Growth and Development

BRs are steroid hormones that are involved in several aspects of plant growth and development (Fang et al. 2019; Lin 2020). BRs control a range of physiological and developmental processes in plants, including floral transition, vascular development, cell type elongation, stem cell maintenance, expansion, and cell division (Lv et al. 2018; Wang et al. 2020). Additionally, they are involved in a variety of processes, including the elongation of the hypocotyl (Oh et al. 2014), the growth of the roots (Hacham et al. 2011), the growth of the shoots (Khan et al. 2013), the development of the pollen tube, the germination of seeds, the germination and development of pollen (Vogler et al. 2014), the differentiation of treachery elements, the formation and differentiation of xylem, photomorphogenesis and plant reproduction (Ahammed et al. 2020), and senescence (Yin et al. 2002). During seed germination, BRs can trigger the cell cycle, govern the advancement of the cell cycle, manage the proliferation of leaf cells, and cause excessive growth in plants cultivated hydroponically (González-García et al. 2011; Arteca and Arteca 2001). According to Ackerman-Lavert and Savaldi-Goldstein (2020), BRs also control the formation of stomata and the responses to biotic and abiotic stress. Furthermore, according to Makarevitch et al. (2012), BR is essential for controlling both male and female

Table 1.1 Role of BRs in promoting plant tolerance to abiotic stresses

Environmental stress	Plant	Response	References
	Zea mays	Stimulated functional components of antioxidative defense system and ultimately reduced oxidative damage	Rattan et al. (2020)
	Allium sativum	BR-mediated lignin accumulation possibly played an important role in garlic adaption to salt stress	Kong et al. (2021)
Salt stress	*Solanum lycopersicum*	SlBRI1, a major receptor of BR signaling, regulated plant growth and development as well as stress resistance	Wang et al. (2023)
	Solanum tuberosum	Improved IAA, ABA levels and activities of CAT, POX, SOD and APX	Ahanger et al. (2020)
Cold stress	*Zea mays*	Enlarged ability of dehydroascorbate reductase, monodehydroascorbate reductase also the ratios of GSH/GSSG and AsA/DHA	Sun et al. (2020)
	Cucumis sativus	Accelerated recovery of photosynthetic apparatus by balancing the electron portioning, carboxylation and redox homeostasis	Jiang et al. (2013a, b)
	Helianthus annuus	Improved protein synthesis in addition to acid inverter action	Filová (2014)
	S. lycopersicum	Positively regulated chilling tolerance via ABA biosynthesis	An et al. (2023)
Heat stress	*Glycine max*	Significantly improved the capacity of antioxidant defense systems to protect photosynthetic apparatus under heat stress, thereby effectively alleviated heat stress-induced growth inhibition	Wang et al. (2022)
	Brassica juncea	Significantly improved the growth and physiological parameters with an enhanced antioxidant defense	Neha et al. (2022)
	Oryza sativa	Modulated ethylene synthesis and antioxidant metabolism	Nazir et al. (2023)

fertility in crops. According to Jaillais and Vert (2012), BR contributes to etiolation and encourages the lengthening of stigma, plant architecture, thermotolerance, proton transport, tiller number, leaf angle, and leaf size. Furthermore, crop yields may be increased by the exogenous application of BR or by changes to their production and signalling. Additionally, BRs regulate a number of genes in plants. Chen et al. (2017) discovered that BRs stimulated the expression of WRKY46, WRKY54, and WRKY70 genes, which were demonstrated to have beneficial effects on BR-regulated plant development in *A. thaliana*. The histone lysine methyltransferase SDG8 is implicated in BR-regulated gene expression, according to a different study. In *A. thaliana*, the knockout mutant sdg8 exhibited a decreased growth phenotype with weakened BR responses (Wang et al. 2014). BR transcriptionally modulates the genes and pathways that control the development of the seed and ovule in *A. thaliana*, hence controlling seed development and influencing seed size, weight, and quantity (Jiang et al. 2013a, b). Additionally, under mild nitrogen deprivation, BRs control the root nitrogen foraging response in A. thaliana. When there is a slight nitrogen deficit, the primary root elongates due to the action of the Brassinosteroid signalling kinase 3 (BSK3) genes (Jia et al. 2019). In rice, BR controls both plant architecture and grain output (Fujioka and Yokota 2003). The height of rice plants is influenced by the gene products of BRD1 and D11, which are involved in the manufacture of BR (Tanabe et al. 2005). A C-6 oxidase required for rice BR biosynthesis is encoded by the OsDwarf2/OsDwarf1 reduction, which decreases second internode and seed length elongation (Tanabe et al. 2005). By raising the level of antioxidant system (AOS) and energy charge during panicle development, BRs in rice modulate the effects of N fertilization on spikelet development and help to increase spikelet growth (Zhang et al. 2019). Male sterility in maize is brought on by insufficient BR biosynthesis, which inhibits the formation of anther and pollen (Makarevitch et al. 2012). Furthermore, in monocotyledonous wheat (*Triticum aestivum* L.), an exogenous treatment of EBR postpones the transition from vegetative to generative state. According to Symons et al. (2006), BRs application play significant role in the ripening of grape berry. Berries treated with EBR were shown to have significantly higher levels of soluble sugars and sucrose synthase (sucrolytic) (Xu et al. 2015). Jujube fruit degradation reduced after treatment with BR, most likely as a result of its capacity to delay senescence and promote disease resistance (Zhu et al. 2010). By adjusting plant development and enhancing plant performance through interactions with plant growth regulators or other plant hormones, BRs can take part in physiological processes in response to stress (Trevisan et al. 2020). Numerous developmental processes are impacted by the interruption of BR signalling, including pollen and seed formation (Ye et al. 2010) and flowering time (Domagalska et al. 2010). Dwarfed plants that lack boron (BR) have modified stomatal development, decreased male fertility, shortened hypocotyls, petioles, and internodes, downward-curling leaves, and delayed flowering (Gudesblat et al. 2012). Additionally, because of the decreased lamina inclination, BR-deficient plants also have a compact plant structure. On the other hand, BR shortage might result in reduced germination of seeds, poor stomatal distribution, reduction in tiller number, and seed fertility. Because of their slowed growth, BR-insensitive and -deficient mutants are sometimes referred to as late flowering (Zhao and Li 2012). On the other hand, plants that are over accumulating BRs have taller stems and longer petioles and hypocotyls (Rozhon et al. 2019).

1.6 BRs Against Drought Stress

Among the most harmful abiotic factors is drought stress. Various researches have been conducted on a variety of substances, such as BRs, which have been utilized to strengthen plants' antioxidant defenses against drought stress. In a study, exogenous application of 24-epibrassinolide on tomato plants during applied drought stress reduced H_2O_2 content and lipid peroxidation. Brassinosteroid-treated plants showed increased activity of the antioxidative enzymes peroxidase, superoxide dismutase, catalase, and ascorbate peroxidase. Drought-stressed plants treated with BRs also showed higher concentrations of proline, ascorbate, and carotenoids (Behnamnia et al. 2009). A study on the Chinese medicinal herb *Salvia miltiorrhiza* revealed similar findings. The application of BRs to a genotype of *S. miltiorrhiza* that is vulnerable to drought caused a considerable increase in proline content and antioxidative enzyme activity, whereas MDA content decreased (Zhu et al. 2014). Additionally, abscisic acid contributes to the stress tolerance brought on by BRs. To clarify the function of ABA and BR for stress tolerance, research has been done on polyethylene glycol-mediated drought stress in maize. Plants treated with brassinazole, an inhibitor of BR, exhibited increased levels of oxidative damage, which decreased when treated with abscisic acid and BR. It has been shown that treating stressed plants with BRs increased the expression of genes associated to ABA, strengthening its function in stress tolerance. Additionally, it has been discovered that BRs cause NO production in mesophyll cells in leaves, which activates ABA biosynthesis and increases maize plants' resistance to drought stress (Zhang et al. 2011). By enhancing the water permeability of aquaporins, membrane water channels, BRs also contribute to the maintenance of cellular water content. Brassinosteroid treatment of BR-insensitive (BRI1) mutants in Arabidopsis led to osmotic permeability; nevertheless, BR does not directly influence the plasma membrane (Morillon et al. 2001).

1.7 BRs Against Salt Stress

One of the main causes of osmotic stress, also known as physiological drought, is salinity. Crop output, development, and growth are all adversely impacted. Many plant species, such as Arabidopsis, mustard, eggplant, pepper, cucumber, common bean (*Phaseolus vulgaris*), maize, and black locust (*Robinia pseudoacacia* L.), have been demonstrated to be positively impacted by salinity in BRs (Yuan et al. 2012; Yue et al. 2019). Increased activity of antioxidant enzymes, decreased Na^+ and Cl^- concentrations, and increased K^+ and Ca^{2+} concentrations are signs of greater resistance to salt stress in eggplants induced by EBR treatment. Similarly, in cucumber plants under salt stress, EBR spraying can lower the content of NO_{3-} and NH^{4+} (Yuan et al. 2012). According to Hayat et al. (2012), foliar treatment of HBL in rapeseed can successfully mitigate the detrimental effects of salt stress even 30 and 45 days after sowing. Increased photosynthesis, nitrogen utilization efficiency, and total polyamines are thought to be responsible for cucumber plants' BR-induced higher tolerance to salinity stress (Yuan et al. 2012). Exogenous EBR application (seed soaking and root dipping) in black locust improves the maximum quantum efficiency of PSII under salinity stress, stomatal conductance, transpiration rate, net photosynthetic rate, and chlorophyll content while reducing leaf Na^+ content and membrane leakage (Yue et al. 2019). Plants that are subjected to combined stress can benefit from BR as well. For example, HBL (0.01 μM) can reduce salt- and high temperature-induced combination stress in mung beans, while EBR (1 μM) can reduce combined stress caused by NaCl and $NiCl_2$ in *Brassica juncea* (Ali et al. 2008; Hayat et al. 2012). This significant variance in BR concentrations emphasizes even more the dose-effect relationship between BR and plant species. According to Cui et al. (2011), ubiquitin-conjugating enzymes32 (UBC32) have a part in BR-induced tolerance to salt stress. UBC32 affects the accumulation of the BRI1 receptor in cells and functions as a functional part of the endoplasmic reticulum-associated protein degradation (ERAD) pathway. It also guides the ERAD pathway towards better salt tolerance in Arabidopsis. Furthermore, BR has been linked to the control of DNA methylation, an essential process for salt tolerance. According to Amraee et al. (2019), seed priming with EBR, for example, increases salt tolerance and causes total methylation, indicating a potential role for BR in epigenetic change under salinity stress.

1.8 BRs Against Temperature Stress

Extremely low and high temperatures can both interfere with a plant's ability to operate normally. It has been established that one of the most detrimental stressors on plant growth and agriculture is global warming (Nolan et al. 2020). High temperatures can harm crops by causing things like fruit burns, abscission and senescence, leaf burn, and slowed plant growth (root and shoot), which can lower agricultural output (food and cash). Analyzing the phonotypic changes associated with stress at the molecular level, it was shown that high temperatures cause BES1 and BZR1 to aggregate, which raises PHYTOCHROME INTERACTING FACTOR4 (PIF4) levels (Martínez et al. 2018). Heterodimers of PIF4-BES1 are formed as a result, which lessens the effect of BZR1 on gene transcription and ultimately steers the plants towards thermogenic growth. Conversely, high temperatures and low BRI1 levels have a significant impact on BR signalling, which causes a sharp rise in root growth (Martins et al. 2017). In addition to the endogenous BR signalling cascade's stress-relieving properties, exogenously applied BR also serves the same purposes in plants (Sadura and Janeczko 2018). Photosynthetic processes are the most susceptible to heat stress of all physiological systems (Ahammed and Yu 2016). Researchers have long demonstrated that extremely high temperatures reduce the effectiveness of photosystem II (PSII) and sharply lower the net photosynthetic rate (Ogweno et al. 2008). Some possible instances are given by the tomato, where a pretreatment with epibrassinolide (EBR) can counteract the negative effects of high temperature mentioned above by increasing the amounts of antioxidant enzymes that reduce lipid peroxidation under such stressful circumstances (Ogweno et al. 2008). The ability of BRs to multitask has been demonstrated in the instance of heat-sensitive and heat-tolerant ecotypes of melon under extremely high temperatures, where they aid in the improvement of thermo-tolerance in both genotypes of plants (thermo tolerant and thermo sensitive). According to Zhang et al. (2013), the pretreatment of photosynthetic pigments with EBR significantly increased their concentrations, as well as other metrics as the net carbon dioxide assimilation rate, PSI photochemical activity, stomatal conductance, water-use efficiency, etc. In a different instance using eggplant, exogenous EBR therapy reduced the negative consequences of heat stress by controlling the way that intense heat causes reactive oxygen species (ROS) to aggregate (Wu et al. 2014). When EBR is applied foliarly to wheat at extremely high temperatures, it improves growth, biomass accumulation, photosynthetic efficiency, and, most importantly, antioxidant potential (Hussain et al. 2019). Sonjaroon et al. (2018) observed an intriguing event in heat-stressed rice, wherein the exogenous administration of BRs replicates a molecule known as 7,8-Dihydro-20-hydroxyecdysone (DHECD). During the reproductive stage, these modifications occur, which ultimately lead to an increase in photosynthesis, carbohydrate load, seed setting, and seed weight. Based on the aforementioned examples, it is now evident that BRs have the potential tonic influence on plant photosynthesis and antioxidant function, thereby mitigating the harmful consequences of high temperatures.

In many parts of the world, low temperatures that cause chilling or freezing stress are a significant hindrance to agricultural production, especially for thermophilic plants (Cui et al. 2011). Plants that are subjected to cold stress exhibit a variety of limitations, including as modifications to membrane fluidity, changes in the activities of macromolecules, a drop in the osmotic potential within the cells, and mechanical constraints (Xiong et al. 2002). Plant photosynthetic processes are similarly impacted by cold stress, as evidenced by a decrease in the rate of CO_2 assimilation, photoinhibition at PSI and PSII, and a drop in Benson–Calvin Cycle enzyme activity (Jiang et al. 2013a, b). As with heat stress, cold stress also leads to the formation of ROS, and elevated ROS levels have the potential to harm biomembranes via lipid peroxidation (Chen et al. 2013). Plants have evolved a wide range of ROS scavenging and photo protective mechanisms, respectively, to prevent excessive ROS accumulation and the over reduction of the photosystems that induce photoinhibition (Fang et al. 2019). According to Xiong et al. (2002), plants that have higher ROS scavenging capacity or genes encoding such enzymes are more resilient to cold stress. Exogenous EBR treatment or overexpression of DWRF enhances chilling tolerance by reducing oxidative damage in tomato plants, whereas BR deficiency reduces chilling tolerance by causing protein oxidation and lipid peroxidation. Remarkably, according to Cui et al. (2011), ROS can also function as signals in triggering BR-regulated reactions to cold stress tolerance. According to a recent study (Xia et al. 2018), RBOH1, GLUTAREDOXIN (GRX), and 2-cysteine peroxiredoxin (2-Cys Prx) are involved in a signalling cascade that mediates tomato BR-induced chilling tolerance. According to Zhou et al. (2014), RBOH1 encodes NADPH oxidase, which is in charge of producing ROS in the apoplast, mostly for signalling functions. Apart from ROS, NO is involved in the cold response pathway mediated by BR (Cui et al. 2011). It has been discovered that NO contributes to

BR-induced cold tolerance by acting downstream of H_2O_2. When exposed to cold stress, exogenous application of EBR (0.1 μM) enhances CO_2 assimilation and reduces PSII's photoinhibition. The activation of important enzymes involved in the ascorbate–glutathione (AsA–GSH) cycle and redox homeostasis is what allows the photosynthetic apparatus to recover after BR treatment (Jiang et al. 2013a, b). Grape (*Vitis vinifera*) seedlings treated with EBR exhibit increased freezing tolerance as a result of the treatment's temporal modulation of the AsA-GSH cycle component (Chen et al. 2019). Similarly, using EBR (0.3 μM) topically strengthens grapevines' resistance to chilling stress by enhancing their antioxidative capability, which reduces membrane lipid peroxidation during stressful situations. Under cold stress, EBR pretreatment (1.0 μM) in maize can boost plant height, biomass, and concentrations of protein, sugar, and chlorophyll (Singh et al. 2012).

Under cold stress, plants use an essential strategy known as photoprotection to prevent photoinhibition. It has been demonstrated that BRs have a role in photoprotection in plants that are stressed by cold. Plants that are exposed to low temperatures eventually build up active BRs and activate BZR1, which raises RBOH1 transcript levels and apoplastic H_2O_2 synthesis (Fang et al. 2019). Conversely, a BZR1 mutation or RBOH1 suppression eliminates BR-induced photoprotection, which exacerbates photoinhibition produced by chilling. Specifically, the induction of non-photochemical quenching (NPQ), accumulation of D1 and PSII subunit S (PsbS) proteins, redox signalling, and the PROTON GRADIENT REGULATION5 (PGR5)-dependent cyclic electron fetching (CEF) are all significantly aided by BRs-induced apoplastic H2O2 (Fang et al. 2019). The results of transcriptome analysis indicate that under low temperature conditions, EBR treatment increases the transcript levels of genes related to photosynthesis and chlorophyll biosynthesis, including those that encode the PSII oxygen-evolving enhancer protein, PSI subunit, light-harvesting chlorophyll protein complexes I and II, and ferredoxins (Zhao et al. 2019).

By activating COLD-RESPONSIVE (COR) genes, BRs have been demonstrated to improve freezing tolerance through both independent and C-REPEAT/DEHYDRATION-RESPONSIVE ELEMENT BINDING FACTOR1 (CBF1)-dependent pathways (Eremina et al. 2016). Furthermore, the increase of BZR1 and BES1 in their unphosphorylated (active) forms contributes to BR-mediated better cold tolerance by promoting CBF1 and CBF2 transcription. However, BRASSINOSTEROIDINSENSITIVE2 (BIN2) destabilizes the transcription factor INDUCER OF CBF EXPRESSION1 (ICE1), which in turn negatively affects cold stress responses during chronic cold stress (Ye et al. 2019). According to Nolan et al. (2020), these data imply that BRs can both increase stress tolerance and reduce stress reactions, which are mostly controlled by spatiotemporal regulation. The increased resistance to cold that BRs provide does not just affect whole plants; it also affects harvested plant products like fruits. Research has demonstrated that by prolonging the shelf life under low temperature stress, BR can enhance the postharvest quality of fruits and vegetables (Aghdam and Mohammadkhani 2014). However, compared to the amounts utilized to confer stress tolerance in the intact plant, relatively large quantities of EBR are employed for postharvest management.

For example, chilling stress significantly reduces the quality of fruit in tomatoes; however, 6 μM EBR therapy could reduce the damage caused by chilling in tomato fruits. This is because BR inhibits the activity of lipoxygenase and phospholipase D (Aghdam and Mohammadkhani 2014). Mango fruits treated with 10 μM EBR are protected from cold-induced damage by an increased expression of several proteins, including type II SK2 dehydrin, abscisic acid stress ripening like protein, remorin, and temperature-induced lipocalin (Li et al. 2012). Furthermore, BR lowers the phase transition temperature and raises the degree of instauration of fatty acids in the mango fruits' plasma membrane lipids, which boosts the fruit's fluidity in cold weather (Li et al. 2012).

1.9 Conclusion and Future Prospects

This chapter describes the promising role of BRs in various morphological, physiological, and metabolic processes. BRs are new class of phytohormones that are emerging role in plant science. BRs can moderate processes including photosynthesis, growth, and metabolism. BRs have been shown to help reduce stress-related adverse conditions and have been utilized to strengthen plants' antioxidant defenses against abiotic stresses because crop production is impacted by climate change in a direct, indirect, and socioeconomic way. Recognition and description of key regulatory genes that are responsible for hormonal crosstalk and the underlying signaling also need to be explored.

References

Abhinandan K, Skori L, Stanic M, Hickerson NM, Jamshed M, Samuel MA (2018) Abiotic stress signaling in wheat–an inclusive overview of hormonal interactions during abiotic stress responses in wheat. Front Plant Sci 9:734

Ackerman-Lavert M, Savaldi-Goldstein S (2020) Growth models from a brassinosteroid perspective. Curr Opin Plant Biol 53:90–97

Aghdam MS, Mohammadkhani N (2014) Enhancement of chilling stress tolerance of tomato fruit by postharvest brassinolide treatment. Food Bioprocess Technol 7:909–914

Ahammed GJ, Yu JQ (eds) (2016) Plant hormones under challenging environmental factors. Springer, Dordrecht, pp 1–21

Ahammed GJ, Li X, Liu A, Chen S (2020) Brassinosteroids in plant tolerance to abiotic stress. J Plant Growth Regul 39:1451–1464

Ahanger MA, Mir RA, Alyemeni MN, Ahmad P (2020) Combined effects of brassinosteroid and kinetin mitigates salinity stress in tomato through the modulation of antioxidant and osmolyte metabolism. Plant Physiol Biochem 147:31–42

Ali B, Hayat S, Fariduddin Q, Ahmad A (2008) 24-Epibrassinolide protects against the stress generated by salinity and nickel in Brassica juncea. Chemosphere 72(9):1387–1392

Amraee L, Rahmani F, Mandoulakani BA (2019) 24-Epibrassinolide alters DNA cytosine methylation of Linum usitatissimum L. under salinity stress. Plant Physiol Biochem 139:478–484

An S, Liu Y, Sang K, Wang T, Yu J, Zhou Y, Xia X (2023) Brassinosteroid signaling positively regulates abscisic acid biosynthesis in response to chilling stress in tomato. J Integr Plant Biol 65(1):10–24

Arteca JM, Arteca RN (2001) Brassinosteroid-induced exaggerated growth in hydroponically grown Arabidopsis plants. Physiol Plant 112(1):104–112
Asseng S, Ewert F, Martre P, Rötter RP, Lobell DB, Cammarano D, Kimball BA, Ottman MJ, Wall GW, White JW, Reynolds MP (2015) Rising temperatures reduce global wheat production. Nat Clim Change 5(2):143–147
Barnabás B, Jäger K, Fehér A (2008) The effect of drought and heat stress on reproductive processes in cereals. Plant Cell Environ 31(1):11–38
Baroowa B, Gogoi N (2014) Biochemical changes in black gram and green gram genotypes after imposition of drought stress. J Food Legumes 27(4):350–353
Behnamnia LM, Kalantari KM, Rezanejad F (2009) Exogenous application of brassinosteroid alleviates drought-induced oxidative stress in Lycopersicon esculentum. Gen Appl Plant Physiol 35:22–34
Bonan GB, Doney SC (2018) Climate, ecosystems, and planetary futures: the challenge to predict life in earth system models. Science 359(6375):eaam8328
Boyer JS (1982) Plant productivity and environment. Science 218(4571):443–448
Challinor AJ, Wheeler TR, Craufurd PQ, Ferro CA, Stephenson DB (2007) Adaptation of crops to climate change through genotypic responses to mean and extreme temperatures. Agric Ecosyst Environ 119(1–2):190–204
Chen S, Jin W, Liu A, Zhang S, Liu D, Wang F, Lin X, He C (2013) Arbuscular mycorrhizal fungi (AMF) increase growth and secondary metabolism in cucumber subjected to low temperature stress. Sci Hortic 160:222–229
Chen J, Nolan TM, Ye H, Zhang M, Tong H, Xin P, Chu J, Chu C, Li Z, Yin Y (2017) Arabidopsis WRKY46, WRKY54, and WRKY70 transcription factors are involved in brassinosteroid-regulated plant growth and drought responses. Plant Cell 29(6):1425–1439
Chen ZY, Wang YT, Pan XB, Xi ZM (2019) Amelioration of cold-induced oxidative stress by exogenous 24-epibrassinolide treatment in grapevine seedlings: toward regulating the ascorbate–glutathione cycle. Sci Hortic 244:379–387
Cui JX, Zhou YH, Ding JG, Xia XJ, Shi KAI, Chen SC, Asami T, Chen Z, Yu JQ (2011) Role of nitric oxide in hydrogen peroxide-dependent induction of abiotic stress tolerance by brassinosteroids in cucumber. Plant Cell Environ 34(2):347–358
de Oliveira ED, Bramley H, Siddique KH, Henty S, Berger J, Palta JA (2012) Can elevated CO2 combined with high temperature ameliorate the effect of terminal drought in wheat? Funct Plant Biol 40(2):160–171
Divi UK, Rahman T, Krishna P (2016) Gene expression and functional analyses in brassinosteroid-mediated stress tolerance. Plant Biotechnol J 14(1):419–432
Domagalska MA, Sarnowska E, Nagy F, Davis SJ (2010) Genetic analyses of interactions among gibberellin, abscisic acid, and brassinosteroids in the control of flowering time in Arabidopsis thaliana. PloS One 5(11):e14012
Easterling WE, Aggarwal PK, Batima P, Brander KM, Erda L, Howden SM, Kirilenko A, Morton J, Soussana JF, Schmidhuber J, Tubiello FN (2007) Food, fibre and forest products. Climate Change 2007:273–313
Eremina M, Unterholzner SJ, Rathnayake AI, Castellanos M, Khan M, Kugler KG, May ST, Mayer KF, Rozhon W, Poppenberger B (2016) Brassinosteroids participate in the control of basal and acquired freezing tolerance of plants. Proc Natl Acad Sci U S A 113(40):E5982–E5991
Espeland EK, Kettenring KM (2018) Strategic plant choices can alleviate climate change impacts: a review. J Environ Manag 222:316–324
Fang P, Yan M, Chi C, Wang M, Zhou Y, Zhou J, Shi K, Xia X, Foyer CH, Yu J (2019) Brassinosteroids act as a positive regulator of photoprotection in response to chilling stress. Plant Physiol 180(4):2061–2076
Faostat FAO (2017). http://www.fao.org/faostat/en/#data.QC. Accessed January 2018
Filová A (2014) Production of secondary metabolities in plant tissue cultures. Res J Agric Sci 46(1):236–245
Fujioka S, Yokota T (2003) Biosynthesis and metabolism of brassinosteroids. Annu Rev Plant Biol 54(1):137–164

Garrity DP, O'Toole JC (1994) Screening rice for drought resistance at the reproductive phase. Field Crop Res 39(2–3):99–110
Gong M, Chen SN, Song YQ, Li ZG (1997) Effect of calcium and calmodulin on intrinsic heat tolerance in relation to antioxidant systems in maize seedlings. Funct Plant Biol 24(3):371–379
González-García MP, Vilarrasa-Blasi J, Zhiponova M, Divol F, Mora-García S, Russinova E, Caño-Delgado AI (2011) Brassinosteroids control meristem size by promoting cell cycle progression in Arabidopsis roots. Development 138(5):849–859
Gudesblat GE, Schneider-Pizoń J, Betti C, Mayerhofer J, Vanhoutte I, Van Dongen W, Boeren S, Zhiponova M, De Vries S, Jonak C, Russinova E (2012) SPEECHLESS integrates brassinosteroid and stomata signalling pathways. Nat Cell Biol 14(5):548–554
Hacham Y, Holland N, Butterfield C, Ubeda-Tomas S, Bennett MJ, Chory J, Savaldi-Goldstein S (2011) Brassinosteroid perception in the epidermis controls root meristem size. Development 138(5):839–848
Hatfield JL, Prueger JH (2015) Temperature extremes: effect on plant growth and development. Weather Clim Extrem 10:4–10
Hayat S, Maheshwari P, Wani AS, Irfan M, Alyemeni MN, Ahmad A (2012) Comparative effect of 28 homobrassinolide and salicylic acid in the amelioration of NaCl stress in *Brassica juncea* L. Plant Physiol Biochem 53:61–68. https://doi.org/10.1016/j.plaphy.2012.01.011
Hellin J, Bellon MR, Hearne SJ (2014) Maize landraces and adaptation to climate change in Mexico. J Crop Improv 28(4):484–501
Hussain M, Khan TA, Yusuf M, Fariduddin Q (2019) Silicon-mediated role of 24-epibrassinolide in wheat under high-temperature stress. Environ Sci Pollut Res 26:17163–17172
Intergovernmental Panel on Climate Change (IPCC) (2000) Special report on emissions scenarios; A special report of working group III of the Intergovernmental Panel on Climate Change. Cambridge University Press, Cambridge
Intergovernmental Panel on Climate Change (IPCC) (2007) Climate change: impacts, adaptation, and vulnerability; Contribution of working group II to the fourth assessment report of the Intergovernmental Panel on Climate Change. Cambridge University Press, Cambridge
Jaillais Y, Vert G (2012) Brassinosteroids, gibberellins and light-mediated signalling are the three-way controls of plant sprouting. Nat Cell Biol 14(8):788–790
Jia Z, Giehl RF, Meyer RC, Altmann T, von Wirén N (2019) Natural variation of BSK3 tunes brassinosteroid signaling to regulate root foraging under low nitrogen. Nat Commun 10(1):2378
Jiang WB, Huang HY, Hu YW, Zhu SW, Wang ZY, Lin WH (2013a) Brassinosteroid regulates seed size and shape in Arabidopsis. Plant Physiol 162(4):1965–1977
Jiang YP, Huang LF, Cheng F, Zhou YH, Xia XJ, Mao WH, Shi K, Yu JQ (2013b) Brassinosteroids accelerate recovery of photosynthetic apparatus from cold stress by balancing the electron partitioning, carboxylation and redox homeostasis in cucumber. Physiol Plant 148(1):133–145
Kang Y, Khan S, Ma X (2009) Climate change impacts on crop yield, crop water productivity and food security—a review. Prog Nat Sci 19(12):1665–1674
Khan M, Rozhon W, Bigeard J, Pflieger D, Husar S, Pitzschke A, Teige M, Jonak C, Hirt H, Poppenberger B (2013) Brassinosteroid-regulated GSK3/Shaggy-like kinases phosphorylate mitogen-activated protein (MAP) kinase kinases, which control stomata development in Arabidopsis thaliana. J Biol Chem 288(11):7519–7527
Kitano H (2002) Systems biology: a brief overview. Science 295(5560):1662–1664
Kjellström E, Nikulin G, Strandberg G, Christensen OB, Jacob D, Keuler K, Lenderink G, van Meijgaard E, Schär C, Somot S, Sørland SL (2018) European climate change at global mean temperature increases of 1.5 and 2°C above pre-industrial conditions as simulated by the EURO-CORDEX regional climate models. Earth Syst Dynam 9(2):459–478
Kong Q, Mostafa HH, Yang W, Wang J, Nuerawuti M, Wang Y, Song J, Zhang X, Ma L, Wang H, Li X (2021) Comparative transcriptome profiling reveals that brassinosteroid-mediated lignification plays an important role in garlic adaption to salt stress. Plant Physiol Biochem 158:34–42
Kour J, Kohli SK, Khanna K, Bakshi P, Sharma P, Singh AD, Ibrahim M, Devi K, Sharma N, Ohri P, Skalicky M (2021) Brassinosteroid signaling, crosstalk and, physiological functions in plants under heavy metal stress. Front Plant Sci 12:608061

Li B, Zhang C, Cao B, Qin G, Wang W, Tian S (2012) Brassinolide enhances cold stress tolerance of fruit by regulating plasma membrane proteins and lipids. Amino Acids 43:2469–2480
Lin WH (2020) Designed manipulation of the brassinosteroid signal to enhance crop yield. Front Plant Sci 11:854
Lobell DB, Bänziger M, Magorokosho C, Vivek B (2011) Nonlinear heat effects on African maize as evidenced by historical yield trials. Nat Clim Change 1(1):42–45
Lv B, Tian H, Zhang F, Liu J, Lu S, Bai M, Li C, Ding Z (2018) Brassinosteroids regulate root growth by controlling reactive oxygen species homeostasis and dual effect on ethylene synthesis in Arabidopsis. PLoS Genet 14(1):e1007144
Makarevitch I, Thompson A, Muehlbauer GJ, Springer NM (2012) Brd1 gene in maize encodes a brassinosteroid C-6 oxidase. PLoS One 7(1):e30798
Maleki A, Naderi A, Naseri R, Fathi A, Bahamin S, Maleki R (2013) Physiological performance of soybean cultivars under drought stress. Bull Environ Pharmacol Life Sci 2(6):38–44
Martínez C, Espinosa-Ruíz A, de Lucas M, Bernardo-García S, Franco-Zorrilla JM, Prat S (2018) PIF 4-induced BR synthesis is critical to diurnal and thermomorphogenic growth. EMBO J 37(23):e99552
Martins S, Montiel-Jorda A, Cayrel A, Huguet S, Roux CPL, Ljung K, Vert G (2017) Brassinosteroid signaling-dependent root responses to prolonged elevated ambient temperature. Nat Commun 8(1):309
Morillon R, Catterou M, Sangwan RS, Sangwan BS, Lassalles JP (2001) Brassinolide may control aquaporin activities in Arabidopsis thaliana. Planta 212:199–204
Mubarik MS, Khan SH, Sajjad M, Raza A, Hafeez MB, Yasmeen T, Rizwan M, Ali S, Arif MS (2021) A manipulative interplay between positive and negative regulators of phytohormones: a way forward for improving drought tolerance in plants. Physiol Plant 172(2):1269–1290
Nazir F, Jahan B, Kumari S, Iqbal N, Albaqami M, Sofo A, Khan MIR (2023) Brassinosteroid modulates ethylene synthesis and antioxidant metabolism to protect rice (Oryza sativa) against heat stress-induced inhibition of source–sink capacity and photosynthetic and growth attributes. J Plant Physiol 289:154096
Neha T, Mohapatra S, Sirhindi G, Dogra V (2022) Seed priming with brassinolides improves growth and reinforces antioxidative defenses under normal and heat stress conditions in seedlings of Brassica juncea. Physiol Plant 174(6):e13814
Nolan TM, Vukašinović N, Liu D, Russinova E, Yin Y (2020) Brassinosteroids: multidimensional regulators of plant growth, development, and stress responses. Plant Cell 32(2):295–318
Ogweno JO, Song XS, Shi K, Hu WH, Mao WH, Zhou YH, Yu JQ, Nogués S (2008) Brassinosteroids alleviate heat-induced inhibition of photosynthesis by increasing carboxylation efficiency and enhancing antioxidant systems in Lycopersicon esculentum. J Plant Growth Regul 27:49–57
Oh E, Zhu JY, Bai MY, Arenhart RA, Sun Y, Wang ZY (2014) Cell elongation is regulated through a central circuit of interacting transcription factors in the Arabidopsis hypocotyl. Elife 3:e03031
Olesen JE, Bindi M (2002) Consequences of climate change for European agricultural productivity, land use and policy. Eur J Agron 16(4):239–262
Olesen JE, Trnka M, Kersebaum KC, Skjelvåg AO, Seguin B, Peltonen-Sainio P, Rossi F, Kozyra J, Micale F (2011) Impacts and adaptation of European crop production systems to climate change. Eur J Agron 34(2):96–112
Otto IM, Reckien D, Reyer CP, Marcus R, Le Masson V, Jones L, Norton A, Serdeczny O (2017) Social vulnerability to climate change: a review of concepts and evidence. Reg Environ Chang 17:1651–1662
Pereira A (2016) Plant abiotic stress challenges from the changing environment. Front Plant Sci 7:1123
Pradhan GP, Prasad PV, Fritz AK, Kirkham MB, Gill BS (2012) Effects of drought and high temperature stress on synthetic hexaploid wheat. Funct Plant Biol 39(3):190–198
Rattan A, Kapoor D, Kapoor N, Bhardwaj R, Sharma A (2020) Brassinosteroids regulate functional components of antioxidative defense system in salt stressed maize seedlings. J Plant Growth Regul 39:1465–1475

Raza A, Mehmood SS, Tabassum J, Batool R (2019) Targeting plant hormones to develop abiotic stress resistance in wheat. In: Wheat production in changing environments: responses, adaptation and tolerance, pp 557–577
Rozhon W, Akter S, Fernandez A, Poppenberger B (2019) Inhibitors of brassinosteroid biosynthesis and signal transduction. Molecules 24(23):4372
Sadura I, Janeczko A (2018) Physiological and molecular mechanisms of brassinosteroid-induced tolerance to high and low temperature in plants. Biol Plant 62(4):601–616
Saini HS, Aspinall D (1982) Abnormal sporogenesis in wheat (Triticum aestivum L.) induced by short periods of high temperature. Ann Bot 49(6):835–846
Salehi-Lisar SY, Bakhshayeshan-Agdam H (2016) Drought stress in plants: causes, consequences, and tolerance. In: Drought stress tolerance in plants, vol 1. Physiology and biochemistry, pp 1–16
Scheben A, Yuan Y, Edwards D (2016) Advances in genomics for adapting crops to climate change. Curr Plant Biol 6:2–10
Schlenker W, Roberts MJ (2009) Nonlinear temperature effects indicate severe damages to US crop yields under climate change. Proc Natl Acad Sci U S A 106(37):15594–15598
Schmidhuber J, Tubiello FN (2007) Global food security under climate change. Proc Natl Acad Sci U S A 104(50):19703–19708
Sheoran IS, Saini HS (1996) Drought-induced male sterility in rice: changes in carbohydrate levels and enzyme activities associated with the inhibition of starch accumulation in pollen. Sex Plant Reprod 9:161–169
Singh I, Kumar U, Singh SK, Gupta C, Singh M, Kushwaha SR (2012) Physiological and biochemical effect of 24-epibrassinoslide on cold tolerance in maize seedlings. Physiol Mol Biol Plants 18:229–236
Singh P, Basu S, Kumar G (2018) Polyamines metabolism: a way ahead for abiotic stress tolerance in crop plants. In: Biochemical, physiological and molecular avenues for combating abiotic stress tolerance in plants. Academic, pp 39–55
Sonjaroon W, Jutamanee K, Khamsuk O, Thussagunpanit J, Kaveeta L, Suksamrarn A (2018) Impact of brassinosteroid mimic on photosynthesis, carbohydrate content and rice seed set at reproductive stage under heat stress. Agric Nat Resourc 52(3):234–240
Sun Y, He Y, Irfan AR, Liu X, Yu Q, Zhang Q, Yang D (2020) Exogenous brassinolide enhances the growth and cold resistance of maize (Zea mays L.) seedlings under chilling stress. Agronomy 10(4):488
Symons GM, Davies C, Shavrukov Y, Dry IB, Reid JB, Thomas MR (2006) Grapes on steroids. Brassinosteroids are involved in grape berry ripening. Plant Physiol 140(1):150–158
Tack J, Barkley A, Nalley LL (2015) Effect of warming temperatures on US wheat yields. Proc Natl Acad Sci U S A 112(22):6931–6936
Tanabe S, Ashikari M, Fujioka S, Takatsuto S, Yoshida S, Yano M, Yoshimura A, Kitano H, Matsuoka M, Fujisawa Y, Kato H (2005) A novel cytochrome P450 is implicated in brassinosteroid biosynthesis via the characterization of a rice dwarf mutant, dwarf11, with reduced seed length. Plant Cell 17(3):776–790
Thornton PK, Ericksen PJ, Herrero M, Challinor AJ (2014) Climate variability and vulnerability to climate change: a review. Glob Chang Biol 20(11):3313–3328
Trevisan S, Forestan C, Brojanigo S, Quaggiotti S, Varotto S (2020) Brassinosteroid application affects the growth and gravitropic response of maize by regulating gene expression in the roots, shoots and leaves. Plant Growth Regul 92:117–130
Van Velthuizen H (2007) Mapping biophysical factors that influence agricultural production and rural vulnerability, no. 11. Food and Agriculture Organization
Vogler F, Schmalzl C, Englhart M, Bircheneder M, Sprunck S (2014) Brassinosteroids promote Arabidopsis pollen germination and growth. Plant Reprod 27:153–167
Wang Z, Huang B (2004) Physiological recovery of Kentucky bluegrass from simultaneous drought and heat stress. Crop Sci 44(5):1729–1736

Wang X, Chen J, Xie Z, Liu S, Nolan T, Ye H, Zhang M, Guo H, Schnable PS, Li Z, Yin Y (2014) Histone lysine methyltransferase SDG8 is involved in brassinosteroid-regulated gene expression in Arabidopsis thaliana. Mol Plant 7(8):1303–1315

Wang Q, Yu F, Xie Q (2020) Balancing growth and adaptation to stress: crosstalk between brassinosteroid and abscisic acid signaling. Plant Cell Environ 43(10):2325–2335

Wang W, Xie Y, Liu C, Jiang H (2022) The exogenous application of brassinosteroids confers tolerance to heat stress by increasing antioxidant capacity in soybeans. Agriculture 12(8):1095

Wang S, Liu D, Jin T, Du J, Tian A, Huang Q, Khan M, Nie S, Zhang S, Deng R, Zhang M (2023) Manipulation of brassinosteroid signaling via overexpression of its major receptor SlBRI1 regulates salt tolerance in tomato. Sci Hortic 322:112457

Wei Z, Li J (2016) Brassinosteroids regulate root growth, development, and symbiosis. Mol Plant 9(1):86–100

Williams ME (2011) Brassinosteroids: plant steroid hormones. Teaching tools in plant biology: lecture notes. Plant Cell (online), https://doi.org/10.1105/tpc.110

Winkel T, Renno JF, Payne WA (1997) Effect of the timing of water deficit on growth, phenology and yield of pearl millet (Pennisetum glaucum (L.) R. Br.) grown in Sahelian conditions. J Exp Bot 48(5):1001–1009

Wu X, Yao X, Chen J, Zhu Z, Zhang H, Zha D (2014) Brassinosteroids protect photosynthesis and antioxidant system of eggplant seedlings from high-temperature stress. Acta Physiol Plant 36(2):251–261

Xia XJ, Fang PP, Guo X, Qian XJ, Zhou J, Shi K, Zhou YH, Yu JQ (2018) Brassinosteroid-mediated apoplastic H2O2-glutaredoxin 12/14 cascade regulates antioxidant capacity in response to chilling in tomato. Plant Cell Environ 41(5):1052–1064

Xiong L, Schumaker KS, Zhu JK (2002) Cell signaling during cold, drought, and salt stress. Plant Cell 14(suppl_1):S165–S183

Xu ZZ, Zhou GS (2006) Combined effects of water stress and high temperature on photosynthesis, nitrogen metabolism and lipid peroxidation of a perennial grass Leymus chinensis. Planta 224:1080–1090

Xu F, Xi ZM, Zhang H, Zhang CJ, Zhang ZW (2015) Brassinosteroids are involved in controlling sugar unloading in Vitis vinifera 'Cabernet Sauvignon' berries during véraison. Plant Physiol Biochem 94:197–208

Ye Q, Zhu W, Li L, Zhang S, Yin Y, Ma H, Wang X (2010) Brassinosteroids control male fertility by regulating the expression of key genes involved in Arabidopsis anther and pollen development. Proc Natl Acad Sci U S A 107(13):6100–6105

Ye K, Li H, Ding Y, Shi Y, Song C, Gong Z, Yang S (2019) BRASSINOSTEROID-INSENSITIVE2 negatively regulates the stability of transcription factor ICE1 in response to cold stress in Arabidopsis. Plant Cell 31(11):2682–2696

Yin Y, Wang ZY, Mora-Garcia S, Li J, Yoshida S, Asami T, Chory J (2002) BES1 accumulates in the nucleus in response to brassinosteroids to regulate gene expression and promote stem elongation. Cell 109(2):181–191

Yuan L, Yuan Y, Du J, Sun J, Guo S (2012) Effects of 24-epibrassinolide on nitrogen metabolism in cucumber seedlings under Ca (NO3) 2 stress. Plant Physiol Biochem 61:29–35

Yue J, You Y, Zhang L, Fu Z, Wang J, Zhang J, Guy RD (2019) Exogenous 24-epibrassinolide alleviates effects of salt stress on chloroplasts and photosynthesis in Robinia pseudoacacia L. seedlings. J Plant Growth Regul 38:669–682

Zandalinas SI, Mittler R, Balfagón D, Arbona V, Gómez-Cadenas A (2018) Plant adaptations to the combination of drought and high temperatures. Physiol Plant 162(1):2–12

Zeid F, El Shihy O, Ibrahim F (2019) Exogenous application of 24-epibrassinolide and spermine improves growth and yield of drought stressed wheat (Triticum aestivum L.). Middle East J Appl Sci 9:1063–1080

Zhang A, Zhang J, Zhang J, Ye N, Zhang H, Tan M, Jiang M (2011) Nitric oxide mediates brassinosteroid-induced ABA biosynthesis involved in oxidative stress tolerance in maize leaves. Plant Cell Physiol 52(1):181–192

Zhang YP, Zhu XH, Ding HD, Yang SJ, Chen YY (2013) Foliar application of 24-epibrassinolide alleviates high-temperature-induced inhibition of photosynthesis in seedlings of two melon cultivars. Photosynthetica 51:341–349
Zhang W, Zhu K, Wang Z, Zhang H, Gu J, Liu L, Yang J, Zhang J (2019) Brassinosteroids function in spikelet differentiation and degeneration in rice. J Integr Plant Biol 61(8):943–963
Zhao B, Li J (2012) Regulation of brassinosteroid biosynthesis and inactivation. J Integr Plant Biol 54(10):746–759
Zhao C, Liu B, Piao S, Wang X, Lobell DB, Huang Y, Huang M, Yao Y, Bassu S, Ciais P, Durand JL (2017) Temperature increase reduces global yields of major crops in four independent estimates. Proc Natl Acad Sci U S A 114(35):9326–9331
Zhao M, Yuan L, Wang J, Xie S, Zheng Y, Nie L, Zhu S, Hou J, Chen G, Wang C (2019) Transcriptome analysis reveals a positive effect of brassinosteroids on the photosynthetic capacity of wucai under low temperature. BMC Genomics 20(1):1–19
Zhou J, Wang J, Li X, Xia XJ, Zhou YH, Shi K, Chen Z, Yu JQ (2014) H2O2 mediates the crosstalk of brassinosteroid and abscisic acid in tomato responses to heat and oxidative stresses. J Exp Bot 65(15):4371–4383
Zhu Z, Zhang Z, Qin G, Tian S (2010) Effects of brassinosteroids on postharvest disease and senescence of jujube fruit in storage. Postharvest Biol Technol 56(1):50–55
Zhu J, Lu P, Jiang Y, Wang M, Zhang L (2014) Effects of brassinosteroid on antioxidant system in Salvia miltiorrhiza under drought stress. J Res Agric Anim Sci 2:1–6

Salicylic Acid: Food, Functions, and Future

2

Maninder Meenu, Kiran Khandare, Manali Singh, Samuel Kenyanya, Kanu Priya Sharma, and Monika Garg

Abstract

Salicylic acid (SA), a ubiquitous plant metabolite with remarkable hydrogen bonding characteristics, has a rich past and a wide range of applications. The medicinal benefits of SA are well-established, having roots in the use of the world's most popular drug, aspirin, and in the ancient willow bark remedies of Mesopotamia. This chapter discusses the background and specifics of foods high in salicylates and the intricate network of actions exhibited by SA has been explored. The use of SA in pharmaceutical and agri-food businesses has been meticulously studied. Through in-depth analysis, we uncover the mechanisms underlying SA's impact on important fruit quality parameters such as such as respiration rate, weight loss, softening, chilling injury, color, and decay have been explored. The numerous biological activities of SA and foods high in it have been briefly discussed.

M. Meenu (✉) · M. Garg (✉)
Department of Agri-Biotechnology, National Agri-Food Biotechnology Institute (NABI), Mohali, Punjab, India

K. Khandare
Division of Chemical Engineering, Center of Innovative and Applied Bioprocessing (CIAB), Mohali, Punjab, India

M. Singh
Microbial Type Culture Collection and Gene Bank (MTCC), Institute of Microbial Technology (IMTECH), Chandigarh, India

S. Kenyanya
Department of Biomedical Sciences, Kisii University, Kisii, Kenya

K. P. Sharma
Department of Biochemistry, Punjab Agricultural University (PAU), Ludhiana, Punjab, India

© The Author(s), under exclusive license to Springer Nature Singapore Pte Ltd. 2024

M. Faizan, S. Hayat (eds.), *Plant Growth Regulators: Resilience for Sustainable Agriculture*, https://doi.org/10.1007/978-981-97-2918-0_2

Keywords
Salicylic acid · Derivatives · Food · Health benefits · Post-harvest quality · Pre-harvest quality

2.1 Introduction

Salicylic acid (SA), also called as 2-hydroxybenzoic acid, is a constituent of many plants. It has very simple phenolic structure with an aromatic ring which has been generated from β-oxidation of cinnamic acid. It is an example of intermolecular of chelate, hydrogen bonds. It shows hydrogen bonds linked to the carboxyl groups and forms dimer (Fig. 2.1). Cochran discovered the crystal structure of SA for the first time in 1953. Later, a crystal structure of SA determined by neutron diffraction was reported by Bacon and Jude, while Sundaralingam and Jesen reported an additional refinement of the X-ray structure. The presence of intermolecular OH…O=C hydrogen bond in SA molecule it is expected the formation of long chains of interacting hydrogen bonds in acid-base system. On further investigation we found the different derivatives of SA have different functions based on its structure relationship. A derivative called 5-ASA is used to trap the free radicals in metabolism. Olsalazine is used to

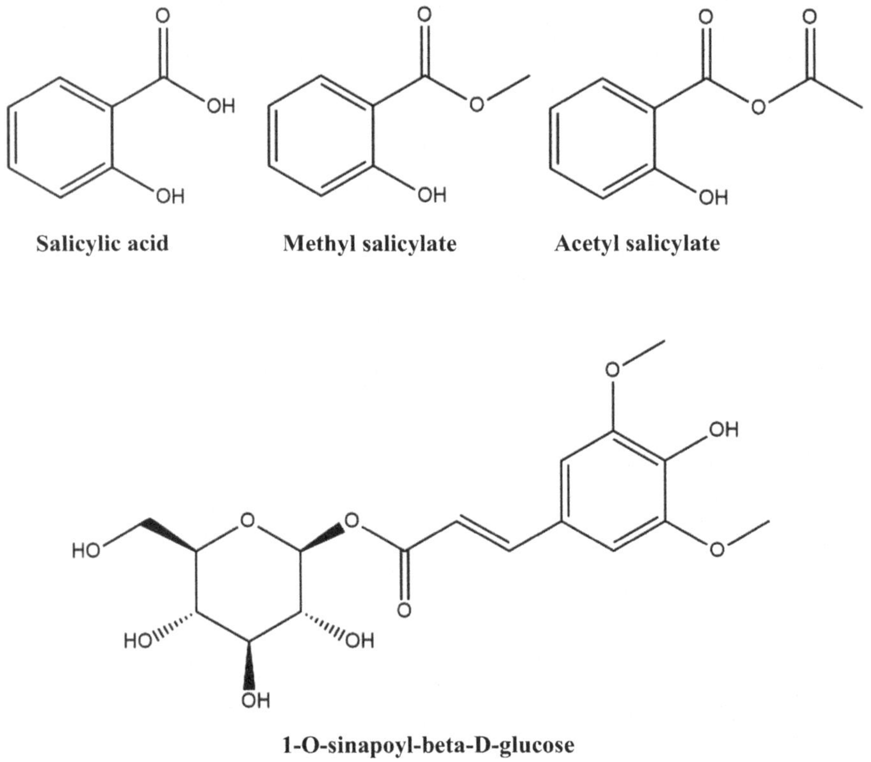

Fig. 2.1 Structure of salicylic acid and derivatives

treat inflammatory disorders. Methyl salicylate (MeSA) is used as flavoring agent in various food and beverages. Acetylsalicylic acid (ASA) is the most widely used drug in the world useful in fever and in treatment of pain. There are several other derivatives which are useful in the different diseases such as Mesalazine, Aspirin, Diflunisal, Sulfasalazine, and Balsalazide (Ekinci et al. 2011).

2.1.1 History

There are several evidence that showed the SA has significant effects on mammalian cell lines. Several plants with high SA content such as myrtle, willow, poplar, and meadowsweet are used in the treatment of human maladies. Since SA is predominantly a plant product and is thought not to occur in the tissues of vertebrates, its therapeutic capabilities have been viewed as "adventitious." Willow trees are presumably not very concerned with curing human illnesses, and their ability to produce leaves containing salicylates may have been beneficial to their survival for a short while prior to the development of synthetic salicylates. Dried myrtle leaves can be infused to treat back issues and rheumatism, according to the Ebers papyrus. Additionally, ancient records mention the Babylonians, Assyrians, and Chinese using willow bark and leaves for medicinal purposes. The Royal Society was notified in 1763 by the Reverend Edward Stone that willow bark contained compounds that successfully reduced "ague" symptoms (probably malarial fever). During the first part of the 1800s, SA, the active component, was extracted and refined from willow and meadowsweet. It was chemically synthesized with the carboxylation of sodium phenoxide in 1863 (Kolbe 1860). In 1876 it was reported that it shows the antipyretic activity and anti-inflammatory activity by Thomas John MacLagan. Due to its significant effect in reduction of pain and fever, it becomes most widely used drug in 1899 under the trademark "Aspirin" (Jack 1997). However, there is still a little more to discuss on the "natural history" of SA before summarizing these functions and making any observations. Its connection to the complex of chemically related substances that build up in willow and other salicaceous plants comes first, followed by the way it works to treat certain human ailments.

2.1.2 Food Rich in Salicylates (Especially Indian Food) and Allowed Dietary Intake

As mentioned in Fig. 2.2, various foods including vegetables, fruits, beverages, spices, and herbs contain significant amount of SA.

A letter published in the Lancet in 1903 contained the first report on SA content of various foods. The salicylate content of conservative food items has been reported in a number of publications at highly different values. Foods containing structurally comparable phenolic acids may react to artifactually enhance absorbance, which would cause the salicylate amount to be overestimated. It could be the reason where similar food shows different amount of salicylate. In addition to this, poor

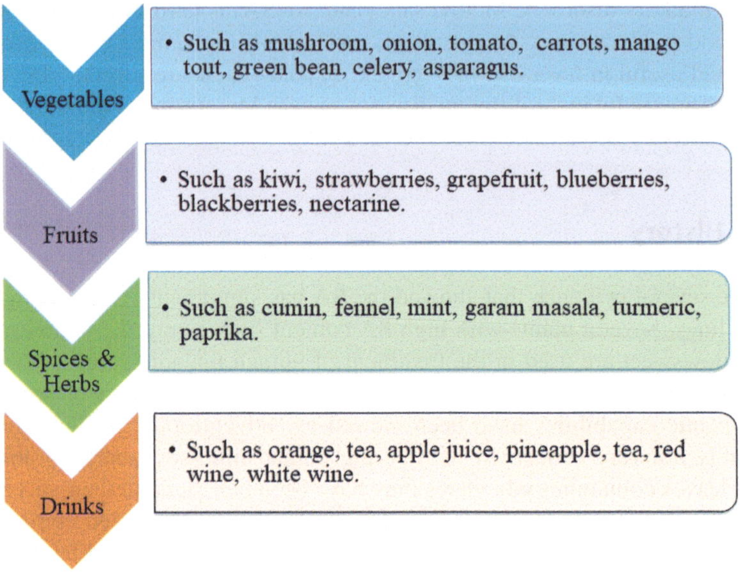

Fig. 2.2 Different foods rich in salicylic acid

estimation of compounds may happen depending on the extraction methodology. For instance, five different brands of orange juice varied in their salicylate level from 0.47 to 3.02 mg L^{-1}. Similar to other phytochemicals, a wide range of other factors, including as plant species, seasonality, growing circumstances, storage, and cooking, are likely to have an impact on the salicylate content of primary food products. SA is present in significant amount in plants as mentioned in Table 2.1 (Venema et al. 1996; Herrmann 1989).

2.1.3 Applications of Salicylic Acid in Agri-food and Pharma Sector

The early ripening of fruits and ethylene production have a direct relationship. In general appearance of ethylene release is considered as intimation of fruit ripening. SA treatment (0.8 mmol L^{-1}) strongly reduces the production of ethylene and delay fruit ripening. Moreover it also decreases the production of O_2 rate and respiration which enhances the shelf life of fruit (Mo et al. 2008). Aids in Strawberry's physical structure is exceedingly delicate, and its respiratory rhythm is non-climacteric. It is highly susceptible to fungal infections, particularly those caused by *Rhizopus stolonifera, Penicillium* sp., and *Botrytis cinerea*, and it has a high metabolic postharvest activity. In order to manage fungal diseases in postharvest fruits and vegetables, heat shock techniques have been employed. The production of antioxidant enzymes and enhancement of metabolism is the ultimate aim of heat shock treatments which involves the inactivation of oxygen radicals. This technology is now replaced by

Table 2.1 Sources of salicylic acid and its content

S. no.	Source	Concentration (mg/100 g)
1.	Asafoetida	3.8
2.	Garlic	5.6
3.	Ginger	3.5
4.	Fenugreek	0.1
5.	Allium sepa	0.80
6.	Apple	0.55
7.	Blackberries	0.81
8.	Green apple	0.55
9.	Gala melon	0.62
10.	Cinnamon	64.2
11.	Cloves	2.5
12.	Pepper seeds	9
13.	Tomato	0.13
14.	Onion	0.80
15.	Carrots	0.16
16.	Strawberry	0.61
17.	Black cumin	1629.4
18.	Cumin	29.76
19.	Asparagus	1.29
20.	Grapefruit	0.44

using SA as it is also an alternative to enhance the antioxidant enzymes and lowers the metabolism in fruits (Coltro et al. 2014). SA also reported to cause thinning of cell wall and decreased cytoplasmic matrix in *R. stolonifer.* Alike benzoic acid, SA also retards some metabolic pathways in microorganisms by inhibiting major enzymes in microbes and retard their growth which could be one of the possible strategies to keep food away from microbial attack (Kong et al. 2019).

2.1.4 Action Mechanism

SA is easily taken up from the dietary matrix by monocarboxylic acid transporters, H+ and Na+ transport systems. Many proteins involved in metabolism, transport, protein folding, and cytoskeletal control are modulated by SA. It is still unclear whether it has biological significance (Konishi et al. 2006). SA when exposed to cells in low concentrations regulates the expression of transcription factors that modulate oxidative stress, like antioxidant response factor. SA inhibits the production of prostaglandins that promote inflammation and cancer by changing how important transcription factors bind to the promoter region of genes (Wu et al. 1991). Additionally, salicylate at micromolar doses inhibits the transcriptional activation of additional genes implicated in inflammatory responses, like iNOS. This would also reduce the exposure of cells to capable detrimental peroxynitrite. That's how the inflammatory genes promoter can be suppressed through diet.

2.2 Function of Salicylic Acid in Food

Vital physiological processes, such as enhancing the plant's response to biotic and abiotic stress, boosting the plant's resistance to System Acquired Resistance, or both, SA is essential to the growth and development of the plant. In order to survive a range of stresses, these physiological activities are achieved by either increasing or modifying the internal endogenous signaling. SA can also create conjugate bonds with specific amino acids, such as proline and arginine, that strengthens the plant's tolerance to stress and preserves systemic acquired resistance. A wide range of important physiological functions are also performed by the SA, such as stimulating protein synthesis, postponing fruit ripening, germination of seeds, immunological response, nutrition transfer, CO_2 representation, ion absorption, flowering, and stomata movement (Wang et al. 2022). As opposed to ABA's work, which causes leaves to fall, it also stops ethylene gas from being represented. It also has a big impact on increasing metabolic rates, which help the plant conserve energy by taking different paths and adjusting the amount of amino and nucleic acids it produces. Moreover, SA and its derivative, aspirin, have been used for therapeutic and disease prevention, particularly in oxidative stress and inflammation. Plant-based foods' salicylates may contribute to disease prevention, and some argue "salicylic acid deficiency" should be considered a necessary vitamin (Duthie and Wood 2011).

Salicylates are emerging as health supporters rather than just plant defenders. Natural and synthetic phenolics target a variety of health conditions, including inflammation, cancer, and others. Salicylates are becoming more than just plant protectors; they are becoming beneficial for health. These adaptable phenolics, both natural and synthetic, work against a variety of illnesses and show promise in the fight against inflammation, neurological disorders, and cancer. The following section provides a brief overview of salicylates health benefits.

2.2.1 Application of Postharvest Salicylic Acid to Enhance Quantity and Quality of Agro-food Products

There are different methods of postharvest application of SA such as postharvest dipping, fumigating, and postharvest as mentioned below.

2.2.1.1 Postharvest Dipping

This method involves immersing fruits in a salicylate solution for a specific duration, leave them to naturally dry at room temperature. The efficacy of SA treatment on fruits depends on the interplay of concentration (ranging from 0.8 μmol L^{-1} to 5.0 mmol L^{-1}) and time (2 min–6 h), exhibiting variability across fruit varieties. Notably, fruits with inedible peels, such as bananas, chestnuts, and pineapples, exhibit greater tolerance for higher concentrations and prolonged treatment durations in comparison to fruits that can be consumed directly like peaches, grapes, and cherries.

2.2.1.2 Coating

Comparable to dipping, coating involves sub-merging fruits in a blend of SA and coating materials, with chitosan being a commonly utilized coating material (Ehteshami et al. 2020). Chitosan, when combined with SA, undergoes various treatments, such as mixing with a plasticizer like glycerol for coating fruits post-SA dipping. Another effective coating method entails applying a chitosan solution having SA to fruits using a brush with fine bristle. While application of salicylic acid (SA) and its derivatives after harvesting proves beneficial for fruit storage, challenges exist, including the limited water solubility of SA and the off-putting odor of chitosan coating, attributed to acetic acid, which can negatively impact consumer satisfaction.

2.2.1.3 Postharvest Fumigation

Postharvest fumigating primarily utilizes methyl salicylate (MeSA) due to its ready volatility. In this treatment, salicylate solutions are placed on filter paper in a container, facilitating quick filling of the container with volatile MeSA. Fruits are then enclosed within the container and left at room temperature for 12–24 h. MeSA fumigation has demonstrated effectiveness in enhancing the quality of storage fruits like pomegranates (Sayyari et al. 2011), sweet cherries (Giménez et al. 2016), and blood oranges (Habibi et al. 2020). Care must be taken to avoid direct contact between MeSA and fruit, and fumigation time should be meticulously controlled to prevent irreversible damage to the fruit.

The quality of fruit is intricately linked to its ripening and senescence processes during storage, encompassing factors such as weight loss, fruit softening, ethylene production, color alteration, and respiration rate (Gao et al. 2020). Numerous studies have highlighted the significant involvement of SA and its derivatives in regulating the fruit physiological metabolism to uphold and preserve fruit quality (Amiri et al. 2021; Giménez et al. 2016; Hanif et al. 2020; Valverde et al. 2015). In the following sections, impact of SA treatment on various biochemical and physiological and biochemical parameters of plant and plant-products is discussed in detail.

2.2.1.4 Ethylene Production and Respiration Rate

Ethylene, an essential plant hormone, has a substantial impact in the senescence and ripening of fruits. Respiratory metabolism, a vital physiological process influenced by ethylene, supplies the energy required for diverse metabolic functions. (Kumar et al. 2013; Zhang et al. 2020). The onset of ripening and senescence in climacteric fruits is marked during the peak of ethylene production and the onset of a respiration peak, a phenomenon similar to that observed in non-climacteric fruits occurs (Zhang et al. 2020). The application of SA has been shown to postpone delaying the increase in ethylene and inhibiting the respiration rate can impede fruit senescence in tomatoes (Kumar et al. 2013), grapes (Champa et al. 2015), asparagus (Wei et al. 2011), and plums (Luo et al. 2011).

The reduction in the expression levels of genes associated with SA induces a decline in ethylene production in tomatoes by affecting enzymes like 1-aminocyclo propane-1-carboxylic acid oxidase (ACO) and 1-aminocyclopropane-1-carboxylic acid synthase (ACS), which are involved in ethylene production (Kumar et al. 2013). SA can also impede ethylene synthesis by decreasing the enzyme activities of ACS and ACO, preventing the conversion of 1-aminocyclopropane-1-carboxylic acid (ACC) to ethylene (Kumar et al. 2013). Moreover, the reduction in the respiration rate influenced by SA and its derivatives is contingent on the suppression of ethylene synthesis and the closure of stomata (Champa et al. 2015). Additionally, Cooper et al. (2007) reported that the combined treatment of 1-methylcyclopropene (1-MCP) and MeSA proves more effective in inhibiting ethylene production than a single 1-MCP treatment.

2.2.1.5 Weight Loss

Fruits lose weight physiologically because of water loss from transpiration and continuous respiration even after harvest. SA has been proposed as a potential and successful treatment for the reduction of senescence-related enzyme activity (Hanif et al. 2020). By doing this, the pace at which fruit respires while being stored is decreased, hence reducing the physiological weight loss (PLW). Fruits that have lost water during storage shrink and wither, losing not just quality and look but also weight. Because SA inhibits respiration and the development of ethylene, postharvest administration of 2 mM of the acid proved successful in lowering the PLW in guava (Madhav et al. 2021). In another study, guavas treated with 200 ppm of SA were kept in cold storage; Kaur et al. (2019) found that there was less deterioration and less PLW in guava. Awad and Al-Qurashi (2021) suggested treating postharvest fruits with SA solution at concentrations of 2 mM or 0.2 mM was successful in lowering the PLW, which enhanced the quality of "Sensation" mangoes under ambient settings by preventing hydrolytic enzymes and boosting the fruit's antioxidant system.

In comparison to the untreated control group, the application of SA after harvest lowered the PWL in fruit (Hajilou and Fakhimrezaei 2013). SA prevented respiration by blocking the ethylene synthesis. SA may decrease PLW and slow down respiration by blocking stomata. At Tabriz University in Iran, the experiment was conducted on "Asgar-Abad," a well-known commercial cultivar of Iranian apricot. In mandarin orange budded on "Rough Lemon" rootstock, Haider et al. in 2020 found that postharvest administration of SA was considerably beneficial in minimizing the physiological loss in weight when compared to control (Haider et al. 2020). The University of Agriculture in Faisalabad, Pakistan, was the site of the study. Fruits coated with 2 mM SA had significantly lower peel-level wax (PLW) after 4, 8, 12, and 16 days of storage, according to research by Hazarika et al. (2019). It was also discovered that SA's closure of stomata reduces the weight loss and respiration rate of fruits. The study was conducted at Mizoram University in India's Department of Horticulture, Aromatic and Medicinal Plants. According to, SA treatments showed the least amount of weight loss after 40 days of storage (1.8%). Weight loss was significantly reduced by the SA treatments in comparison to the control group. The use of SA, MeSA, and acetylsalicylic acid (ASA) has been demonstrated to mitigate the weight loss of various fruits, including plums (Davarynejad et al. 2015), cherries (García-Pastor et al. 2020), oranges (Amiri et al. 2021), and pomegranates (García-Pastor et al. 2020). Additionally, Koyuncu et al. 2019 observed that the application of putrescine and oxalic acid (OA) significantly reduced the weight loss of pomegranates compared to a singular SA treatment.

2.2.1.6 Fruit Softening

The softening of fruit is accompanied by an accelerated ripening process, which is closely tied to the degradation of the cell wall. Fruit cell walls consist of pectin, cellulose, and lignin (Koyuncu et al. 2019). Pectin degradation leads to the breakdown of cellulose and hemicelluloses, a process regulated by cell wall enzymes, including polygalacturonase (PG), pectin methylesterase (PME), and cellulase (CEL) (Koyuncu et al. 2019). The synthesis of PG is triggered by ethylene, indicating a connection between cell wall degradation and ethylene production (Kumar et al. 2013). SA has the ability to postpone fruit softening by inhibiting the activities of PG, PME, and CEL, thereby preserving the integrity of the cell wall (Kumar et al. 2013; Valero et al. 2011). Additionally, SA can enhance the water-holding capacity of fruit by establishing a more crosslinked pectin network (Batool et al. 2022). Similarly, the applications of ASA and MeSA have been shown to delay the softening process in various fruits, including pomegranates (García-Pastor et al. 2020), sweet cherries (Giménez et al. 2016), tomatoes (Changwal et al. 2021), and apricots (Fan et al. 2021). Kaur et al. 2019 reported that postharvest application of SA was proven helpful in preserving higher firmness on guava. In guava, postharvest SA treatment efficiently maintains fruit firmness, while SA immersion delayed fruit softening because it controlled the development of soluble pectin and decrease of insoluble pectin, as indicated by Kaushik et al. (2021).

2.2.1.7 Effect of Salicylic Acid on Chilling Injury

According to a study by Madhav et al., postharvest administration of SA on the ripening phase helps to regulate the ethylene evolution and senescence further resulting in decreased chilling injury of the fruits and less generation of free radicals in guava. According to Mosa et al. (2021), guava treated with SA acquires chilling resistance through the activation of heat shock proteins. According to Sayyari, the best postharvest SA treatment for pomegranate ascorbic acid levels to be maintained and for minimizing chilling harm was 2 mM (Sayyari et al. 2009).

2.2.1.8 Color Change

The alteration of fruit skin color is a common occurrence during postharvest storage, primarily determined by the degradation of phytochrome (Batool et al. 2022; Koyuncu et al. 2019). SA and its derivatives have the capacity to impede the color transformation and flesh browning of fruit. Lycopene, a potent antioxidant, is closely associated with the color transition in ripening tomatoes (Amiri et al. 2021). SA can postpone chlorophyll degradation and the synthesis rate of lycopene by inhibiting ethylene and respiration rates, thereby preserving the color of tomato fruit (Kumar et al. 2013). The application of SA, ASA, and MeSA can restrain the color change of pomegranate fruit during storage (García-Pastor et al. 2020). Additionally, García-Pastor et al. (2020) suggested that the use of salicylates resulted in a deeper red color by increasing anthocyanin concentration, enhancing the profitability of pomegranates in the international market. Sayyari et al. (2009) in The Division of Food Science and Postharvest Technology, IARI, New Delhi, India, found that post-harvest application of SA or SSA considerably postponed the ripening process of guava fruits, most likely by inhibiting the production of ethylene. According to Lo'ay et al., fruit treated with SA at a concentration of 2 mM experiences a less decline in hue value across all fruit development stages (Lo'ay and Dawood 2019). The effects of SA, which inhibits ethylene biosynthesis and lowers respiration rate to delay fruit ripening, are responsible for the observed outcomes. When SA is applied after harvest, guava cultivars take longer to become yellow, which indicates a pause in their respiratory activities. The inhibitory effect SA on ethylene synthesis and the vegetable wax formulation's barrier qualities on gaseous exchange, which further resulted in reduced activity of enzymes like ACC synthase and ACC oxidase, were the main causes of the ripening process delay. The effects of SA treatment on guava fruits have also been documented, as have studies on wax-coated guava by Madhav et al. and Lo'ay et al.

2.2.1.9 Decay Rate

The occurrence of decay poses a significant challenge, arising from either fruit senescence or disease during storage. The postharvest application of SA has been demonstrated to mitigate decay incidence, thereby extending the shelf life of fruits such as oranges (Amiri et al. 2021), apricots (Batool et al. 2022), bananas (Xu et al. 2019), pomegranates (Koyuncu et al. 2019), and grapes (Gomes et al. 2021). Moreover, the combination of SA with acetic acid proves more effective in reducing decay incidence compared to SA treatment alone (Amiri et al. 2021).

2.2.1.10 Soluble Solid Content and Titratable Acid

Soluble solid content (SSC) serves as a crucial indicator of sweetness, making a significant contribution to consumer acceptability (Fan et al. 2021). The decline in SSC during storage primarily stems from respiration metabolism (Amiri et al. 2021; Hazarika and Marak 2022). Starch and other insoluble carbohydrates undergo hydrolysis into soluble sugars during fruit ripening, leading to an increase in the respiration rate and SSC (Davarynejad et al. 2015). Kumar et al. observed that SA treatment slowed the rise in SSC in tomato fruit, as SA reduced metabolic activity and hindered the hydrolysis of complex carbohydrates into SSC. Xu et al. (2019) noted that the combination of SA and 1-MCP effectively restrained the increase in SSC compared to 1-MCP alone in bananas.

Titratable acid is positively correlated with organic acid content, a major respiratory substrate (Koyuncu et al. 2019). The decrease in titratable acid during fruit storage results from the pyruvate decarboxylation reaction occurring in the ripening process (Hazarika and Marak 2022). The application of SA or MeSA has been reported to reduce the consumption of organic acid by suppressing the respiration rate and delaying the ripening process in certain fruits, such as oranges (Amiri et al. 2021), sweet cherries (Giménez et al. 2016), and apricots (Fan et al. 2021). However, lower acidity has also been observed in SA-treated grapes compared to the control

(Hazarika and Marak 2022). The SSC/acidity ratio is considered an indicator of fruit taste (Hazarika and Marak 2022). The increase in SSC/acidity during storage results from a decline in acidity and an increase in SSC in fruit (Champa et al. 2015). SA has the ability to suppress the deviation of SSC/acidity from its initial value, presumably due to decreased respiration and metabolic activities in grape fruit (Champa et al. 2015). However, Kumar et al. (2013) indicated a decrease in the SSC/acidity ratio in litchi fruit treated with SA.

2.2.1.11 Flavor

The flavor of fruit is a result of the intricate interplay between aroma, sweetness, and sourness (Yang et al. 2020). Fruit aroma, crucial for consumer acceptability and commercial value, is often lost during postharvest cold storage. Various volatile organic compounds (VOCs), including C6 aldehydes, terpenes, alcohols, esters, and lactones, contribute to fruit aroma (Amiri et al. 2021). Linoleic and linolenic acids can undergo conversion to C6 aldehydes through the actions of lipoxygenase (LOX) and hydroperoxide lyase (HPL), followed by transformation into C6 alcohols by alcohol dehydrogenase (ADH). Eventually, alcohol acyltransferase (AAT) catalyzes the formation of volatile esters (Wang et al. 2017; Yang et al. 2020). The presence of fruity-note volatile esters and lactones in peach fruit increased with the application of SA, enhancing the expression levels of PpLOX, PpHPL, PpADH, and PpAAT during cold storage (Yang et al. 2020). Wang et al. (2017) demonstrated that MeSA fumigation reduced C6 alcohol generation in tomato fruit, attributed to the down-regulated transcript level of ADH, followed by suppressed ADH activity in the metabolic pathway. However, e-nose analysis revealed no difference in fragrance profile between SA-treated fruit and the control. Further study is needed to confirm the effects of SA and its derivatives on volatiles.

The sweetness of fruit is contingent on the soluble sugars content, while sourness is often attributed to organic acids. The metabolism of sugars and organic acids is linked to glycolysis and the tricarboxylic acid cycle (Amiri et al. 2021). Key sugars in fruit, such as sucrose, glucose, fructose, and sorbitol, are regulated by enzymes like sucrose synthase (SS), neutral invertase (NINV), and sucrose phosphate synthase (SPS) (Amiri et al. 2021; Zhao et al. 2021). As illustrated in Fig. 2.2b, sorbitol can be converted to fructose through the action of sorbitol dehydrogenase (Amiri et al. 2021).

2.2.1.12 Impact of Salicylic Acid and Its Derivatives on Phenolic Compounds and Antioxidant Capacity of Fruit

The presence of reactive oxygen species (ROS), such as the hydroxyl radical, superoxide, and hydrogen peroxide, during fruit ripening poses a risk of oxidative damage to membrane lipids, leading to cell death and harm to fruit tissue (Champa et al. 2015). Phenolic compounds, known for their antioxidant activity, play a dual role by scavenging free radicals, reducing oxidative damage to fruits, and contributing to the maintenance of fruit flavor and quality. The application of SA and its derivatives has been observed to preserve phenolics and enhance antioxidant capacity, thereby reducing oxidative damage. This protective effect can extend the shelf life and postpone senescence in various fruits, including sweet cherry (Valverde et al. 2015; Zhang et al. 2020), pomegranate (García-Pastor et al. 2020), lemon (Amiri et al. 2021), and plum (Davarynejad et al. 2015).

Phenolic compounds, encompassing flavonoids, anthocyanins, and phenolic acids, are vital secondary metabolites in fruit (Oraei et al. 2019). Key flavonoids found in fruits include quercetin, rutin, epicatechin, and catechin, while primary phenolic acids consist of caffeic acid, chlorogenic acid, neochlorogenic acid, and gallic acid (Amiri et al. 2021). Flavonoids, particularly, play a significant role in the antioxidant activity of fruits, and the vibrant colors of grapes are attributed to anthocyanins synthesized through the flavonoid pathway (Devi et al. 2018; Gomes et al. 2021).

The biosynthesis of phenolics involves the phenylpropanoid pathway. Phenylalanine ammonialyase (PAL) initiates the conversion of phenylalanine to cinnamic acid, which is further transformed into phenolic acid (Oraei et al. 2019). Cinnamate 4-hydroxylase (C4H), 4-coumarate-coenzyme A ligase (4CL), and chalcone synthase (CHS) contribute to the conversion of cinnamic acid to chalcone (Amiri et al. 2021). Chalcone isomerase (CHI) catalyzes the formation of flavanone, which can be subsequently converted into flavone, flavonol, and anthocyanin. The activities of PAL, C4H, and 4CL can be augmented by SA, leading to an increase in phenolic content in fruits such as oranges (Amiri et al. 2021) and grapes (Blanch et al. 2020). Moreover, SA or MeSA has been observed to inhibit the degradation of anthocyanins caused by polyphenol oxidase (PPO) enzymes in blood oranges (Habibi et al. 2020).

The utilization of SA or MeSA has been documented to enhance the antioxidant capacities of apricots (Fan et al. 2021), asparagus (Wei et al. 2011), and grapes (Gomes et al. 2021) during storage. Following a 12-day storage period, SA demonstrated an increase in DPPH free radical scavenging capacity, ABTS. Generally, both enzymatic and non-enzymatic antioxidants, such as ascorbic acid and phenolic compounds, can counteract the detrimental effects of ROS on cell structure (Amiri et al. 2021; Gomes et al. 2021). ROS can be effectively scavenged by antioxidant enzymes, including superoxide dismutase (SOD), catalase (CAT), peroxidase (POD), glutathione reductase (GR), and ascorbate peroxidase (APX), activation of which can be induced by SA, MeSA, or ASA treatments. This leads to heightened antioxidant activity in sweet cherries (Giménez et al. 2016; Zhang et al. 2020), pomegranates (García-Pastor et al. 2020), citrus fruits (Zhou et al. 2018), papayas (Hanif et al. 2020), and bananas (Xu et al. 2019). Additionally, the browning of fruit peel, attributed to phenolic oxidation catalyzed by the PPO enzyme, can be inhibited by SA (Kumar et al. 2013). Moreover, ascorbic acid (AA) plays a crucial role in neutralizing radicals to impede oxidative reactions in plant tissues (Huang et al. 2017; Sangprayoon et al. 2019). SA application can enhance antioxidant ability by inhibiting ascorbic acid oxidase (AAO) enzymes, positively influencing the ascorbate-glutathione cycle, resulting in elevated AA content in orange fruits (Amiri et al. 2021; Hanif et al. 2020; Wei et al. 2011). Other reports have indicated that SA, MeSA, or ASA treatments can lead to a high content of dehydroascorbic acid (DHA) in fruit, oxidized from AA by AAO in pomegranate fruit (García-Pastor et al. 2020; Hazarika and Marak 2022). Several studies have shown a positive correlation between total phenolic compounds (TPC) and total antioxidant activity (TAA) (Amiri et al. 2021; Wei et al. 2011). MeSA treatment has been observed to decrease the lipophilic antioxidant capacity of apricots by reducing carotenoid content and increase hydrophilic antioxidant capacity by elevating phenolic content, resulting in high TAA (Amiri et al. 2021).

According to research done by Lo'ay et al., using SA postharvest was successful in slightly lowering the total phenol content to inhibit polyphenol oxidase (PPO) and improve fruit color quality during ripening (2011). Postharvest application of SA found to reduce total phenolic content of guava during storage, according to research by Adhav et al. Fruits treated with SA may lose less phenols since PPO takes longer to oxidize phenolic compounds. Khademi and Ershadi (2013) previously observed similar results in peach. The phenolic content of apricots treated with 3 mM SA decreased after storage (Hajilou and Fakhimrezaei 2013). SA treatment resulted in the creation of all important phenolics, the synthesis of new polyphenols, and the stimulation of phenylalanine ammonia lyase activity in apricot. The prominent commercial cultivar of Iranian apricot known as "Asgar-Abad" was the subject of a study carried out at Tabriz University. In 2013 Khademi et al. discovered that fruits treated with SA at high concentrations of 2 and 4 mM had the greatest levels of TPC when compared to controls (Khademi and Ershadi 2013).

Fruits treated with SA at 1 mM were found to contain more TPC than controls during the first 2 weeks of storage, but after that point there was no noticeable change. Following storage, total phenols exhibited a declining trend, with the SA-treated fruits seeing a notably faster rate of drop than the control ones (SA at a concentration of 4 mM). The loss of phenolic compounds during storage has been connected to numerous enzymatic and non-enzymatic activities. Peroxidase (POD) and polyphenoloxidase are the two primary enzymes that oxidize phenolic substances (PPO). According to Tareen et al. (2012), SA applied after harvest significantly contributed to the preservation of peach fruits' increased ascorbic acid content. The study was carried out in the Postharvest Laboratory of the Department of Horticulture, Pir Mehar Ali Shah, Arid Agriculture University Rawalpindi, Pakistan, on peach (Prunus persica L. Batsch) fruits of cv. "Flordaking" engrafted on wild rootstock (Tareen et al. 2012).

2.2.2 Biological Activity/Health Benefits of Salicyclic Acid and Food Rich in Salicyclic Acid

As a pharmacological agent, salicylates are crucial to human health. Aspirin (acetyl salicyclic acid) is the most extensively used and recognized salicylate. Apart from its antipyretic and analgesic properties, it can also help in prevention of strokes, heart attacks, and even certain type of cancer (Fig. 2.3). This wonder drug also boasts anti-platelet, anti-diabetic, anti-inflammatory and neuroprotective properties (Anthony et al. 2000; Thun et al. 2012; Shirakawa et al. 2016). Combining aspirin with a salicylate-rich diet can further enhance its power, potentially preventing colorectal cancer and certain types of inflammatory diseases (Paterson and Lawrence 2001). Colorectal cancers have a high level of cyclooxygenase-2 (COX-2) expression. Cyclooxygenase-1 (COX-1) and cyclooxygenase-2 (COX-2) are key enzymes in the arachidonic acid cascade, catalyzing the conversion of arachidonic acid to prostaglandin H2 (PGH2), a precursor to various prostaglandins and thromboxanes. PGH2 plays a pivotal role in mediating inflammatory responses (Eberhart et al. 1994) and platelet aggregation leading to thrombosis (Preston et al. 1981; Smith et al. 1996). Aspirin exerts its anti-inflammatory and antithrombotic effects by irreversibly inhibiting both COX-1 and COX-2, thereby reducing the biosynthesis of PGH2 and downstream prostanoid mediators. This mechanism underlies the diverse clinical benefits of aspirin, including its use in preventing cardiovascular events and colorectal cancer (Kopp and Ghosh 1994).

Aspirin's therapeutic repertoire encompasses a variety of protein targets in addition to its well-established job of inhibiting cyclooxygenases (COX), indicating a potentially wider range of effects on human health. Studies have identified interactions with ribosomal S6 kinase 2 (RSK2), implicated in cell proliferation and potentially contributing to aspirin's anti-cancer properties (Ostrowska 1996). Additionally, aspirin's inhibition of cathepsin A (CTSA) might explain its anti-inflammatory effects, given CTSA's role in inflammation and tissue breakdown (Stevenson et al. 1999). Furthermore, modulation of nuclear factor-kappa B (NF-kB), a pivotal player in the immune system and inflammation, by aspirin could contribute to its anti-inflammatory and anti-cancer activities (Dachineni et al. 2016). The interaction with cyclin-dependent kinase 2 (CDK2), involved in cell cycle regulation, suggests potential anti-cancer effects of aspirin (Ostrowska 1996). Salicylates, both natural and synthetic, are beneficial for a range of health issues because of their capacity to target specific human proteins. Certain naturally occurring salicylates, such as amorfrutin, grifolic acid, and ginkgolic acid, have pharmacological roles in human disease and multitarget inhibitory activities. Amorfrutins are naturally occurring salicylates that are extracted from *Glycyrrhiza foetida* (Isah 2015) and *Amorpha*

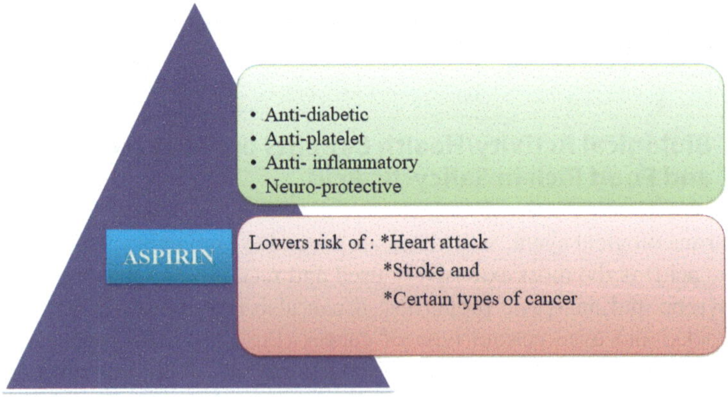

Fig. 2.3 Role of aspirin in human health

fruticosa (Mitscher et al. 1981). The naturally occurring salicylate grifolic acid is derived from grifolin, which is isolated from the fungus *Albatrellus confluens'* fruiting bodies (Zhao et al. 2018). Grifolin induces apoptosis and stops cell growth, which exhibit anticancer activity in various cancer types. Ginkgolic acid, also known as 2-hydroxy-6-alkylbenzoic acid, is found in Ginkgo biloba leaves and fruits (Weidner et al. 2016). Because of their anticancer qualities and other pharmacological roles, ginkgolic acids may be a viable target for drug development in human disease (Ye et al. 2005; Jin et al. 2007; Luo et al. 2015; Wu and Li 2017).

Due to the presence of salicylic acid (SA), bismuth subsalicylate (SA and a bismuth salt) has anti-inflammatory properties and relieves stomach pain by acting as an antacid functions as an antiseptic. Like other hydroxyl acids, SA is a crucial component of many skincare products. It can be used to treat keratosis pilaris, calluses, and acanthosis nigricans (Madan and Levitt 2014). It can also prevent dandruff, ichthyosis, psoriasis, ringworm, warts, and corns. Due to its ability to exfoliate the stratum corneum, it has also been used for many years to treat various skin conditions. In treatment of conditions like melasma, freckles, photodamage, and lentigines, SA possesses exfoliating properties (Arif 2015).

For thousands of years, salicylates (SA and its derivative) have been utilized in medicine. The Ebers papyrus (an Egyptian compilation of therapeutic remedies) specifies a dried myrtle leaves tea for back problems and rheumatism. Salicylate-rich plants like medeowsweet, willow, and poplar were used for their medicinal properties long before salicylates were identified. Hippocrates of Cos who lived in the first century BC suggested an extract of willow bark as a remedy because of its analgesic properties. The corpus Hippocratian, written around 300 BC, mentions using this extract to ease the pain of childbirth and to alleviate fever (Vlot et al. 2009). In the early 1800s, salicylic acid, the active component, was extracted and refined from willow and meadowsweet. Dr. Thomas John MacLaggan, a Scottish physician, published research in 1876 that provided the first definitive evidence of salicylic acid's analgesic, antipyretic, and anti-inflammatory properties. Moreover, salicylates can be found in a variety of foods, including fruits, vegetables and drinks and herbs, and spices (Robertson and Kermode 1982; Venema et al. 1996).

2.3 Conclusion

SA transcends its initial guise as a mere aspirin, weaving a complex tapestry at the intersection of food, science, and well-being. It presents a compelling picture of a molecule with much to offer due to its impact on fruit quality, hidden health benefits, and widespread presence in a variety of food sources. Moreover, we gain a deeper understanding of the delicate balance between nature and human intervention as we continue to peel back its layers, inspiring further exploration and responsible use of this multifaceted wonder.

References

Amiri S, Nicknam Z, Radi M, Sayadi M, Bagheri F, Karimi Khorrami N, Abedi E (2021) Postharvest quality of orange fruit as influenced by salicylic acid, acetic acid, and carboxymethyl cellulose coating. J Food Meas Charact 15(5):3912–3930

Anthony JC, Breitner JCS, Zandi PP, Meyer MR, Jurasova I, Norton MC, Stone SV (2000) Reduced prevalence of AD in users of NSAIDs and H2 receptor antagonists: the Cache County study. Neurology 54:2066–2071

Arif T (2015) Salicylic acid as a peeling agent: a comprehensive review. Clin Cosmet Investig Dermatol 8:455–461

Awad MA, Al-Qurashi AD (2021) Postharvest salicylic acid and melatonin dipping delay ripening and improve quality of 'sensation' mangoes. Philipp Agric Sci 104:34–44

Batool M, Bashir O, Amin T, Wani SM, Masoodi FA, Jan N, Bhat SA, Gul A (2022) Effect of oxalic acid and salicylic acid treatments on the post-harvest life of temperate grown apricot varieties (Prunus armeniaca) during controlled atmosphere storage. Food Sci Technol Int 28(7):557–569

Blanch GP, Gómez-Jiménez MC, del Castillo MLR (2020) Exogenous salicylic acid improves phenolic content and antioxidant activity in table grapes. Plant Foods Hum Nutr 75(2):177–183

Champa WAH, Gill MIS, Mahajan BVC, Arora NK (2015) Preharvest salicylic acid treatments to improve quality and postharvest life of table grapes (Vitis vinifera L.) cv. flame seedless. J Food Sci 52(6):3607–3616

Changwal C, Shukla T, Hussain Z, Singh N, Kar A, Singh VP, Abdin MZ, Arora A (2021) Regulation of postharvest tomato fruit ripening by endogenous salicylic acid. Front Plant Sci 12:663943. https://doi.org/10.3389/fpls.2021.663943. PMID: 34163503; PMCID: PMC8216237

Coltro S, Broetto L, Rotilli MCC, de Moraes AJ, Barp FK, Braga GC (2014) Heat shock and salicylic acid on postharvest preservation of organic strawberries. Rev Ceres 61:306–312

Cooper D, Doucet L, Pratt M (2007) Understanding in multinational organizations. J Organ Behav 28(3):303–325

Dachineni R, Guoqiang A, Kumar DR, Sadhu SS, Tummala H, Bhat GJ (2016) Cyclin A2 and CDK2 as novel targets of aspirin and salicylic acid: a potential role in cancer prevention. Mol Cancer Res 14:241–252

Davarynejad GH, Zarei M, Nasrabadi ME, Ardakani E (2015) Effects of salicylic acid and putrescine on storability, quality attributes and antioxidant activity of plum cv. 'Santa Rosa'. J Food Sci 52(4):2053–2062. https://doi.org/10.1007/s13197-013-1232-3

Devi J, Bhatia S, Alam MS, Dhillon TS (2018) Effect of calcium and salicylic acid on quality retention in relation to antioxidative enzymes in radish stored under refrigerated conditions. J Food Sci 55(3):1116–1126

Duthie GG, Wood AD (2011) Natural salicylates: foods, functions and disease prevention. In: Food and function, vol. 2(9), pp 515–520. https://doi.org/10.1039/c1fo10128e

Eberhart CE, Coffey RJ, Radhika A, Giardiello FM, Ferrenbach S, DuBois RN (1994) Up-regulation of cyclooxygenase 2 gene expression in human colorectal adenomas and adenocarcinomas. Gastroenterology 107(4):1183–1188

Ehteshami S, Abdollahi F, Ramezanian A, Mirzaalian Dastjerdi A (2020) Maintenance of quality and bioactive compounds in pomegranate fruit (Punica granatum L.) by combined application of organic acids and chitosan edible coating. J Food Biochem 44(9):1–11

Ekinci D, Şentürk M, Küfrevioğlu Öİ (2011) Salicylic acid derivatives: synthesis, features and usage as therapeutic tools. Expert Opin Ther Pat 21(12):1831–1841

Fan X, Du Z, Cui X, Ji W, Ma J, Li X, Wang X, Zhao H, Liu B, Guo F, Gong H (2021) Preharvest methyl salicylate treatment enhance the chilling tolerance and improve the postharvest quality of apricot during low temperature storage. Postharvest Biol Technol 177:111535

Gao J, Zhang Y, Li Z, Liu M (2020) Role of ethylene response factors (ERFs) in fruit ripening. Food Qual Saf 4(1):15–20

García-Pastor ME, Giménez MJ, Zapata PJ, Guillén F, Valverde JM, Serrano M, Valero D (2020) Preharvest application of methyl salicylate, acetyl salicylic acid and salicylic acid alleviated disease caused by Botrytis cinerea through stimulation of antioxidant system in table grapes. Int J Food Microbiol 334:108807

Giménez MJ, Valverde JM, Valero D, Zapata PJ, Castillo S, M. (2016) Serrano postharvest methyl salicylate treatments delay ripening and maintain quality attributes and antioxidant compounds of "Early Lory" sweet cherry. Postharvest Biol Technol 117:102–109

Gomes EP, Vanz Borges C, Monteiro GC, Filiol Belin MA, Minatel IO, Pimentel Junior A, Tecchio MA, Lima GPP (2021) Preharvest salicylic acid treatments improve phenolic compounds and biogenic amines in 'Niagara Rosada' table grape. Postharvest Biol Technol 176:111505

Habibi F, Ramezanian A, Guillén F, Serrano M, Valero D (2020) Blood oranges maintain bioactive compounds and nutritional quality by postharvest treatments with γ-aminobutyric acid, methyl jasmonate or methyl salicylate during cold storage. Food Chem 306:125634

Haider SA et al (2020) Effects of salicylic acid on postharvest fruit quality of "kinnow" mandarin under cold storage. Sci Hortic 259:108–843

Hajilou J, Fakhimrezaei S (2013) Effects of post-harvest calcium chloride or salicylic acid treatments on the shelf-life and quality of apricot fruit. J Hortic Sci Biotechnol 88(5):600–601

Hanif A, Ahmad S, Shahzad S, Liaquat M, Anwar R (2020) Postharvest application of salicylic acid reduced decay and enhanced storage life of papaya fruit during cold storage. J Food Sci 14(6):3078–3088

Hazarika TK, Marak T (2022) Salicylic acid and oxalic acid in enhancing the quality and extending the shelf life of grape cv. Thompson seedless. Food Sci Technol Int 28(6):463–475

Hazarika TK, Lalrinfeli L, Lalchhanmawia J, Mandal D (2019) Alteration of quality attributes and shelf-life in strawberry (Fragaria × ananassa) fruits during storage as influenced by edible coatings. Indian J Agric Sci 89(1):28–34

Herrmann K (1989) Occurrence and content of hydroxycinnamic and hydroxybenzoic acid compounds in foods. Crit Rev Food Sci Nutr 28(4):315–347

Huang Z, Li J, Zhang J, Gao Y, Hui G (2017) Physicochemical properties enhancement of Chinese kiwi fruit (Actinidia chinensis Planch) via chitosan coating enriched with salicylic acid treatment. J Food Sci 11(1):184–191

Isah T (2015) Rethinking Ginkgo biloba L.: medicinal uses and conservation. Pharmacogn Rev 9:140–148

Jack DB (1997) One hundred years of aspirin. Lancet 350(9075):437–439

Jin S, Pang RP, Shen JN, Huang G, Wang J, Zhou JG (2007) Grifolin induces apoptosis via inhibition of PI3K/AKT signalling pathway in human osteosarcoma cells. Apoptosis 12:1317–1326

Kaur M, Singh G, Akshi M (2019) Post-harvest treatment for improvement of shelf life in guava (Psidium guajava L.) cultivar 'Allahabad safeda' under ambient storage condition. Plant Arch 19:3005–3010

Kaushik L, Kumar R, Reddy DK, Kaushik P (2021) Effect of pre-harvest calcium chloride and salicylic acid spray on morphological and biochemical traits of guava (Psidium guajava). https://doi.org/10.1101/2021.03.04.433865

Khademi Z, Ershadi A (2013) Postharvest application of salicylic acid improves storability of peach (Prunuspersicacv elberta) fruits. Int J Agric Crop Sci 5(6):651

Kolbe H (1860) Ueber Synthese der Salicylsäure. Justus Liebigs Ann Chem 113(1):125–127

Kong J, Zhang Y, Ju J, Xie Y, Guo Y, Cheng Y, Qian H, Quek SY, Yao W (2019) Antifungal effects of thymol and salicylic acid on cell membrane and mitochondria of Rhizopus stolonifer and their application in postharvest preservation of tomatoes. Food Chem 285:380–388

Konishi Y, Zhao Z, Shimizu M (2006) Phenolic acids are absorbed from the rat stomach with different absorption rates. J Agric Food Chem 54(20):7539–7543

Kopp E, Ghosh S (1994) Inhibition of NF-kappa B by sodium salicylate and aspirin. Science 265:956–959

Koyuncu MA, Erbas D, Onursal CE, Secmen T, Guneyli A, Uzumcu SS (2019) Postharvest treatments of salicylic acid, oxalic acid and putrescine influences bioactive compounds and quality of pomegranate during controlled atmosphere storage. J Food Sci Technol 56(1):350–359

Kumar D, Mishra DS, Chakraborty B, Kumar P (2013) Pericarp browning and quality management of litchi fruit by antioxidants and salicylic acid during ambient storage. J Food Sci Technol 50(4):797–802

Lo'ay AA, Dawood HD (2019) Chilling injury, fruit color maturity stages, and antioxidant enzyme activities of lemon "baladi CV" fruits under cold storage stress. Sci Hortic 257:108676. https://doi.org/10.1016/j.scienta.2019.108676

Luo Z, Chen C, Xie J (2011) Effect of salicylic acid treatment on alleviating postharvest chilling injury of "Qingnai" plum fruit. Postharvest Biol Technol 62(2):115–120

Luo X, Yang L, Xiao L, Xia X, Dong X, Zhong J, Liu Y, Li N, Chen L, Hongde L (2015) Grifolin directly targets ERK1/2 to epigenetically suppress cancer cell metastasis. Oncotarget 6:42704–42716

Madan RK, Levitt J (2014) A review of toxicity from topical salicylic acid preparations. J Am Acad Dermatol 70(4):788–792

Madhav JV, Sethi S, Sharma RR, Nagaraja A, Arora A, Varghese E (2021) Influence of bilayer coating of salicylic acid and edible wax on chilling injury and functional attributes of guava. J Food Process Preserv 45(7):1–8

Mitscher LA, Park YH, AlShamma A, Hudson P, Haas T (1981) Amorfrutin A and B, bibenzyl antimicrobial agents from *Amorpha fruticosa*. Phytochemistry 20:781–785

Mo Y, Gong D, Liang G, Han R, Xie J, Li W (2008) Enhanced preservation effects of sugar apple fruits by salicylic acid treatment during post-harvest storage. J Sci Food Agric 88(15):2693–2699

Mosa AWF, Sas-Paszt L, Górnik K, Ali HM, Salem MZ (2021) Biostimulants & growth. In: Bioresources 16(4)

Oraei M, Panahirad S, Zaare-Nahandi F, Gohari G (2019) Pre-véraison treatment of salicylic acid to enhance anthocyanin content of grape (Vitis vinifera L.) berries. J Sci Food Agric 99(13):5946–5952

Ostrowska H (1996) Inhibition of human platelet cathepsin A by non-steroidal anti-inflammatory drugs-in vitro study. Pol J Pharmacol 48:113–116

Paterson JR, Lawrence JR (2001) Salicylic acid: a link between aspirin, diet and the prevention of colorectal cancer. Q J Med 94:445–448

Preston FE, Whipps S, Jackson CA, French AJ, Wyld PJ, Stoddard CJ (1981) Inhibition of prostacyclin and platelet thromboxane A2 after low-dose aspirin. N Engl J Med 304(2):76–79

Robertson GL, Kermode WJ (1982) J Sci Food Agric 32:833

Sangprayoon P, Supapvanich S, Youryon P, Wongs-Aree C, Boonyaritthongchai P (2019) Efficiency of salicylic acid or methyl jasmonate immersions on internal browning alleviation and physicochemical quality of Queen pineapple cv. "Sawi" fruit during cold storage. J Food Biochem 43(12):1–11

Sayyari M, Babalar M, Kalantari S, Serrano M, Valero D (2009) Effect of salicylic acid treatment on reducing chilling injury in stored pomegranates. Postharvest Biol Technol 53(3):152–154

Sayyari M, Babalar M, Kalantari S, Martínez-Romero D, Guillén F, Serrano M, Valero D (2011) Vapour treatments with methyl salicylate or methyl jasmonate alleviated chilling injury and enhanced antioxidant potential during postharvest storage of pomegranates. Food Chem 124(3):964–970

Shirakawa K, Wang L, Man N, Maksimoska J, Sorum AW, Lim HW, Lee IS, Shimazu T, Newman JC, Schroder S, Melanie O, Marmorstein R, Meier J, Nimer S, Verdin E (2016) Salicylate, diflunisal and their metabolites inhibit CBP/p300 and exhibit anticancer activity. Elife 5:e11156

Smith WL, Garavito RM, DeWitt DL (1996) Prostaglandin endoperoxide H synthases (cyclooxygenases)-1 and-2. J Biol Chem 271(52):33157–33160

Stevenson MA, Zhao MJ, Asea A, Coleman CN, Calderwood SK (1999) Salicylic acid and aspirin inhibit the activity of RSK2 kinase and repress RSK2-dependent transcription of cyclic AMP response element binding protein- and NF-kappa B-responsive genes. J Immunol 163:5608–5616

Tareen MJ, Abbasi NA, Hafiz IA (2012) Effect of salicylic acid treatments on storage life of peach fruits cv. 'flordaking'. Pak J Bot 44(1):119–124

Thun MJ, Jacobs EJ, Patrono C (2012) The role of aspirin in cancer prevention. Nat Rev Clin Oncol 9:259–267

Valero D, Díaz-Mula HM, Zapata PJ, Castillo S, Guillén F, Martínez-Romero D, Serrano M (2011) Postharvest treatments with salicylic acid, acetylsalicylic acid or oxalic acid delayed ripening and enhanced bioactive compounds and antioxidant capacity in Sweet cherry. J Agric Food Chem 59(10):5483–5489

Valverde JM, Giménez MJ, Guillén F, Valero D, Martínez-Romero D, Serrano M (2015) Methyl salicylate treatments of sweet cherry trees increase antioxidant systems in fruit at harvest and during storage. Postharvest Biol Technol 109:106–113

Venema DP, Hollman PCH, Janssen PLTM, Katan MB (1996) J Agric Food Chem 44:1762

Vlot AC, Dempsey DA, Klessig DF (2009) Salicylic acid, a multifaceted hormone to combat disease. Annu Rev Phytopathol 47:177–206

Wang L, Li X, Bai J, Luo H, Jin C, Hui J, Yu Z (2017) Residual impact of methyl salicylate fumigation at the breaker stage on C6 volatile biopathway in red tomato fruit. J Food Process Preserv 41(6):1–8

Wang J, Lewis D, Shi R, McGhie T, Wang L (2022) The colour variations of flowers in wild Paeonia delavayi plants are determined by four classes of plant pigments. N Z J Crop Hortic Sci 50(1):69–84

Wei Y, Liu Z, Su Y, Liu D, Ye X (2011) Effect of salicylic acid treatment on postharvest quality, antioxidant activities, and free polyamines of Asparagus. J Food Sci 76(2):126–132

Weidner M, Rousseau RJ, Micikas C, Fischer A, Plauth SJ, Wowro K, Siems G, Hetterling M, Kliem FC, Schroeder S, Sauer S (2016) Amorfrutin C induces apoptosis and inhibits proliferation in colon cancer cells through targeting mitochondria. J Nat Prod 79:2–12

Wu Z, Li Y (2017) Grifolin exhibits anti-cancer activity by inhibiting the development and invasion of gastric tumor cells. Oncotarget 8:21454–21460

Wu KK, Sanduja R, Tsai A-L, Ferhanoglu B, Loose-Mitchell DS (1991) Aspirin inhibits interleukin 1-induced prostaglandin H synthase expression in cultured endothelial cells. Proc Natl Acad Sci U S A 88(6):2384–2387

Xu F, Liu Y, Xu J, Fu L (2019) Influence of 1-methylcyclopropene (1-MCP) combined with salicylic acid (SA) treatment on the postharvest physiology and quality of bananas. J Food Process Preserv 43(3):1–7

Yang C, Duan W, Xie K, Ren C, Zhu C, Chen K, Zhang B (2020) Effect of salicylic acid treatment on sensory quality, flavor-related chemicals and gene expression in peach fruit after cold storage. Postharvest Biol Technol 161:111089

Ye M, Liu JK, Lu ZX, Zhao Y, Liu SF, Li L, Tan M, Weng XX, Li W, Cao Y (2005) Grifolin, a potential antitumor natural product from the mushroom *Albatrellus confluens*, inhibits tumor cell growth by inducing apoptosis in vitro. FEBS Lett 579:3437–3443

Zhang W, Cao J, Fan X, Jiang W (2020) Applications of nitric oxide and melatonin in improving postharvest fruit quality and the separate and crosstalk biochemical mechanisms. Trends Food Sci Technol 99:531–541

Zhao YF, Jiang F, Liang XY, Wei LL, Zhao YY, Ma Q, Hu YS, Su X (2018) Grifolic acid causes osteosarcoma cell death in vitro and in tumor-bearing mice. Biomed Pharmacother 103:1035–1042

Zhao Y, Song C, Brummell DA, Qi S, Lin Q, Bi J, Duan Y (2021) Salicylic acid treatment mitigates chilling injury in peach fruit by regulation of sucrose metabolism and soluble sugar content. Food Chem 358:129867

Zhou Y, Ma J, Xie J, Deng L, Yao S, Zeng K (2018) Transcriptomic and biochemical analysis of highlighted induction of phenylpropanoid pathway metabolism of citrus fruit in response to salicylic acid, Pichia membranaefaciens and oligochitosan. Postharvest Biol Technol 142:81–92

Nitric Oxide: A Key Bioactive Regulator of Plant Tolerance Mechanism Under Metal Induced Oxidative Stress

3

Lata Sharma, Suman Parihar, and Gyan Singh Shekhawat

Abstract

The redox-related signalling molecule nitric oxide (NO) is considered a crucial regulator of various physiological mechanisms and response to abiotic stress in plants. NO is produced by various pathways, such as oxidative, reductive and non enzymatic, in plants. It has a significant role in plant tolerance mechanisms against heavy metal (HM) stress that is most dominant abiotic stresses that affects growth and production. The common attribute of heavy metal toxicity is the increased production and accumulation of ROS (reactive oxygen species) that causes oxidative stress. These ROS disturb the normal cellular functioning by interacting with and damaging essential components of plant cell, such as lipids, protein, enzymes, and DNA. NO regulates the antioxidant enzymes activity and transcription factors (TFs) of genes regarded to stress tolerance, especially through post-transcriptional modifications (PTMs) such as S-nitrosylation and tyr-nitration. NO activates MAP-kinase and Ca^{2+}-dependent signalling to induce HM stress tolerance mechanisms. Over the past 20 years, nitric oxide has received a lot of attention as a molecule that can reduce the toxicity of heavy metals through both endogenous production and exogenous delivery. This chapter addressed various aspects of NO, such as biosynthesis, interactions with other molecules, signalling, and its participation in the alleviation of heavy metal toxicity.

Keywords

Nitric oxide · Heavy metal toxicity · Signalling · Reactive oxygen species · S-nitrosylation · Antioxidant enzymes · MAP-kinase

L. Sharma · S. Parihar · G. S. Shekhawat (✉)
Plant Biotechnology and Molecular Biology Laboratory, Department of Botany (UGC-CAS), Jai Narain Vyas University, Jodhpur, Rajasthan, India

© The Author(s), under exclusive license to Springer Nature Singapore Pte Ltd. 2024

M. Faizan, S. Hayat (eds.), *Plant Growth Regulators: Resilience for Sustainable Agriculture*, https://doi.org/10.1007/978-981-97-2918-0_3

Abbreviations

bZIP	Basic leucine zipper
cADPR	Cyclic ADP-ribose
GSH	Gluthathione
GSNO	S-nitrosoglutathione
HM	Heavy metal
MAPK	Mitogen-activated protein kinase
MYB	Myeloblastosis protein
NO	Nitric oxide
NOS	Nitric oxide synthase
NR/NiR	Nitrate/ Nitrite reductase
PTMs	Post transcription modification
RNS	Reactive nitrogen species
ROS	dReactive oxygen species
TF	Transcriptional factors

3.1 Introduction

Initially, Joseph Priestley described nitric oxide (NO) as a highly poisonous compound in 1772, as it is part of the pollutant gases released by automobile exhaust, industrial waste and cigarette smoke (Saxena and Shekhawat 2013). Later, nitric oxide was discovered to be a crucial signalling molecule that regulates the molecular, biochemical, and physiological responses of both stressed and non-stressed plants and animals. Klepper made the initial discovery of it in plants (Khan et al. 2023b). However, the evaluation of NO in plants was accelerated after the identification of its vital roles in mammals, such as (1) NO's role as endothelium relaxing factor in the cardiovascular system, (2) NO's role as an important cytotoxical agent of immune system, and (3) NO's role as a nervous system signalling molecule (Crane et al. 2010; Tewari et al. 2021). In plants, NO plays vital functions, namely growth and development, seed germination, formation of roots, stomata closure, programmed cell death, and response against environmental stresses (Khator et al. 2021).

Gaseous NO is a bio-active, lipophilic, diffusive free radical and ubiquitous signalling molecule. It plays dual functions: cytoprotectant and cytotoxic, mainly depends on its concentration alone with its localization in cells of plant. It participates in signal transduction pathways at nano-molar concentrations (1.0 nmol L^{-1}) (Tewari et al. 2021). NO can exist in three interchangeable redox active forms, such as nitroxyl radical (NO·), nitroxyl anion (NO^{-}) and nitrosonium cation (NO^{+}). It is a small diatomic molecule that can migrate easily from both the hydrophilic and hydrophobic regions of a cell; these characteristics make it an excellent intracellular and intercellular messenger (Khan et al. 2023b). In plants, NO regulates fundamental cellular processes through paracrine signalling manner (Wendehenne et al. 2001). NO can indirectly react with nucleic acids, lipids and proteins to produce its derivatives, such as GSNO (S-nitrosoglutathione), $ONOO^-$ (peroxynitrite), SNOs (S-nitrosothiols) and NO_2 (nitrogen dioxide), collectively known as RNS (reactive nitrogen species) (Corpas et al. 2011). S-nitrosoglutathione is produced by the S-nitrosylation reaction of nitric oxide with gluthathione, which has been documented to serve as a mobile NO reservoir in plants (Foster et al. 2009; Airaki et al. 2011). While peroxynitrite ($ONOO^-$) is produced through the reaction of NO with the superoxide anion (O_2^-), NO and its derivatives are involved in several interactions, such as (1) in redox-based post-translational modifications (PTM), (2) nitration of tyrosine and unsaturated fatty acids, (3) binding with metal centers of proteins, and (4) S-nitrosylation of thiol (–SH) groups (Arora et al. 2016). Furthermore, GSNO mediates the signalling via specific post-translational modification (PTM) of proteins which are redox-sensitive; in this modification S-nitrothiol is formed through a reaction of trans-nitrosylation from GSNO to Cys-NO. According to various studies, NO acts as an antioxidant, chain breaker, or potential inhibitor of ROS under stress induced by heavy metal (Khator et al. 2021).

Plants constantly cope with various stresses, including heavy metals, temperature (low and high), drought, salinity, etc., that primarily interrupt their growth and consequently exacerbates the world's food insecurity (Khan et al. 2021). All abiotic stresses produce reactive oxygen species, which causes oxidative stress in plants (Arora et al. 2016). The accumulation of ROS molecules is responsible for the peroxidation of lipid molecules, the inactivation of proteins, and the degradation of nucleic acids. In addition of ROS generation and accumulation, abiotic stress factors also disrupt the antioxidant defense system, which also leads to oxidative damage (Nabi et al. 2019). Moreover, different physiological and metabolic pathways, including photosynthesis, respiration and protein synthesis, are affected. Plants use different signalling processes and molecules, like NO, to combat the ROS generation. NO counteracts oxidative stress by two possible mechanisms: either by directly interact with reactive oxygen species, such as superoxide (O_2^-), to generate peroxynitrite ($ONOO^-$, which is less harmful than peroxides) or by functioning as a signalling molecule, thereby modulating the expression level of genes (Misra et al. 2011).

3.2 Biosynthesis of Nitric Oxide

Nitric oxide (NO) is a multifaceted bioactive molecule that is an important regulator of various plant physiological functions, stress tolerance mechanisms, and disease resistance mechanisms. The synthesis of NO was initially reported in legumes, specifically soybean, through the application of synthetic inhibitor herbicides under dark and anaerobic environments (Klepper 1991). NO is biosynthesized through various reductive and oxidative mechanisms which involve enzymatic or non-enzymatic reactions. Overall, NO production is accelerated by NOS (nitric oxide synthase), NR (nitrate reductase), NiR (nitrite reductase), other enzymes, and non-enzymatic compounds (Fig. 3.1).

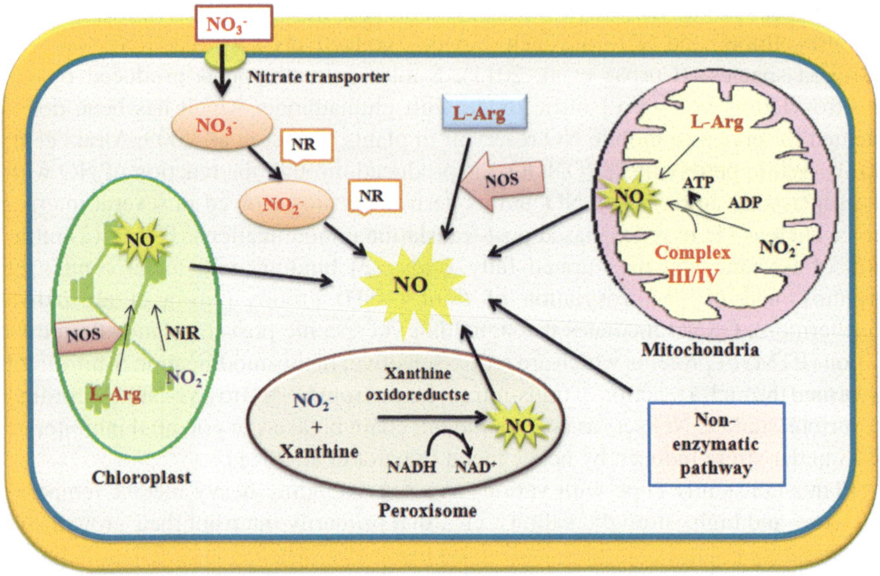

Fig. 3.1 Biosynthesis of nitric oxide through different pathways

3.2.1 NO Biosynthesis Through Nitric Oxide Synthase (NOS)

The synthesis of NO is still controversial in plant as compare to in animal. In mammals, nitric oxide is produced from three isozymes of nitric oxide synthase (NOS), namely (1) nNOS or NOS-I (neuronal NOS), (2) iNOS or NOS-II (inducible NOS), and (3) eNOS or NOS-III (endothelial NOS) (Crane et al. 2010). NOS catalyzes the NO production from L-Arg to L-citrulline, which is NADPH-dependent reaction (Misra et al. 2010). In higher plant systems, NOS-like functional enzyme is not isolated and characterized (Corpas et al. 2001). However, according to several biochemical and molecular investigations, a NOS-like sequence has been observed in the genomes of *Ostreococcus tauri* and *Ostreococcus lucimarinus* green algae (Foresi et al. 2010); additional research on *Arabidopsis thaliana* reveals that NOS-like enzyme is present in the peroxisomes of its root tips (Corpas and Barroso 2014; Tewari et al. 2021). Besides catalyzing the synthesis of NO, NOS is also involved in the production of two additional molecules in plants: hydroxylamine (Rumer et al. 2009) and polyamines (Tun et al. 2006), which is activated by stresses and reactive oxygen species.

3.2.2 NO Biosynthesis Through Nitrate/Nitrite Reductase (NR/NiR) and Other Enzymes

Nitrate reductase (NR) is another major enzyme that catalyzes the NADPH-dependent reduction of NO_3 (nitrate) to NO_2 (nitrite) in the plant cell's cytoplasm

(Fig. 3.1). Some in vivo and in vitro analyses have established that nitrate reductase also catalyzes the reduction of NO_2 to NO and its derivative peroxynitrite ($ONOO^-$) (Fancy et al. 2017). There are some inhibitors of nitrate reductase in plants, namely, sodium azide, tungstate, and potassium cyanide, which inhibits the NO production (Sang et al. 2008; Fancy et al. 2017). Furthermore, NR activity is affected by various factors like concentrations of NO_3^- and NO_2^-, cellular pH and post-translational modification etc. At low cellular pH, the NO biosynthesis is stimulated due to reduced activity of plastidial NiR and nitrite accumulation patterns. Other reductive ways for NO biosynthesis are mitochondrial ETC (electron transport chain) and peroxisomal xanthine dehydrogenase/oxidase (Fig. 3.1) (Gupta et al. 2011). In contrast, NO is produced in different possible oxidative ways, such as hydroxylamine-mediated and polyamines-mediated (Tun et al. 2006).

3.2.3 NO Biosynthesis Through Non-enzymatic Mechanism

Beside the enzymatic biosynthesis of NO, it can also be synthesized by non-enzymatic mechanisms. Under elevated nitrate concentrations and acidic pH, non-enzymatic NO_2-dependent nitric oxide synthesis was observed in the apoplast region of aleurone layer in barley seed (Astier et al. 2018). It has been proposed that the regulation of germination of barley seed is depended on NO production in the apoplast region of seed (Bethke et al. 2004; Astier et al. 2018). Ascorbic acid and few phenolic compounds are some reducing agents, that are essential for enhancement of non-enzymatic NO production (Allagulova et al. 2023). Ascorbic acid reduces nitrite to yield NO and dehydroascorbic acid, and this reduction reaction takes place in chloroplast and apoplastic spaces. Furthermore, carotenoids are also included in NO production through the light regulated reduction of NO_2^- (nitrite) (Bethke et al. 2004). Hydrogen peroxide (H_2O_2) and salicylic acid (SA) trigger the rapid generation of NO in companion cell of *Vicia faba,* while in the phloem tissue, its synthesis depends on the Ca^{2+} concentration and NOS activity (Gaupels et al. 2008).

3.3 Heavy Metal Toxicity

Heavy metal (HM) toxicity is one of the major abiotic stresses and became the most serious risk factor to crop productivity. Heavy metals are a set of 52 metals that include cadmium, arsenic, lead, nickel, copper, manganese, mercury, and cobalt (Boyd and Rajakaruna 2013; Riyazuddin et al. 2021). There are various causes of increasing HM contamination in the environment, such as the long-term use of sewage sludge applications, phosphate fertilizers, industrial waste, and bad watering (Wei et al. 2020). All plant tissue and life stages, from seed germination to senescence, are negatively affected by HM toxicity; these effects are more prominent during germination of seed and root growth (Riyazuddin et al. 2021). Plants show various symptoms under HM stress such as chlorosis, stunted growth, browning of roots, and sometimes even death. Heavy metal toxicity induces excessive methylglyoxyl (MG) and reactive oxygen species (ROS) production, which causes oxidative stress (Hossain et al. 2009) (Fig. 3.2). There are various reactive oxygen species, such as superoxide radicals (O_2^-) and hydroxyl radicals ($OH^·$) which are in radical form, hydrogen peroxide (H_2O_2) and singlet oxygen ($^1O_2$) which are non-radical form. These ROS are unstable, extremely reactive, and immediately react with many other molecules to produce more new free radicals because they have unpaired electrons, which have the tendency to pair and form two stable electron bonds (Foyer and Halliwell 1976). MG is also a highly reactive αβ-dicarbonyl aldehyde and is formed naturally by non-enzymatic mechanisms from glycolysis and from photosynthesis intermediates (Hossain et al. 2012). Under stress conditions, MG production is induced because of increased glycolysis rate, which results in numerous destructive effects like protein degradation and the inactivation of antioxidant defense system (Hossain et al. 2012). Various investigations suggest that HMs disrupt the cellular equilibrium in five possible ways: (1) by changing the redox potential to high oxidized values (Foyer and Noctor 2016); (2) by directly generating reactive oxygen species from the Haber-Weiss cycle and Fenton-like reactions (Sharma and Dietz 2009); (3) by increasing NADPH oxidases activity (Shahid et al. 2014). (4) by consuming GSH that promotes the production of phytochelatins (PCs), which enhance chelation and sequestration of heavy metals through

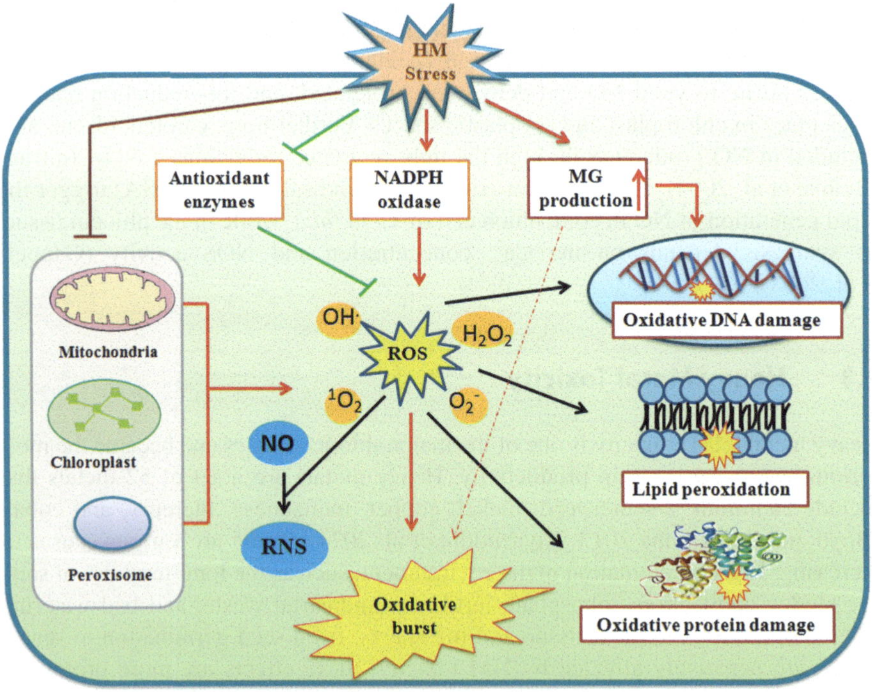

Fig. 3.2 Oxidative burst induced by heavy metal stress in plant cell

ABC transporters (Song et al. 2014); and (5) by inhibition of antioxidant enzymes by removing crucial cations from the certain sites and focusing on their –SH groups (Shahid et al. 2014).

3.4 Mechanisms of HM Stress-Tolerance in Plants

Plants govern tolerance against HMs through controlled and inter-related complex web of physiological and molecular processes. There are different mechanisms evolved in different species of plant to resist high concentration of HMs, and even more than one mechanism could be operated within one plant species. Two defense mechanisms are adopted by plants known as avoidance and tolerance, to overcome toxicity in addition to maintain cellular concentration of HMs (Yan et al. 2020). At the extracellular level, plants use avoidance as a first line of defense to prevent heavy metals from entering their tissues via root cells. It involves several processes for avoidance for example adsorption, precipitation, immobilization, and exclusion of metals. Heavy metals are immobilized through root adsorption or by forming stable complexes with variety of exudates (amino acids and organic acids) in the rhizosphere (Dalvi and Bhalerao 2013). Tolerance is the second defense line at the intracellular level that is the potential of plants to deal with HMs, which are accumulated through different processes for instance inactivation, chelation, and compartmentalization (Yan et al. 2020). Proteins, phytochelatins, chaperones, metallothioneins, some amino acids and organic acids play significant role in HM chelation and compartmentalization/sequestration.

Plants have an innate immune system that involves antioxidants such as enzymatic and non-enzymatic which are significant for HM-stress tolerance. Enzymatic antioxidants involve ascorbate peroxidase (APx), superoxide dismutase (SOD), monodehydro-ascorbate reductase (MDHAR), glutathione reductase (GR), dehydroascorbate reductase (DHAR), catalase (CAT) and glutathione peroxidase (GPx) (Saxena and Shekhawat 2013), whereas antioxidants which are non-enzymatic involve ascorbic acid (AsA), glutathione (GSH), α-tocopherol and proline; these are water soluble (Hossain et al. 2012). According to cellular requirements, these antioxidants regulate the ROS concentration, while both GSH and AsA act as co-factors for enzymatic antioxidants as well as they can directly reduce ROS and modulate the expression level of genes associated to both biotic and abiotic stress (Singh et al. 2016).

3.5 NO Mediated Signalling in HM Stress Tolerance Mechanisms

Heavy metal contaminated soil is the most challenging issues. When heavy metal concentrations exceed a certain threshold in plants, it causes negative effects such as oxidative damage (Hossain et al. 2012). NO produced due to stress induced by HMs in plants can trigger many kinds of signalling cascades that strengthens the tolerance mechanisms. NO plays a significant part in various physiological mechanisms in plants. NO transmits its signal through some signalling proteins that are activated by NO and its derivative-mediated post-translational modifications (PTMs), such as S-nitrosylation and tyrosine nitration (Begara-Morales et al. 2016). It induces transcriptional changes in numerous genes that are associated with transduction of signal, basic metabolisms, transportation, defense and cell death, production, and degeneration of ROS (Palmieri et al. 2008).

3.5.1 Through Crosstalk Between NO and ROS

Under the abiotic stress, overproduction of ROS stimulates high NO synthesis through the NR, which indicates the relation between reactive nitrogen species and reactive oxygen species (Khan et al. 2023a). Both ROS and NO are crucial biological messengers and play significant role in response to diverse abiotic and biotic stresses. The application of H_2O_2 promotes the production of NO and subsequent accumulation which induce the antioxidant activity, reduce ROS overproduction, and re-establish redox balance in Arabidopsis (Hasanuzzaman et al. 2018). Conversely, the elimination of H_2O_2 results in a depletion of NO production (Bright et al. 2006). Therefore, despite being phytotoxic, both H_2O_2 and NO are crucial for the transmission of signals. In the response to stress, NO interacts with O_2^- (superoxide anion) to produce peroxynitrite ($ONOO^-$) that is involved in activation of different many enzymes and protein for further signalling. It causes the nitration of Tyr- (tyrosine/Y) residue of a protein, which is a type of post-translational modification (PTM) (Liu et al. 2018) (Fig. 3.3). S-nitrosylation regulates H_2O_2 level at the cellular level through prevent the CAT and glycolate oxidase activity in peroxisome (Ortega-Galisteo et al. 2012).

3.5.2 Activation of MAP-Kinase Cascade and Ca^{2+} Signalling

It has been observed that mitogen-activated protein kinases (MAP-Kinases) cascades, regulation of genes at transcriptional level associated with stress-response, and changes in Ca^{2+} levels are all part of the HM stress response in plants (Thapa et al. 2012). Both MAP-Kinases and calcium ions (Ca^{2+}) are the downstream targets of NO. To increase free cytosolic Ca^{2+} levels in cells, NO modulate intracellular Ca^{2+}-permeable channels through cGMP, cADPR, and protein kinases (Arasimowicz and Floryszak-Wieczorek 2007). cGMP produced by the enzymatic action of sGC(guanylate cyclase) after its binding with NO (Besson-Bard et al. 2008; Verma et al. 2020) Ca^{2+} is an important signalling molecule that reduces the chromium toxicity through inhibiting Cr^{2+} transportation from root to shoot in rice seedlings (Mukta et al. 2019). Furthermore, Ca^{2+} affects the ROS level produced under arsenic stress (Rahman et al. 2015) as well as activates CIPK and MAP-kinase signalling and stress tolerance under copper stress in rice seedlings (Wilkins et al. 2016). Different Ca-binding proteins, namely, CaM (calmodulins), CBLs (Calcineurin

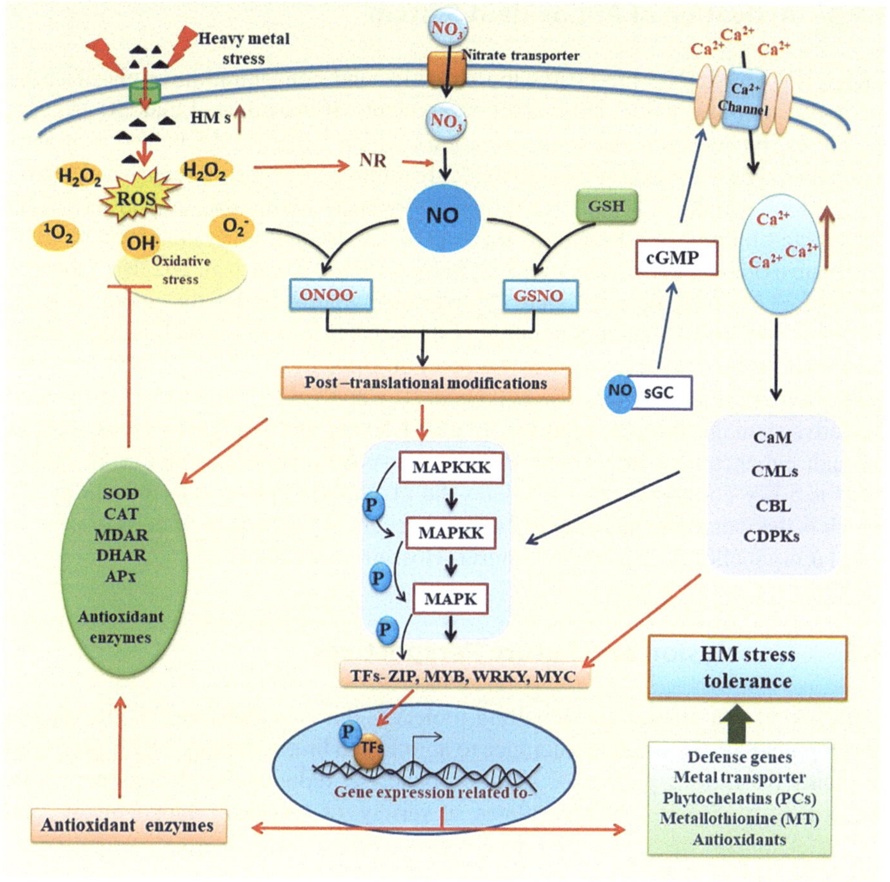

Fig. 3.3 Nitric oxide signalling through activation of MAP-kinase cascade, Ca^{2+} -signalling, and antioxidant system, in response of heavy metal stress tolerance in plant

B-like), CMLs (CaM-like proteins), and CDPKs (calcium ion-dependent protein kinases), also trigger Ca^{2+}-signalling pathway (Kaur et al. 2021).

MAP-kinases are activated through NO, ROS, and some plant hormones such as abscisic acid (ABA), salicylic acid (SA), ethylene (ET), and auxin (IAA) under HM stress (Li et al. 2022). The MAP-kinase cascade, which activates MYB, WRKY, bZIP, and MYC transcriptional factors through phosphorylation (Li et al. 2022), is a sequence of phosphorylation events from MAPKKKs via MAPKKs to MAPK (Jonak et al. 2002). These transcription factors induce the expression level of genes related to defense, metal transporters, phytochelatin (PC), metallothionine (MT), and antioxidants by binding specific cis-regulatory elements and act as a key component that regulate heavy metal tolerance and detoxification mechanisms in plants (Li et al. 2022) (Fig. 3.3).

3.5.3 Activation of Antioxidant System

Under HM stress, NO modulates the activities and concentration levels of enzymatic and non-enzymatic antioxidant components. It is evaluated that NO triggers the PTMs (tyr-nitration and S-nitrosylation), which influence the behavior of antioxidant enzymes for example superoxide dismutase (SOD), different peroxiredoxins (Prxs), catalase (CAT) enzymes associated with the Asa-GSH cycle (ascorbate-glutathione) (Begara-Morales et al. 2016), and other enzymes, namely, glycolate oxidase and hydroxy pyruvate reductase etc. (Khan et al. 2023b) (Table 3.1 and Fig. 3.3). It is reported that Cd-tolerance improved by the application of NO-donor through enhancing the total GSH and AsA content, which reduces damages induced by oxidative stress in the roots of *Madicago truncatula* seedlings (Xu et al. 2009). In sunflower (*Helianthus annuus* L.), applied NO attenuates Cd-induced oxidative damage through recovery of GSH levels and CAT activity as well as through enhancement in activity of APx and AsA content (Laspina et al. 2005). GSH is a key component that regulates the glyoxalase system (Gly I and Gly II), which is the metabolic pathway of MG that occurs in the cytosol of cells or even in cellular organelles such as mitochondria (Hossain et al. 2012).

3.6 Conclusion and Future Perspectives

Nitric oxide is a ubiquitous signalling molecule that is significant for growth and development as well as plant tolerance to abiotic and biotic stresses. NO is produced in plants by various pathways, such as oxidative, reductive, and non-enzymatic. According to reports, NO regulates a variety of physical, biochemical, and

Table 3.1 Post-translational modifications of enzymes induced by nitric oxide

Enzymes	NO-induced PTMs	Plants	References
Hydroxy pyruvate reductase	S-nitrosylation/ tyrosine nitration	*Pisum sativum* and *Arabidopsis*	Puyaubert et al. (2014); Khan et al. (2023b)
Glycolate oxidase	S-nitrosylation/ tyrosine nitration	*Pisum sativum*, citrus and *Arabidopsis*	Puyaubert et al. (2014); Khan et al. (2023b)
Catalase (CAT)	S-nitrosylation/ tyrosine nitration	*Arabidopsis* and *Capsicum annuum*	Chaki et al. (2015); Khan et al. (2023b)
Monodehydro-ascorbate reductase (MDAR)	S-nitrosylation/ tyrosine nitration	*Pisum sativum*	Begara-Morales et al. (2015)
Dehydro-ascorbate reductase (DHAR)	S-nitrosylation	*Arabidopsis* and *Solanum tuberosum*	Puyaubert et al. (2014)
Ascorbate peroxidase (APx)	S-nitrosylation/ tyrosine nitration	*Pisum sativum* and *Arabidopsis*	Begara-Morales et al. (2014)
Superoxide dismutase (SOD)	Tyrosine nitration	*Arabidopsis*	Holzmeister et al. (2015)

molecular processes in response to varying environmental conditions, thereby mitigating the detrimental effects of environmental challenges on plants. As well, NO participates in various developmental stages from germination to vegetative growth. Under HM stress, nitric oxide can activate the antioxidant system, which includes non-enzymatic and enzymatic antioxidants, to reduce oxidative damage. The consequences of NO are depended on its concentration, signalling, and interaction with different molecules. Interactivity of NO with ROS, phyto-hormones, and enzymes modulates plant functions under stress. NO activates MAP-kinase and Ca^{2+} dependent signalling, which regulate the genes expression, which are related to defense, metal transporter, phytochelatin (PC), metallothionine (MT), and antioxidants. NO transmits its action mainly through post-transcriptional modifications (PTMs), such as tyrosine nitration and S-nitrosylation of the target enzymes and proteins. Thus, NO is essential in plant tolerance mechanism under HM stress and other abiotic stress. Still, the precise signalling and mechanisms of interaction with other enzymes and molecules are unclear. It is imperative to evolve strategies for practical application of NO to enhance growth, tolerance to various stresses and productivity of crop, which will be facilitated by an understanding of NO functioning in the highly complicated metabolism of plants. These strategies have significant socioeconomic advantages under changing climatic conditions.

Acknowledgement The authors wish to thank the University Grants Commission, New Delhi, for providing financial assistance through the Center for Advanced Study programme.

References

Airaki M, Sánchez-Moreno L, Leterrier M, Barroso JB, Palma JM, Corpas FJ (2011) Detection and quantification of S-nitrosoglutathione (GSNO) in pepper (*Capsicum annuum* L.) plant organs by LC-ES/MS. Plant Cell Physiol 52(11):2006–2015

Allagulova CR, Lubyanova AR, Avalbaev AM (2023) Multiple ways of nitric oxide production in plants and its functional activity under abiotic stress conditions. Int J Mol Sci 24(14):11637

Arasimowicz M, Floryszak-Wieczorek J (2007) Nitric oxide as a bioactive signalling molecule in plant stress responses. Plant Sci 172(5):876–887

Arora D, Jain P, Singh N, Kaur H, Bhatla SC (2016) Mechanisms of nitric oxide crosstalk with reactive oxygen species scavenging enzymes during abiotic stress tolerance in plants. Free Radic Res 50(3):291–303

Astier J, Gross I, Durner J (2018) Nitric oxide production in plants: an update. J Exp Bot 69(14):3401–3411

Begara-Morales JC, Sanchez-Calvo B, Chaki M, Valderrama R, Mata-Perez C, Lopez-Jaramillo J, Barroso JB (2014) Dual regulation of cytosolic ascorbate peroxidase (APX) by tyrosine nitration and S-nitrosylation. J Exp Bot 65(2):527–538

Begara-Morales JC, Sanchez-Calvo B, Chaki M, Mata-Perez C, Valderrama R, Padilla MN, Barroso JB (2015) Differential molecular response of monodehydroascorbate reductase and glutathione reductase by nitration and S-nitrosylation. J Exp Bot 66(19):5983–5996

Begara-Morales JC, Sanchez-Calvo B, Chaki M, Valderrama R, Mata-Perez C, Padilla MN, Barroso JB (2016) Antioxidant systems are regulated by nitric oxide-mediated post-translational modifications (NO-PTMs). Front Plant Sci 7:152

Besson-Bard A, Courtois C, Gauthier A, Dahan J, Dobrowolska G, Jeandroz S, Wendehenne D (2008) Nitric oxide in plants: production and cross-talk with Ca2+ signaling. Mol Plant 1(2):218–228

Bethke PC, Badger MR, Jones RL (2004) Apoplastic synthesis of nitric oxide by plant tissues. Plant Cell 16(2):332–341

Boyd RS, Rajakaruna N (2013) Heavy metal tolerance. Oxford University Press, Oxford

Bright J, Desikan R, Hancock JT, Weir IS, Neill SJ (2006) ABA-induced NO generation and stomatal closure in *Arabidopsis* are dependent on H2O2 synthesis. Plant J 45(1):113–122

Chaki M, Alvarez de Morales P, Ruiz C, Begara-Morales JC, Barroso JB, Corpas FJ, Palma JM (2015) Ripening of pepper (*Capsicum annuum*) fruit is characterized by an enhancement of protein tyrosine nitration. Ann Bot 116(4):637–647

Corpas FJ, Barroso JB (2014) Functional implications of peroxisomal nitric oxide (NO) in plants. Front Plant Sci 5:97

Corpas FJ, Barroso JB, del Río LA (2001) Peroxisomes as a source of reactive oxygen species and nitric oxide signal molecules in plant cells. Trends Plant Sci 6(4):145–150

Corpas FJ, Leterrier M, Valderrama R, Airaki M, Chaki M, Palma JM, Barroso JB (2011) Nitric oxide imbalance provokes a nitrosative response in plants under abiotic stress. Plant Sci 181(5):604–611

Crane BR, Sudhamsu J, Patel BA (2010) Bacterial nitric oxide synthases. Annu Rev Biochem 79:445–470

Dalvi AA, Bhalerao SA (2013) Response of plants towards heavy metal toxicity: an overview of avoidance, tolerance and uptake mechanism. Ann Plant Sci 2(9):362–368

Fancy NN, Bahlmann AK, Loake GJ (2017) Nitric oxide function in plant abiotic stress. Plant Cell Environ 40(4):462–472

Foresi N, Correa-Aragunde N, Parisi G, Calo G, Salerno G, Lamattina L (2010) Characterization of a nitric oxide synthase from the plant kingdom: NO generation from the green alga *Ostreococcus tauri* is light irradiance and growth phase dependent. Plant Cell 22(11):3816–3830

Foster MW, Hess DT, Stamler JS (2009) Protein S-nitrosylation in health and disease: a current perspective. Trends Mol Med 15(9):391–404

Foyer CH, Halliwell B (1976) The presence of glutathione and glutathione reductase in chloroplasts: a proposed role in ascorbic acid metabolism. Planta 133:21–25

Foyer CH, Noctor G (2016) Stress-triggered redox signalling: what's in pROSpect? Plant Cell Environ 39(5):951–964

Gaupels F, Furch AC, Will T, Mur LA, Kogel KH, Van Bel AJ (2008) Nitric oxide generation in *Vicia faba* phloem cells reveals them to be sensitive detectors as well as possible systemic transducers of stress signals. New Phytol 178(3):634–646

Gupta KJ, Fernie AR, Kaiser WM, van Dongen JT (2011) On the origins of nitric oxide. Trends Plant Sci 16(3):160–168

Hasanuzzaman M, Nahar K, Rahman A, Inafuku M, Oku H, Fujita M (2018) Exogenous nitric oxide donor and arginine provide protection against short-term drought stress in wheat seedlings. Physiol Mol Biol Plants 24:993–1004

Holzmeister C, Gaupels F, Geerlof A, Sarioglu H, Sattler M, Durner J, Lindermayr C (2015) Differential inhibition of *Arabidopsis* superoxide dismutases by peroxynitrite-mediated tyrosine nitration. J Exp Bot 66(3):989–999

Hossain MA, Hossain MZ, Fujita M (2009) Stress-induced changes of methylglyoxal level and glyoxalase I activity in pumpkin seedlings and cDNA cloning of glyoxalase I gene. Aust J Crop Sci 3(2):53

Hossain MA, Piyatida P, da Silva JAT, Fujita M (2012) Molecular mechanism of heavy metal toxicity and tolerance in plants: central role of glutathione in detoxification of reactive oxygen species and methylglyoxal and in heavy metal chelation. J Bot 2012:872875

Jonak C, Okresz L, Bogre L, Hirt H (2002) Complexity, cross talk and integration of plant MAP kinase signalling. Curr Opin Plant Biol 5(5):415–424

Kaur R, Das S, Bansal S, Singh G, Sardar S, Dhar H, Ram H (2021) Heavy metal stress in rice: uptake, transport, signaling, and tolerance mechanisms. Physiol Plant 173(1):430–448

Khan MK, Pandey A, Hamurcu M, Avsaroglu ZZ, Ozbek M, Omay AH, Gezgin S (2021) Variability in physiological traits reveals boron toxicity tolerance in *Aegilops* species. Front Plant Sci 12:736614

Khan M, Ali S, Al Azzawi TNI, Saqib S, Ullah F, Ayaz A, Zaman W (2023a) The key roles of ROS and RNS as a signaling molecule in plant–microbe interactions. Antioxidants (Basel) 12(2):268

Khan M, Ali S, Al Azzawi TNI, Yun BW (2023b) Nitric oxide acts as a key signaling molecule in plant development under stressful conditions. Int J Mol Sci 24(5):4782

Khator K, Saxena I, Shekhawat GS (2021) Nitric oxide induced Cd tolerance and phytoremediation potential of *B. juncea* by the modulation of antioxidant defense system and ROS detoxification. Biometals 34:15–32

Klepper L (1991) NOx evolution by soybean leaves treated with salicylic acid and selected derivatives. Pestic Biochem Physiol 39(1):43–48

Laspina NV, Groppa MD, Tomaro ML, Benavides MP (2005) Nitric oxide protects sunflower leaves against Cd-induced oxidative stress. Plant Sci 169(2):323–330

Li S, Han X, Lu Z, Qiu W, Yu M, Li H, Zhuo R (2022) MAPK cascades and transcriptional factors: regulation of heavy metal tolerance in plants. Int J Mol Sci 23(8):4463

Liu D, Liu M, Liu XL, Cheng XG, Liang ZW (2018) Silicon priming created an enhanced tolerance in alfalfa (*Medicago sativa* L.) seedlings in response to high alkaline stress. Front Plant Sci 9:716

Misra AN, Misra M, Singh R (2010) Nitric oxide biochemistry, mode of action and signaling in plants. J Med Plants Res 4(25):2729–2739

Misra AN, Misra M, Singh R (2011) Nitric oxide ameliorates stress responses in plants. Plant Soil Environ 57(3):95–100

Mukta RH, Khatun MR, Nazmul Huda AKM (2019) Calcium induces phytochelatin accumulation to cope with chromium toxicity in rice (*Oryza sativa* L.). J Plant Interact 14(1):295–302

Nabi RBS, Tayade R, Hussain A, Kulkarni KP, Imran QM, Mun BG, Yun BW (2019) Nitric oxide regulates plant responses to drought, salinity, and heavy metal stress. Environ Exp Bot 161:120–133

Ortega-Galisteo AP, Rodríguez-Serrano M, Pazmino DM, Gupta DK, Sandalio LM, Romero-Puertas MC (2012) S-Nitrosylated proteins in pea (*Pisum sativum* L.) leaf peroxisomes: changes under abiotic stress. J Exp Bot 63(5):2089–2103

Palmieri MC, Sell S, Huang X, Scherf M, Werner T, Durner J, Lindermayr C (2008) Nitric oxide-responsive genes and promoters in *Arabidopsis thaliana*: a bioinformatics approach. J Exp Bot 59(2):177–186

Puyaubert J, Fares A, Rézé N, Peltier JB, Baudouin E (2014) Identification of endogenously S-nitrosylated proteins in *Arabidopsis* plantlets: effect of cold stress on cysteine nitrosylation level. Plant Sci 215:150–156

Rahman A, Mostofa MG, Alam MM, Nahar K, Hasanuzzaman M, Fujita M (2015) Calcium mitigates arsenic toxicity in rice seedlings by reducing arsenic uptake and modulating the antioxidant defense and glyoxalase systems and stress markers. Biomed Res Int 2015:340812

Riyazuddin R, Nisha N, Ejaz B, Khan MIR, Kumar M, Ramteke PW, Gupta R (2021) A comprehensive review on the heavy metal toxicity and sequestration in plants. Biomolecules 12(1):43

Rumer S, Gupta KJ, Kaiser WM (2009) Plant cells oxidize hydroxylamines to NO. J Exp Bot 60(7):2065–2072

Sang J, Jiang M, Lin F, Xu S, Zhang A, Tan M (2008) Nitric oxide reduces hydrogen peroxide accumulation involved in water stress-induced subcellular anti-oxidant defense in maize plants. J Integr Plant Biol 50(2):231–243

Saxena I, Shekhawat GS (2013) Nitric oxide (NO) in alleviation of heavy metal induced phytotoxicity and its role in protein nitration. Nitric Oxide 32:13–20

Shahid M, Pourrut B, Dumat C, Nadeem M, Aslam M, Pinelli E (2014) Heavy-metal-induced reactive oxygen species: phytotoxicity and physicochemical changes in plants. Rev Environ Contam Toxicol 232:1–44

Sharma SS, Dietz KJ (2009) The relationship between metal toxicity and cellular redox imbalance. Trends Plant Sci 14(1):43–50

Singh S, Parihar P, Singh R, Singh VP, Prasad SM (2016) Heavy metal tolerance in plants: role of transcriptomics, proteomics, metabolomics, and ionomics. Front Plant Sci 6:1143

Song WY, Mendoza-Cozatl DG, Lee Y, Schroeder JI, Ahn SN, Lee HS, Martinoia E (2014) Phytochelatin–metal (loid) transport into vacuoles shows different substrate preferences in barley and *Arabidopsis*. Plant Cell Environ 37(5):1192–1201

Tewari RK, Horemans N, Watanabe M (2021) Evidence for a role of nitric oxide in iron homeostasis in plants. J Exp Bot 72(4):990–1006

Thapa G, Sadhukhan A, Panda SK, Sahoo L (2012) Molecular mechanistic model of plant heavy metal tolerance. Biometals 25:489–505

Tun NN, Santa-Catarina C, Begum T, Silveira V, Handro W, Floh EIS, Scherer GF (2006) Polyamines induce rapid biosynthesis of nitric oxide (NO) in *Arabidopsis thaliana* seedlings. Plant Cell Physiol 47(3):346–354

Verma N, Tiwari S, Singh VP, Prasad SM (2020) Nitric oxide in plants: an ancient molecule with new tasks. Plant Growth Regul 90:1–13

Wei L, Zhang J, Wang C, Liao W (2020) Recent progress in the knowledge on the alleviating effect of nitric oxide on heavy metal stress in plants. Plant Physiol Biochem 147:161–171

Wendehenne D, Pugin A, Klessig DF, Durner J (2001) Nitric oxide: comparative synthesis and signaling in animal and plant cells. Trends Plant Sci 6(4):177–183

Wilkins KA, Matthus E, Swarbreck SM, Davies JM (2016) Calcium-mediated abiotic stress signaling in roots. Front Plant Sci 7:1296

Xu J, Yin H, Li X (2009) Protective effects of proline against cadmium toxicity in micropropagated hyperaccumulator, *Solanum nigrum* L. Plant Cell Rep 28:325–333

Yan A, Wang Y, Tan SN, MohdYusof ML, Ghosh S, Chen Z (2020) Phytoremediation: a promising approach for revegetation of heavy metal-polluted land. Front Plant Sci 11:359

Recent Advances in the Role of Nitric Oxide and Cross-Talk with Signalling Molecules Under Environmental Stress in Plants

4

Prasann Kumar and Padmanabh Dwivedi

Abstract

Nitric oxide (NO) plays a crucial role as a signalling molecule in coordinating the plant's response to various stressors. The intricate role of nitric oxide (NO) and its cross-talk with other signalling molecules under environmental stress conditions has been emphasised in recent advancements in plant physiology and molecular biology. Nitric oxide (NO), frequently functioning as a signalling molecule, is crucial in regulating various physiological processes, including stomatal closure, photosynthesis, antioxidant defence, and programmed cell death. Consequently, NO plays a significant role in modulating the plant's adaptive responses to stress. Moreover, recent findings indicate a substantial interplay between nitric oxide (NO) and various signalling molecules, such as reactive oxygen species (ROS), phytohormones (e.g., abscisic acid, salicylic acid, jasmonic acid, and ethylene), calcium ions, and secondary messengers like cyclic nucleotides. The complex interactions described above play a significant role in adjusting plant stress responses, allowing plants to develop a suitable defence mechanism customised to the particular environmental obstacle. Comprehending the molecular mechanisms that underlie signalling pathways mediated by nitric oxide (NO) and their interplay with other signalling molecules is of utmost importance in developing stress-tolerant crops that exhibit enhanced resilience to unfavourable environmental circumstances. Furthermore, investigating the regulatory networks associated with nitric oxide (NO) in stress responses shows

P. Kumar
Department of Agronomy, School of Agriculture, Lovely Professional University, Phagwara, India

P. Dwivedi (✉)
Department of Plant Physiology, Institute of Agricultural Sciences, Banaras Hindu University, Varanasi, India
e-mail: pdwivedi25@bhu.ac.in

© The Author(s), under exclusive license to Springer Nature Singapore Pte Ltd. 2024

M. Faizan, S. Hayat (eds.), *Plant Growth Regulators: Resilience for Sustainable Agriculture*, https://doi.org/10.1007/978-981-97-2918-0_4

potential for developing innovative approaches to promote sustainable agriculture and effectively manage ecological systems in the context of climate change. This chapter focuses on recent progress in understanding the function of nitric oxide and its complex interactions with signalling molecules in plants experiencing environmental stress. Future research must investigate the intricacies of signalling networks mediated by nitric oxide (NO) to fully exploit the capabilities of plant stress tolerance mechanisms and guarantee global food security in the face of evolving environmental conditions.

Keywords

Nitric oxide · Abiotic stress · Plants · Phytohormones · Reactive oxygen species · Environment · Zero hunger · No poverty

4.1 Introduction

As immobile organisms, plants face various environmental stresses that can significantly impact their growth, development, and overall health. Environmental stressors such as drought, salinity, extreme temperatures, and pollution activate intricate signalling networks within plant cells (Altaf et al. 2023a). These networks coordinate various responses that enable the plants to survive and adapt. Nitric oxide (NO) is a versatile and multifaceted molecule crucial in modulating plant responses to environmental challenges within intricate signalling pathways (Singh et al. 2023d). Recent developments in plant molecular biology and physiology have provided insight into the essential function of nitric oxide and its complex interaction with different signalling molecules during periods of environmental stress (Pramanik et al. 2023). Nitric oxide, a gaseous free radical, was first recognized as a signalling molecule in animals, where it plays a vital role in various physiological processes. Researchers did not discover the presence and significance of NO in plants until later (Kaya et al. 2023a). Over the past 20 years, extensive research has clarified the diverse functions of nitric oxide in plant biology, encompassing processes such as seed germination and senescence (Lone et al. 2023). Nitric oxide controls various physiological processes, such as stomata movement, root growth, and reactions to non-living and living environmental factors (Mahiwal et al. 2024). Plants face numerous environmental stresses that endanger their overall health and survival. Abiotic stresses, such as drought and salinity, and biotic stresses, caused by pathogens and pests, initiate complex signalling cascades. To address these challenges, plants employ various molecular and biochemical mechanisms to improve their ability to adapt and survive. Recent research has been primarily concerned with understanding the communication between stress-responsive pathways and the signalling molecules that coordinate these complex processes. Plants adjust their physiological and molecular responses to deal with unfavourable conditions like drought and salinity, known as abiotic stress. Nitric oxide has been recognized as a pivotal mediator in these responses (Sodhi and Saxena 2023). It plays a role in regulating antioxidant systems, maintaining ion balance, and activating stress-responsive genes. The intricate interaction between NO and other signalling molecules, such as reactive oxygen species (ROS) and phytohormones like abscisic acid (ABA), emphasizes the complex nature of the plant's stress response network. Reactive oxygen species, produced from cellular metabolism, serve two functions in plants—signalling molecules and potential inducers of oxidative stress (Feng et al. 2023). The interaction between nitric oxide and reactive oxygen species (ROS) is a captivating field of research, with recent findings indicating complex regulatory mechanisms. NO has been linked to regulating reactive oxygen species (ROS) levels, thereby impacting the cellular redox. In specific stress signalling pathways, hydrogen peroxide (H_2O_2), a component of ROS, can act either upstream or downstream of NO. Comprehending the subtle distinctions of this intercommunication is vital for deciphering the complexities of plant stress reactions (He et al. 2023a, b). Phytohormones are crucial in controlling plant growth and development, and their contribution to stress responses is widely recognized. Nitric oxide interacts with diverse phytohormones, such as abscisic acid, jasmonic acid, and salicylic acid, forming a complex network of signalling pathways. Recent research has revealed information about the communication between NO and phytohormones during periods of stress. This research emphasizes how these two substances can work together or against each other to regulate how plants respond to stress. Recent technological advancements, such as advanced imaging techniques, genetically encoded biosensors, and omics approaches, have revolutionized the study of nitric oxide signalling in plants (Mitra et al. 2023). The real-time visualization of nitric oxide dynamics and high-throughput omics analyses have given researchers powerful tools to examine the complexities of NO-mediated signalling pathways in environmental stress. Technological advancements have expedited the rate of exploration and enhanced our comprehension of the function of NO in plant stress reactions. Although significant improvements have been made in comprehending the role of nitric oxide and its interaction with signalling molecules during periods of environmental stress, several obstacles remain (Seabra and Tortella 2023). Further investigation is needed to unravel the precise molecular targets of NO, understand the context-dependent nature of NO signalling, and develop strategies to manipulate NO levels for enhanced stress tolerance. Furthermore, it is essential to incorporate the discoveries made in model plants into crop species to apply research findings to agriculture effectively. Recent progress in investigating nitric oxide and its interaction with signalling molecules during environmental stress has revealed the complex regulatory mechanisms that plants utilize to adapt to difficult conditions (Devi et al. 2023). The diverse functions of nitric oxide, its interaction with reactive oxygen species (ROS) and plant hormones, and the incorporation of advanced technologies have significantly advanced our comprehension of plant stress responses. This emerging discipline offers excellent potential for uncovering fundamental aspects of plant biology and devising methods to improve crop resilience in response to a shifting climate and growing environmental challenges (Basit et al. 2023).

4.2 Function of Nitric Oxide in Plant Growth and Development

Nitric oxide, a diminutive gaseous molecule, has surfaced as a pivotal signalling constituent in plant biology, exerting influence over diverse facets of growth and development. Formerly predominantly linked to its function in stress reactions, NO is presently acknowledged as a versatile controller that coordinates crucial mechanisms throughout the life cycle of plants. Nitric oxide plays a wide range of roles and is closely connected to essential cellular processes, from the initial growth of a seed to its ageing and decline. There are various and complex roles of nitric oxide in the growth and development of plants (Singh et al. 2023a).

4.2.1 Regulation of Seed Germination and Dormancy

A plant's life cycle commences with its seed's germination, a process closely controlled by environmental stimuli and internal communication networks. Nitric oxide is believed to play a role in maintaining the seed germination process and dormancy release. Research has shown that the external application of nitric oxide NO or substances that release NO can enhance the function of seed germination. This indicates a direct relationship between NO levels and the shift from dormancy to germination. Nitric oxide NO seems to regulate the function of enzymes responsible for breaking down compounds associated with inactivity, thereby promoting the emergence of the embryonic plant (Jindal and Seth 2023).

4.2.2 Study of the Growth and Structure of Roots

The complex system of roots beneath the soil surface is crucial for the absorption of nutrients, the uptake of water, and the overall stability of plants. Nitric oxide is essential in determining the growth and structure of roots. NO influences cell division and elongation in the root meristem, thereby contributing to the growth of the primary source. In addition, nitric oxide plays a role in controlling the process of lateral root development, allowing the plant to adjust to different soil conditions. An effective root system development is essential for the plant's capacity to explore its surroundings and maximize the acquisition of resources (Singh et al. 2023a, b, c, d).

4.2.3 Control of Stomatal Aperture and the Process of Gas Exchange

Stomata, which are tiny openings found on the surfaces of leaves and stems, are vital in controlling the exchange of gases between the plant and its surroundings. Nitric oxide plays a role in regulating stomatal aperture, affecting the process of opening and closing these pores. Nitric oxide-mediated signalling pathways play a role in influencing the behaviour of guard cells, which in turn affects stomatal conductance and transpiration rates. The plant's ability to regulate gas exchange is crucial for maintaining the ideal water balance and promoting efficient photosynthesis (Madhu et al. 2024).

4.2.4 Shoot Growth and Elongation Refer to the Process by Which the Shoot of a Plant Increases in Length and Size

Nitric oxide promotes shoot growth and extension, which are crucial for forming above-ground plant structures. NO plays a role in controlling the growth and division of cells in shoot tissues. It regulates the function of enzymes and proteins that affect the movement and development of cell walls, thus affecting the elongation of stem and shoot structures. The precise equilibrium of nitric oxide levels significantly impacts the overall design of the plant, thereby influencing its capacity to harness sunlight and allocate resources efficiently (Khan et al. 2023).

4.2.5 Flowering and Reproductive Development

The shift from vegetative growth to reproductive development, characterized by the emergence of flowers, is a crucial stage in the life cycle of plants. Nitric oxide has been associated with the control of flowering time and the development of floral organs. Research has demonstrated that nitric oxide (NO) can impact the activation of genes related to the flowering process. It is a molecular switch that combines external environmental cues with internal signals. In addition, NO plays a role in regulating the growth of pollen tubes and the process of fertilization, thereby contributing to the successful completion of reproduction (Ahmed et al. 2023).

4.2.6 Senescence and Programmed Cell Death

During plants' natural ageing process, the end of specific tissues or entire organs becomes unavoidable. Nitric oxide plays a crucial role in the control of senescence and programmed cell death (PCD). It functions as a signalling molecule that regulates the expression of genes linked to senescence, impacting the degradation of chlorophyll and the reutilization of nutrients from ageing tissues. Precisely regulating NO levels is essential for maintaining a delicate equilibrium between plant growth and ageing mechanisms (Kaur et al. 2023).

4.2.7 Phytohormone Interactions

Nitric oxide is crucial in a complex signalling network closely linked with various phytohormones. NO modulates the biosynthesis, perception, and downstream signalling of hormones such as auxins, cytokinins, gibberellins, abscisic acid, and ethylene. These interactions facilitate the synchronization of growth and developmental processes, enabling the plant to adapt flexibly to internal and external stimuli (Chauhan et al. 2023).

4.2.8 Adaptation to Living Organisms and Non-living Environmental Factors

Although the impact of nitric oxide on growth and development is significant, its contribution to stress responses should not be disregarded. NO plays a crucial role in the plant's defence against living and non-living stresses. During periods of stress, NO functions as a signalling molecule that triggers defence mechanisms, such as the production of pathogenesis-related proteins and the reinforcement of antioxidant systems (Cuypers et al. 2023). The versatility of NO is emphasized by its ability to function as a molecular switch, redirecting resources from growth and development to stress adaptation as necessary. Nitric oxide plays a wide range of roles in plant growth and development, encompassing all stages of the plant's life cycle (Nawaz et al. 2023). Throughout the entire life cycle of a plant, starting from the initial process of seed germination and continuing until the final stage of senescence, NO plays a crucial role as a flexible regulator, effectively combining external environmental cues with internal biological processes (Naeem et al. 2024). The complex interplay between nitric oxide and different signalling pathways, including those related to phytohormones, highlights the intricacy of plant growth and development. Comprehending the intricate mechanisms of nitric oxide yields valuable knowledge for researchers and presents potential opportunities for manipulating plant growth to benefit agriculture and the environment (He et al. 2023a, b).

4.3 Nitric Oxide Synthesis in Plants

The synthesis of nitric oxide (NO) in plants is an intriguing and intricate process that has a crucial impact on various physiological and developmental functions. Previously known for its role in animal systems, the discovery of NO in plants has expanded the scope of research, uncovering its participation in various processes such as seed germination and defence mechanisms against pathogens(Cao et al. 2023a, b). This study examines the complex operations involved in plant nitric oxide production, focusing on the enzymes and pathways involved, the regulatory agencies controlling its production, and the various roles nitric oxide plays in plant biology (Altaf et al. 2023b). Nitric oxide, a diminutive gaseous molecule, has become a pivotal signalling molecule in plants, facilitating various responses to internal and external stimuli. The production of NO in plants is a tightly controlled process that relies on the function of different enzymes, mainly nitric oxide synthase (NOS)-like proteins, as well as non-enzymatic pathways. Identifying nitric oxide in plants has expanded our comprehension of signalling networks, linking environmental stimuli with complex cellular reactions (Dwivedi and Singh 2014; Wu and Liao 2023).

4.3.1 Pathways for the Synthesis of Nitric Oxide Through Enzymatic Processes

4.3.1.1 Proteins Similar to Nitric Oxide Synthase (NOS)

Although plants do not possess a gene homologous to the traditional NOS gene found in animals, they do have NOS-like proteins that play a role in nitric oxide synthesis. These proteins share structural and functional similarities with animal NOS and are vital in facilitating the conversion of L-arginine to NO. The discovery and analysis of NOS-like proteins have played a crucial role in understanding the enzymatic process of NO production in plants (Ali et al. 2023). The NR pathway refers to the process of nitrate reduction by the enzyme nitrate reductase. Nitrate reductase (NR) is a crucial enzyme in nitrate assimilation, a significant enzymatic pathway for nitric oxide synthesis in plants. Nitrate reductase (NR) converts nitrite into nitric oxide, offering an alternative NO production method. This pathway is especially significant in nitrogen metabolism, highlighting the incorporation of NO synthesis with nutrient assimilation processes (Dilawari et al. 2023).

4.3.1.2 Nitric Oxide Synthesis Without Enzymes

Plants can produce NO through non-enzymatic mechanisms, particularly when they are exposed to stressful conditions, in addition to their enzymatic pathways. Nitric oxide can be generated by reducing nitrite in the presence of different reductants, such as ascorbic acid or metal-containing proteins. Non-enzymatic trails provide a versatile and adaptable method for synthesizing NO, especially when faced with environmental challenges (Altaf et al. 2023c).

4.4 Control of Nitric Oxide Production

4.4.1 Calcium Signaling

Calcium ions (Ca^{2+}) are essential for controlling the synthesis of nitric oxide in plants. Fluctuations in the calcium concentration inside cells, typically caused by external factors or signals related to growth, can trigger the activation of calcium-dependent protein kinases (CDPKs). These CDPKs then regulate the function of enzymes involved in synthesizing nitric oxide. The complex interaction between calcium signalling and NO production emphasizes the intricate regulatory mechanisms that control these processes (Zhong et al. 2023).

4.4.2 Plant Hormones

Phytohormones like abscisic acid (ABA), salicylic acid (SA), and jasmonic acid (JA) are involved in controlling the production of nitric oxide. These signalling molecules function as intermediaries in plant reactions to different stresses and developmental processes, impacting the expression and function of enzymes involved in nitric oxide production. The interaction between phytohormones and NO grants plants a sophisticated and situation-dependent regulatory system (Czékus and Poór 2023).

4.5 The Roles of Nitric Oxide in Plants

4.5.1 The Process of Seed Germination and Dormancy

Nitric oxide is essential for controlling the process of seed germination and dormancy. Research has demonstrated that it improves germination rates and disrupts seed dormancy, suggesting its role in the initial phases of plant development (Seth et al. 2023).

4.5.2 Growth and Structure of Roots

NO exerts regulatory control over cell division, elongation, and the formation of lateral roots in plants. It aids in developing a robust root system, allowing plants to search the soil for water and nutrients effectively (Sharma et al. 2023b).

4.5.3 Control of Stomatal Aperture and Exchange of Gases

Nitric oxide plays a role in controlling the opening and closing of stomata, which affects the exchange of gases and the balance of water in plants. This role is crucial for enhancing the process of photosynthesis and adapting to dynamic environmental conditions (Singh et al. 2024).

4.5.4 Processes of Flowering and Reproduction

The regulation of the transition from vegetative growth to flowering is complex and highly controlled by NO. It exerts an influence on the development of floral organs, the change of pollen tubes, and the process of fertilization, thereby affecting the reproductive success of plants (Yadav et al. 2023).

4.5.5 Ageing and Apoptosis

Nitric oxide plays a role in regulating senescence and programmed cell death. It controls the degradation of chlorophyll and the redistribution of nutrients that occur naturally as plants age (Marothia et al. 2023).

4.5.6 Reaction to Living and Non-living Pressures

During periods of stress, such as when plants are attacked by pathogens or exposed to environmental pressures, they employ nitric oxide as a signalling molecule to activate defence responses. It triggers the activation of defence mechanisms, such as the production of pathogenesis-related proteins and antioxidant systems (Rahman et al. 2023). Plant nitric oxide production is a complex process encompassing both enzymatic and non-enzymatic pathways. The regulatory mechanisms that govern nitric oxide production highlight its versatility as a signalling molecule (Ghorbel and Brini 2023). The various roles of NO in plant growth, development, and stress response emphasize its fundamental significance in plant biology. Potential future research avenues could entail a more thorough investigation into the regulatory networks governing the synthesis of nitric oxide NO, the discovery of supplementary enzymatic pathways, and the utilization of this knowledge in agriculture to enhance crop quality and resistance to stress. Nitric oxide continues to fascinate researchers with its intricate role in the complex coordination of plant life (Sandalio et al. 2023).

4.6 Mechanism of Nitric Oxide in Plant Tolerance to Abiotic Stresses

Nitric oxide, previously regarded exclusively as an atmospheric contaminant, has been recognized as a crucial signalling molecule in plants. It serves as a central player in their reactions to diverse environmental pressures. Abiotic stresses, including drought, salinity, extreme temperatures, and exposure to heavy metals, present substantial obstacles to the growth and development of plants. The complex interaction between nitric oxide and the mechanisms contributing to plant tolerance to these stresses has become a central research focus (Kumari et al. 2022; Sharma et al. 2023a, b).

4.6.1 The Role of Nitric Oxide in the Perception of Stress

4.6.1.1 Production of Nitric Oxide as a Result of Stress

Plants react to abiotic stresses by stimulating nitric oxide production. Different forms of stress induce the activation of enzymes responsible for producing nitric oxide, such as nitrate reductase and nitric oxide synthase-like proteins (Cao et al. 2023a, b).

4.6.1.2 NO as a Stress Signal Transducer

Nitric oxide functions as a secondary messenger in the process of stress signal transduction. It regulates the activity of stress-responsive genes and proteins, triggering adaptive reactions (Ferrante and Francini 2023).

4.6.2 Control of Reactive Oxygen Species (ROS) Concentrations

4.6.2.1 ROS as Signalling Molecules

Abiotic stresses frequently produce reactive oxygen species (ROS) within plant cells. NO plays a role in controlling reactive oxygen species (ROS) levels, thereby preventing oxidative harm and enhancing the ability to withstand stress (Rosenkranz et al. 2023).

4.6.2.2 Interaction Between NO and ROS

Nitric oxide modulates generating and eliminating reactive oxygen species (ROS) by interacting with these molecules. Nitric oxide-mediated antioxidant responses play a role in preserving cellular redox homeostasis during periods of stress (Dhiman et al. 2023).

4.6.3 Regulation of Osmotic and Ionic Homeostasis

4.6.3.1 Water Deficit Stress

During periods of drought, nitrogen oxide plays a crucial role in controlling the closure of stomata, thereby decreasing the amount of water lost. The increased aquaporin activity caused by nitric oxide promotes water movement and helps keep cells hydrated.

4.6.3.2 Salinity Stress

Nitric oxide (NO) regulates ion homeostasis during periods of salinity stress. It controls the activity of ion transporters, thereby mitigating the detrimental consequences of excessive salt buildup in plant cells (Singh et al. 2023b).

4.6.4 Regulation of Antioxidant Defence Mechanisms

4.6.4.1 Enzymatic Antioxidants

Nitric oxide augments the functions of antioxidant enzymes, such as superoxide dismutase (SOD) and catalase. It functions as a signalling molecule, activating pathways that enhance the production of antioxidant defence mechanisms.

4.6.4.2 Non-enzymatic Antioxidants

Nitric oxide affects the build up of non-enzymatic antioxidants, such as ascorbic acid and glutathione. These antioxidants are vital in removing reactive oxygen species (ROS) and reducing oxidative stress (Mariyam et al. 2023).

4.6.5 Phytohormone Interactions

4.6.5.1 Abscisic Acid (ABA)

NO interacts with ABA, a pivotal phytohormone implicated in stress responses. This interaction regulates the process of stomatal closure, thereby improving the efficiency of water usage in plants during periods of stress.

4.6.5.2 Salicylic Acid (SA) and Jasmonic Acid (JA)

Salicylic Acid (SA) and Jasmonic Acid (JA) are two types of acids. The activation of defence mechanisms against biotic and abiotic stresses is not influenced by cross-talk between SA and JA. The interaction between NO and these plant hormones plays a role in developing stress tolerance (Kaya et al. 2023b).

4.6.6 Gene Expression and Signal Transduction Through Nitric Oxide

Genes that do not respond to NO stimulation: Nitric oxide modulates stress-responsive genes' expression, which encodes stress tolerance proteins. NO modulates transcription factors and downstream effectors to coordinate adaptive responses.

4.6.6.1 Protein S-Nitrosylation

Nitric oxide is involved in S-nitrosylation, a post-translational modification affecting proteins. S-nitrosylation modulates protein function, durability, and associations, contributing to stress adaptation (Pal et al. 2023).

4.6.7 The Process of Priming and Its Impact on Memory Recall

4.6.7.1 Priming Induced by Nitric Oxide

Plant exposure to sub-lethal stress levels and the presence of NO primes them to develop enhanced stress tolerance. Priming refers to the prior activation of defence mechanisms, strengthening the plant's capacity to endure subsequent stress events (Ahmed et al. 2023; Feng et al. 2023; Jindal and Seth 2023; Mitra et al. 2023; Sodhi and Saxena 2023).

4.6.7.2 Nitric Oxide-Mediated Memory Responses

Nitric oxide NO triggers memory responses, enabling plants to retain information about past encounters with stress. This memory enhances the speed and strength of the response when faced with future stressful situations.

4.6.8 Prospects and Obstacles for the Future

4.6.8.1 Technological Innovations

Employ cutting-edge technologies, such as omics methodologies and bioinformatics, to decipher the complex networks governed by nitric oxide—advanced instruments for immediately tracking NO fluctuations in plants exposed to various stress conditions (Altaf et al. 2023a; Lone et al. 2023; Seabra and Tortella 2023; Singh et al. 2023a).

4.6.8.2 Developing Nitric Oxide-Responsive Characteristics in Engineering

Utilizing NO-related pathways to cultivate stress-tolerant crop varieties. Challenges and ethical considerations in utilizing genetic engineering to enhance stress tolerance mediated by nitric oxide.

The role of nitric oxide in enabling plants to withstand abiotic stresses is a complex and interconnected system of signalling pathways, molecular responses, and physiological adaptations. The complex interaction between nitric oxide and different cellular processes highlights its crucial function as a signalling molecule in coordinating plant responses to environmental challenges. Advancements in research will lead to a more profound comprehension of these mechanisms, which will benefit the fundamental study of plant biology and facilitate the creation of inventive approaches to improve crop resilience and promote sustainable agriculture in response to evolving environmental conditions (Table 4.1).

4.7 Mechanism of Nitric Oxide in Plant Tolerance to Biotic Stress

Nitric oxide (NO) is crucial in the complex and ever-changing signalling networks that regulate plant responses to biotic stresses. Pathogens and herbivores, which are biotic stressors, present substantial risks to the health and productivity of plants. An essential area of current research is the exploration of the mechanisms by which nitric oxide enhances plant resilience against these challenges. This study investigates the intricate relationship between nitric oxide and different molecular, biochemical, and physiological processes constituting plants' defence mechanisms to respond to biotic stressors (Table 4.2).

Table 4.1 Various mechanisms are activated under nitric oxide (NO) regulation in response to abiotic stress

Mechanism	Description
1. Antioxidant enzyme activation	NO regulates the activity of antioxidant enzymes, such as superoxide dismutase and catalase, helping plants combat oxidative stress induced by abiotic factors
2. Modulation of reactive oxygen species (ROS)	NO interacts with ROS, influencing the balance between reactive oxygen species and antioxidants in plant cells, thereby reducing oxidative damage
3. Enhancement of photosynthesis	NO participates in the regulation of photosynthetic processes, improving the efficiency of photosynthesis under stressful conditions like drought or high salinity
4. Stomatal regulation	NO controls the stomatal aperture, affecting water loss and gas exchange, which is crucial for plants facing challenges like drought and high temperatures
5. Regulation of ion homeostasis	NO plays a role in maintaining ion homeostasis, helping plants cope with salinity stress by regulating the uptake and distribution of ions
6. Osmotic adjustment	NO contributes to osmotic adjustment in plants, aiding in the accumulation of compatible solutes and reducing cellular water loss under conditions like drought
7. Activation of stress-responsive genes	NO induces the expression of genes involved in stress responses, synthesizing proteins that enhance plant tolerance to abiotic stressors
8. Hormonal crosstalk	NO interacts with various plant hormones, such as abscisic acid and auxins, to modulate hormonal signalling pathways involved in stress responses
9. Cell wall modification	NO participates in cell wall modifications, influencing the flexibility and strength of cell walls, which is essential for plant adaptation to environmental challenges
10. Enhancement of nitrogen metabolism	NO regulates nitrogen metabolism, influencing nutrient uptake and assimilation, crucial for plant growth under stressful conditions

Source: Altaf et al. (2023a); Lone et al. (2023); Seabra and Tortella (2023); Singh et al. (2023a)

4.7.1 Absence of Signalling in the Perception of Biotic Stress

4.7.1.1 Identification of Pathogen-Associated Molecular Patterns (PAMPs)

Plants detect potential pathogens by recognizing conserved PAMPs. Nitric oxide is produced in response to identifying pathogen-associated molecular patterns (PAMPs), acting as an initial signalling event in the plant's immune response against pathogens.

4.7.1.2 Effector-Triggered Immunity (ETI)

Pathogens release effector molecules that stimulate a robust immune response called ETI. Nitric oxide production plays a crucial role in the ETI (Effector-Triggered Immunity) process, as it contributes to initiating subsequent defence responses.

Table 4.2 Diverse mechanisms are triggered in response to biotic stress under the regulation of nitric oxide (NO)

Mechanism	Description
1. Induction of defence genes	NO can activate the expression of various defence-related genes, enhancing the production of proteins involved in plant defence mechanisms
2. Antioxidant enzyme activation	NO can modulate the activity of antioxidant enzymes such as catalase, superoxide dismutase, and peroxidase, helping plants to counteract oxidative stress induced by pathogens
3. Regulation of cell signaling	NO acts as a signalling molecule in plant cells, activating various signalling pathways crucial in defence responses against biotic stress
4. Enhancement of phytoalexin production	NO regulates phytoalexin biosynthesis, which are antimicrobial compounds produced by plants in response to pathogen attacks
5. Stomatal regulation	NO has been implicated in the modulation of stomatal closure, influencing the plant's ability to control water loss and pathogen entry through stomatal pores
6. Interaction with reactive oxygen species (ROS)	NO can interact with ROS to form peroxynitrite, which may have antimicrobial properties and contribute to the plant's defence against invading pathogens
7. Modulation of hormone signalling	NO interacts with various plant hormones such as salicylic acid, jasmonic acid, and ethylene, influencing hormonal signalling pathways in plant defence against biotic stress
8. Regulation of programmed cell death (PCD)	NO has been shown to play a role in modulating programmed cell death, a critical aspect of the plant's defence response against pathogen invasion

Sources: Czékus and Poór (2023); Ferrante and Francini (2023); Marothia et al. (2023); Sandalio et al. (2023); Rahman et al. (2023); Yadav et al. (2023)

4.7.2 Interaction Between Reactive Oxygen Species (ROS) and Nitric Oxide

4.7.2.1 ROS as Second Messengers

Biotic stress triggers the production of reactive oxygen species (ROS) in plant cells. Nitric oxide and reactive oxygen species (ROS) interact, creating a dynamic relationship that influences defence signalling pathways.

4.7.2.2 The Function of Nitric Oxide in Reactive Oxygen Species Homeostasis

Nitric oxide regulates reactive oxygen species (ROS) levels, preventing excessive oxidative harm during the plant's defence mechanisms. The intricate equilibrium between nitric oxide (NO) and reactive oxygen species (ROS) guarantees a proficient and regulated defence mechanism (Basit et al. 2023; Devi et al. 2023; Jindal and Seth 2023; Mahiwal et al. 2024; Mitra et al. 2023; Seabra and Tortella 2023; Singh et al. 2023a).

4.7.3 Defense Gene Activation and Expression

4.7.3.1 Transcriptional Regulation Mediated by Nitric Oxide (NO)

Nitric oxide modulates the expression of genes associated with defence mechanisms. Transcription factors influenced by nitric oxide regulate the expression of genes responsible for producing antimicrobial compounds and proteins related to the development of diseases.

4.7.3.2 Induction of Secondary Metabolites

Nitric oxide stimulates the production of secondary metabolites that possess antimicrobial properties, such as phytoalexins. These compounds are essential in preventing the growth of pathogens and restricting the spread of infections.

4.7.4 Control of Phytohormones

4.7.4.1 Salicylic Acid (SA) Signalling

Nitric oxide interacts with the SA signalling pathway, a central regulator of plant defence against biotrophic pathogens. Nitric oxide promotes the accumulation of SA, thereby enhancing the plant's resistance to pathogens.

4.7.4.2 Interactions Between Jasmonic Acid (JA) and Ethylene (ET)

NO regulates the signalling pathways of JA and ET, which play a crucial role in plant defence against necrotrophic pathogens. NO, JA, and ET interaction contributes to a harmonized and efficient defence response (Basit et al. 2023; Devi et al. 2023; Jindal and Seth 2023; Mahiwal et al. 2024; Mitra et al. 2023; Seabra and Tortella 2023; Singh et al. 2023a).

4.7.5 Regulation of Programmed Cell Death (PCD)

4.7.5.1 Nitric Oxide as a Modulator of Programmed Cell Death

Apoptosis is an essential immune response defence mechanism against invading pathogens. Nitric oxide functions as a regulator of programmed cell death, impacting the timing and magnitude of cell demise in reaction to biotic stress.

4.7.5.2 HR-PCD (Hypersensitive Response-Programmed Cell Death)

NO triggers HR-PCD, a specific and swift cell death reaction observed in plant-pathogen interactions. HR-PCD hinders the dissemination of pathogens by sacrificing cells that have been infected.

4.7.6 Facilitating and Recalling Memories

4.7.6.1 Priming Induced by Nitric Oxide

The exposure to NO prepares plants to have a more robust defence response when pathogens attack them in the future. Priming entails stimulating defence mechanisms, resulting in accelerated and enhanced reactions.

4.7.6.2 Nitric Oxide-Mediated Memory Responses

NO triggers memory responses in plants, enabling them to retain information about past encounters with pathogens. This memory facilitates a more effective and targeted defence response upon re-exposure to the same or related pathogens (Cao et al. 2023a, b; Ghorbel and Brini 2023; Pal et al. 2023).

4.7.7 Interactions with Microbial Consortia

4.7.7.1 The Role of Nitric Oxide in Promoting Beneficial Microorganisms

NO engages in plant interactions with symbiotic microorganisms, such as mycorrhizal fungi and rhizobacteria. These interactions facilitate the enhancement of plant health and bolster resistance against pathogens.

4.7.7.2 Cross-Kingdom Signaling

NO plays a crucial role in cross-kingdom signalling, modulating the communication between plants and microbes. Comprehending this communication is crucial for formulating sustainable and ecologically conscious approaches to pathogen management.

The role of nitric oxide in plant resistance to living organisms' stress is a complex and ever-changing process that includes intricate communication networks, gene control, and physiological reactions. As scientific investigation progresses in understanding the intricacies of NO-mediated plant defence, the possibility of utilizing this knowledge for sustainable agriculture becomes more evident. The interaction between NO and different defence mechanisms underscores the adaptability of plants in responding to biotic challenges. Studying these mechanisms not only increases our knowledge of how plants and microbes interact but also provides opportunities for new methods to improve the ability of crops to withstand and reduce the effects of biological stressors on global food security (Ahmed et al. 2023; Altaf et al. 2023b; Cuypers et al. 2023; Kaur et al. 2023; Khan et al. 2023).

4.8 Crosstalk with Signaling Compounds: H_2O_2, H_2S, and Ca^{2+}

The complex signalling networks within plant cells involve a dynamic interaction between different signalling compounds. Nitric oxide is critical in regulating various physiological processes. Nevertheless, the intercommunication between NO and other signalling molecules, including hydrogen peroxide (H_2O_2), hydrogen sulphide (H_2S), and calcium ions (Ca^{2+}), introduces additional levels of intricacy to the control of plant reactions. This study examines the complex interactions and communication mechanisms between NO and other signalling compounds. It aims to understand how these pathways work together to regulate plant growth, development, and responses to the environment.

4.8.1 Hydrogen Peroxide (H_2O_2)

Hydrogen Peroxide (H_2O_2) is a chemical compound of two hydrogen and two oxygen atoms. Crosstalk refers to the communication and interaction between different components or systems. Redox signalling refers explicitly to the network of chemical reactions involved in the transfer of electrons, which play a crucial role in cellular signalling processes. H_2O_2 plays a vital role as a signalling molecule in redox signalling pathways. The interaction between NO and H_2O_2 includes both cooperative and opposing interactions.

4.8.2 ROS Balance and Homeostasis

Nitric oxide NO regulates hydrogen peroxide (H_2O_2) levels, thereby preserving redox homeostasis. The intricate equilibrium between nitric oxide (NO) and hydrogen peroxide (H_2O_2) affects cellular reactions to stress and developmental signals.

4.8.3 Cooperative Roles in Defence Responses

NO and H_2O_2 work together in the plant's defence against pathogens. They cooperatively stimulate the expression of defence genes and trigger programmed cell death (PCD) as part of the hypersensitive response (Sharma et al. 2023a, b; Singh et al. 2024; Rahman et al. 2023).

4.8.4 Hydrogen Sulphide (H_2S) Is a Chemical Compound

4.8.4.1 Hydrogen Sulphide as a Signalling Molecule

Hydrogen sulphide (H_2S) is becoming increasingly important in plant signalling pathways. It interacts with NO to regulate multiple physiological processes, such as stomata closure and root growth.

4.8.4.2 Non-hydrogen Sulphide (H_2S) Interactions in Stress Responses

The exchange of information between NO and H_2S plays a role in controlling stress responses. Both molecules influence antioxidant systems and mitigate oxidative stress in the face of environmental challenges.

4.8.4.3 Role in Seed Germination

Nitric oxide (NO) and hydrogen sulphide (H_2S) work together to regulate seed germination. Their combined and cooperative effect impacts the shift from seed dormancy to germination, guaranteeing the most favourable conditions for plant establishment (Cao et al. 2023a, b; Ghorbel and Brini 2023; Marothia et al. 2023; Sharma et al. 2023b; Rahman et al. 2023; Yadav et al. 2023).

4.8.5 Intercommunication of Calcium Ions (Ca^{2+})

4.8.5.1 Calcium Ions (Ca^{2+}) as a Widely Applicable Second Messenger

Calcium ions (Ca^{2+}) serve as ubiquitous second messengers in plant signalling. The interaction between NO and Ca^{2+} is bidirectional, forming a complex regulatory network.

4.8.5.2 Calcium-Dependent Nitric Oxide Synthesis

Calcium ions control nitric oxide production by activating proteins similar to nitric oxide synthase (NOS). The production of NO, dependent on the presence of Ca^{2+}, connects cellular calcium dynamics with signalling pathways that NO mediates.

4.8.5.3 Calcium-NO Interactions in Stomatal Regulation

NO and calcium ions (Ca^{2+}) work together to control the opening and closing of stomata. The synchronized activity of these signalling molecules regulates the behaviour of guard cells, thereby impacting the process of gas exchange and maintaining water balance.

4.8.6 Integration in Growth and Development

4.8.6.1 Root Architecture and Elongation

The compounds H_2O_2, H_2S, and Ca^{2+} have a combined effect on root development. Their intercommunication influences the processes of cell division, elongation, and lateral root formation, thereby determining the overall structure of the root system (Ahmed et al. 2023; Basit et al. 2023; Devi et al. 2023; Jindal and Seth 2023; Madhu et al. 2024; Singh et al. 2023a, c).

4.8.6.2 Flowering and Reproductive Processes

The regulation of flowering is influenced by the interactions among NO, H_2O_2, H_2S, and Ca^{2+}. Crosstalk impacts the development of floral organs, the growth of pollen tubes, and the achievement of successful fertilization.

4.8.6.3 Senescence and Aging

Nitric oxide (NO), hydrogen peroxide (H_2O_2), hydrogen sulphide (H_2S), and calcium ions (Ca^{2+}) are involved in regulating senescence and ageing. Their collaborative and conflicting actions dictate the timing and advancement of intrinsic ageing in plant tissues.

4.8.7 Environmental Sensing and Adaptive Responses

4.8.7.1 Responses to Abiotic Stress

Negative hydrogen peroxide (H_2O_2), hydrogen sulphide (H_2S), and calcium ions (Ca^{2+}) cooperate in plant reactions to non-living environmental pressures. Their cohesive actions impact the adaptive responses to environmental factors such as drought, salinity, and extreme temperatures.

4.8.7.2 Biotic Stress Interactions

Exchanging information between these signalling compounds is essential for efficient protection against pathogens.

4.8.8 Prospects and Obstacles for the Future

4.8.8.1 Technological Innovations in Signal Monitoring

Employing cutting-edge technologies to monitor signalling dynamics in real-time and employing omics methodologies to unravel the complex molecular networks regulated by NO, H_2O_2, H_2S, and Ca^{2+}.

4.8.8.2 Engineering Strategies for Enhancing Crop Improvement

Altering signalling pathways can be used to enhance crop performance and increase tolerance to environmental stressors. Challenges and ethical considerations in using genetic engineering to improve traits related to crosstalk are taken into consideration.

The interaction between nitric oxide, hydrogen peroxide, hydrogen sulphide, and calcium ions in plant signalling pathways enhances the complexity of cellular communication. The interactions between these signalling compounds, which can enhance or inhibit each other, determine how plants respond to environmental stimuli. This, in turn, affects their growth, development, and ability to tolerate stress. As research progresses in understanding the intricacies of crosstalk, it becomes more apparent that these pathways can be manipulated for agricultural and environmental purposes. The complex interaction among NO, H_2O_2, H_2S, and Ca^{2+} underscores the integrative character of plant signaling networks and presents promising possibilities for future progress in plant biology and crop science (Basit et al. 2023; Devi et al. 2023; Feng et al. 2023; Mahiwal et al. 2024; Mitra et al. 2023; Pramanik et al. 2023; Seabra and Tortella 2023; Singh et al. 2023d).

4.9 Future Direction

Future research should prioritize the investigation of the complex interconnections between nitric oxide and other signalling molecules in various environmental stress conditions. Utilizing advanced systems biology methodologies, such as omics technologies, enables the comprehensive mapping of the entire signalling cascade and the identification of crucial interaction points. Incorporating genomics, transcriptomics, proteomics, and metabolomics will comprehensively comprehend the interactions between NO and other signaling molecules. Discovery and analysis of new signalling molecules that interact with NO during periods of stress. This study investigates the role of these molecules in plant adaptation and stress tolerance, which could be used as targets for genetic engineering or agronomic interventions. Stress-specific crosstalk mechanisms to elucidate the interaction between NO and signaling molecules in various environmental stressors should be explored. The future directions in studying the role of nitric oxide and its interaction with signalling molecules under environmental stress in plants require a comprehensive and integrated approach. The progress in technology and a more profound comprehension of molecular and physiological mechanisms will facilitate the development of inventive methods to bolster plant resistance, optimize agricultural productivity, and tackle the difficulties presented by a shifting environment. These future directions have significant potential for promoting sustainable agriculture and ecosystem management.

4.10 Conclusion

Recent progress in comprehending the function of NO and its intricate interaction with signalling molecules has shed light on the complex network of plant reactions to environmental stress. The extensive participation of NO in facilitating stress adaptation has been a central focus of the scientific investigation, uncovering its crucial function as a signalling molecule in coordinating a wide range of cellular mechanisms. The interaction between NO and other signalling molecules, including hydrogen peroxide (H_2O_2), hydrogen sulphide (H_2S), and calcium ions (Ca^{2+}), has been recognized as a crucial regulatory mechanism that underlies plant responses to various environmental challenges. These interactions create dynamic networks that precisely regulate stress signalling pathways, allowing plants to navigate and adjust to constantly changing environmental conditions. Significantly, the interaction between NO and these signalling molecules goes beyond the stress response, impacting essential cellular processes such as redox homeostasis and hormonal regulation. The interconnectedness of signalling networks and the study of these interactions are crucial for a comprehensive understanding of plant stress physiology. The recent advancements in understanding NO-mediated signalling pathways and their interaction with other molecules lay a strong groundwork for devising novel approaches to improve plant resilience to environmental stress. Deciphering the complexities of these molecular interactions not only enhances our theoretical comprehension but also creates opportunities for the practical implementation of this knowledge in agriculture, thereby aiding the growth of stress-tolerant crop varieties and sustainable agricultural practices. The ongoing research in this field offers the possibility of utilising NO-mediated signalling pathways to reduce the adverse effects of environmental stress on plants. This can contribute to a more resilient and adaptable future in agriculture and ecosystem management.

References

Ahmed I, Kumar A, Bheri M, Srivastava AK, Pandey GK (2023) Glutamate receptor like channels: emerging players in calcium mediated signaling in plants. Int J Biol Macromol 234:123522. https://doi.org/10.1016/j.ijbiomac.2023.123522

Ali S, Tyagi A, Bae H (2023) ROS interplay between plant growth and stress biology: challenges and future perspectives. Plant Physiol Biochem 203:108032. https://doi.org/10.1016/j.plaphy.2023.108032

Altaf MA, Sharma N, Singh J, Samota MK, Sankhyan P, Singh B, Kumar A, Naz S, Lal MK, Tiwari RK, Kumar R (2023a) Mechanistic insights on melatonin-mediated plant growth regulation and hormonal cross-talk process in Solanaceous vegetables. Sci Hortic 308:111570. https://doi.org/10.1016/j.scienta.2022.111570

Altaf MA, Naz S, Kumar R, Sardar H, Nawaz MA, Kumar A, Lal P, Ahmad R, Hayat F, Wani MA, Tiwari RK, Lal MK (2023b) Unraveling the mechanisms of cadmium toxicity in horticultural plants: implications for plant health. S Afr J Bot 163:433–442. https://doi.org/10.1016/j.sajb.2023.10.064

Altaf MA, Naz S, Kumar R, Tiwari RK, Lal MK (2023c) Chapter 10—Role of nitric oxide in regulating physiological and molecular aspects of plants under abiotic stresses. In: Sharma A, Pandey S, Bhardwaj R, Zheng B, Tripathi DK (eds), The role of growth regulators and phytohormones in overcoming environmental stress. Academic, pp 235–248

Basit F, Bhat JA, Alyemeni MN, Shah T, Ahmad P (2023) Nitric oxide mitigates vanadium toxicity in soybean (glycine Max L.) by modulating reactive oxygen species (ROS) and antioxidant system. J Hazard Mater 451:131085. https://doi.org/10.1016/j.jhazmat.2023.131085

Cao Y, Du PH, Shang YW, Ji JH, Tan LQ, Zhang X, Xu JH, Liang BW (2023a) Melatonin and dopamine alleviate waterlogging stress in apples by recruiting beneficial endophytes to enhance the physiological resilience1. J Integr Agric. https://doi.org/10.1016/j.jia.2023.12.012

Cao Y, Ma C, Hao Y, Tan Q, Dhankher OP, White JC, Xing B (2023b) The role of sulfur nutrition in plant response to metal(loid) stress: facilitating biofortification and phytoremediation. J Hazard Mater 443:130283. https://doi.org/10.1016/j.jhazmat.2022.130283

Chauhan J, Prathibha MD, Singh P, Choyal P, Mishra UN, Saha D, Kumar R, Anuragi H, Pandey S, Bose B, Mehta B, Dey P, Dwivedi KK, Gupta NK, Singhal RK (2023) Plant photosynthesis under abiotic stresses: damages, adaptive, and signaling mechanisms. Plant Stress 10:100296. https://doi.org/10.1016/j.stress.2023.100296

Cuypers A, Vanbuel I, Iven V, Kunnen K, Vandionant S, Huybrechts M, Hendrix S (2023) Cadmium-induced oxidative stress responses and acclimation in plants require fine-tuning of redox biology at subcellular level. Free Radic Biol Med 199:81–96. https://doi.org/10.1016/j.freeradbiomed.2023.02.010

Czékus Z, Poór P (2023) The role of nitric oxide in systemic responses of plants. In: Khan MIR, Iqbal N, Poór P, Ferrante A, Pratap Singh V, Tripathi DK, Fotopoulos V (eds), Nitric oxide in developing plant stress resilience. Plant gasotransmitters and molecules with hormonal activity. Academic, pp 217–231

Devi K, Dhiman S, Kour J, Ibrahim M, Bhardwaj T, Sharma N, Madaan I, Sidhu A, Ohri P, Singh AP, Sirhindi G, Mir BA, Bhardwaj R (2023) Chapter 9—Regulatory role of nitric oxide in plants and its crosstalk with phytohormones. In: Khan MIR, Iqbal N, Poór P, Ferrante A, Pratap Singh V, Tripathi DK, Fotopoulos V (eds) Nitric oxide in developing plant stress resilience. Plant gasotransmitters and molecules with hormonal activity. Academic, pp 173–200

Dhiman S, Ibrahim M, Khanna K, Bhardwaj T, Devi K, Sharma I, Arora U, Mir BA, Bhardwaj R (2023) Chapter 16—Cross-talk between ROS, RNS, RCS, and RSS in plants under abiotic stresses. In: Khan MIR, Iqbal N, Poór P, Ferrante A, Pratap Singh V, Tripathi DK, Fotopoulos V (eds) Nitric oxide in developing plant stress resilience. Plant gasotransmitters and molecules with hormonal activity. Academic, pp 305–326

Dilawari R, Mehta S, Singh B, Pandey V (2023) Switching role of hydrogen sulfide in amelioration of metal stress in plant. In: Ghorbanpour M, Adnan Shahid M (eds) Plant stress mitigators. Academic, pp 187–99

Dwivedi P, Singh BN (2014). Nitric oxide in plants: sources, methods of detection and role as signaling agent in plant responses to biotic stresses. In: Singh AL (ed) Advances in crop physiology. Daya Publishing House, New Delhi, pp 261–278

Feng L, Wei L, Liu Y, Ren J, Liao W (2023) Carbon monoxide/heme oxygenase system in plant: roles in abiotic stress response and crosstalk with other signals molecules. Nitric Oxide 138–139:51–63. https://doi.org/10.1016/j.niox.2023.06.005

Ferrante A, Francini A (2023) Cross-talk of NO between RNS, ROS, and RSS under abiotic stress. In: Khan MIR, Iqbal N, Poór P, Ferrante A, Pratap Singh V, Tripathi DK, Fotopoulos V (eds) Nitric oxide in developing plant stress resilience. Plant gasotransmitters and molecules with hormonal activity. Academic, pp 125–34

Ghorbel M, Brini F (2023) Role of brassinosteroids in regulating physiological and molecular aspects of plants under abiotic stress. In: Sharma A, Pandey S, Bhardwaj R, Zheng B, Tripathi DK (eds), The role of growth regulators and phytohormones in overcoming environmental stress. Academic, pp 197–233

He B, Zhang Z, Huang Z, Xirui Duan Y, Wang JC, Li L, He K, Nice EC, He W, Gao W, Shen Z (2023a) Protein persulfidation: rewiring the hydrogen sulfide signaling in cell stress response. Biochem Pharmacol 209:115444. https://doi.org/10.1016/j.bcp.2023.115444

He Y, Zhou Y, Zhou Z, He J, Liu Y, Xiao Y, Long L, Deng O, Xiao H, Shen F, Deng S, Luo L (2023b) Allelopathic effect of pyrogallic acid on cyanobacterium Microcystis aeruginosa: the regulatory role of nitric oxide and its significance for controlling harmful algal blooms (HABs). Sci Total Environ 858:159785. https://doi.org/10.1016/j.scitotenv.2022.159785

Jindal A, Seth CS (2023) Chapter 12—Nitric oxide mediated post-translational modifications and its significance in plants under abiotic stress. In: Khan MIR, Iqbal N, Poór P, Ferrante A, Pratap Singh V, Tripathi DK, Fotopoulos V (eds) Nitric oxide in developing plant stress resilience. Plant gasotransmitters and molecules with hormonal activity. Academic, pp 233–250

Kaur H, Tashima SS, Hussain SJ (2023) Interplay between phytohormones and hydrogen sulfide regulates plant growth and development under normal and abiotic stresses. In: Khan MIR, Singh A, Poór P (eds) Plant hormones in crop improvement. Academic, pp 265–281

Kaya C, Ugurlar F, Ashraf M, Ahmad P (2023a) Salicylic acid interacts with other plant growth regulators and signal molecules in response to stressful environments in plants. Plant Physiol Biochem 196:431

Kaya C, Ugurlar F, Ashraf M, Alam P, Ahmad P (2023b) Nitric oxide and hydrogen sulfide work together to improve tolerance to salinity stress in wheat plants by upraising the AsA-GSH cycle. Plant Physiol Biochem 194:651–663. https://doi.org/10.1016/j.plaphy.2022.11.041
Khan MSS, Ahmed S, Ul Ikram A, Hannan F, Yasin MU, Wang J, Zhao B, Islam F, Chen J (2023) Phytomelatonin: a key regulator of redox and phytohormones signaling against biotic/abiotic stresses. Redox Biol 64:102805. https://doi.org/10.1016/j.redox.2023.102805
Kumari A, Binny S, Singh BN, Dwivedi P (2022) Nitric oxide: a dynamic signaling molecule under plant stress. In: Ahanger MA, Ahmad P (eds) Nitric oxide in plants: a molecule with dual roles. Wiley, Hoboken
Lone R, Hassan N, Bashir B, Rohela GK, Malla NA (2023) Role of growth elicitors and microbes in stress management and sustainable production of sorghum. Plant Stress 9:100179. https://doi.org/10.1016/j.stress.2023.100179
Madhu AS, Kaur A, Singh K, Upadhyay SK (2024) Modulation in gene expression and enzyme activity suggested the roles of monodehydroascorbate reductase in development and stress response in bread wheat. Plant Sci 338:111902. https://doi.org/10.1016/j.plantsci.2023.111902
Mahiwal S, Pahuja S, Pandey GK (2024) Review: structural-functional relationship of WRKY transcription factors: unfolding the role of WRKY in plants. Int J Biol Macromol 257:128769. https://doi.org/10.1016/j.ijbiomac.2023.128769
Mariyam S, Bhardwaj R, Khan NA, Sahi SV, Seth CS (2023) Review on nitric oxide at the forefront of rapid systemic signaling in mitigation of salinity stress in plants: crosstalk with calcium and hydrogen peroxide. Plant Sci 336:111835. https://doi.org/10.1016/j.plantsci.2023.111835
Marothia D, Kaur N, Jhamat C, Sharma I, Pati PK (2023) Plant lectins: classical molecules with emerging roles in stress tolerance. Int J Biol Macromol 244:125272. https://doi.org/10.1016/j.ijbiomac.2023.125272
Mitra M, Agarwal P, Roy S (2023) Plant response to heavy metal stress: an insight into the molecular mechanism of transcriptional regulation. In: Srivastava V, Mishra S, Mehrotra S, Upadhyay SK (eds) Plant transcription factors. Contribution in development, metabolism, and environmental stress. Academic, pp 337–367
Naeem M, Gill SS, Aftab T, Tuteja N (2024) Editorial: crop improvement and plant resilience to abiotic stresses. Plant Sci 339:111958. https://doi.org/10.1016/j.plantsci.2023.111958
Nawaz H, Irum A, Nasim W, Hussain N, Usman M, Alam J (2023) Chapter 1—Hormonal cross-talk mechanisms and plant immunity or defense: an overview. In: Husen A, Zhang W (eds) Hormonal cross-talk, plant defense and development. Plant biology, sustainability and climate change. Academic, pp 1–12
Pal P, Kumar A, Ansari MI (2023) Chapter 7—Role of nitric oxide in regulation of biotic and abiotic stresses tolerance in plants. In: Khan MIR, Iqbal N, Poór P, Ferrante A, Pratap Singh V, Tripathi DK, Fotopoulos V (eds) Nitric oxide in developing plant stress resilience. Plant gasotransmitters and molecules with hormonal activity. Academic, pp 135–155
Pramanik B, Sar P, Bharti R, Gupta RK, Purkayastha S, Sinha S, Chattaraj S, Mitra D (2023) Multifactorial role of nanoparticles in alleviating environmental stresses for sustainable crop production and protection. Plant Physiol Biochem 201:107831. https://doi.org/10.1016/j.plaphy.2023.107831
Rahman U, Shafeeq YL, Hussain S, Hussain B, Khan W-u-D, Riaz L, Ashraf MN, Khaliq MA, Zhenjie D, Cheng H (2023) Role of Phytohormones in heavy metal tolerance in plants: a review. Ecol Indic 146:109844. https://doi.org/10.1016/j.ecolind.2022.109844
Rosenkranz M, Shi H, Ballauff J, Schnitzler JP, Polle A (2023) Chapter 8—Reactive oxygen species (ROS) in mycorrhizal fungi and symbiotic interactions with plants. In: Van Breusegem F, Mittler R (eds) Advances in botanical research. Oxidative stress response in plants. vol 105. Academic, pp 239–275
Sandalio LM, Collado-Arenal AM, Romero-Puertas MC (2023) Deciphering peroxisomal reactive species interactome and redox signalling networks. Free Radic Biol Med 197:58–70. https://doi.org/10.1016/j.freeradbiomed.2023.01.014
Seabra AB, Tortella GR (2023) Chapter 9—Biochemical role of gasotransmitters on plant growth, development, biomass, and tolerance under as stress. In: Singh VP, Tripathi DK, Fotopoulos

V, Lemos Batista B, Tortella Fuentes G, Da Silva Lobato AK (eds) The role of gasotransmitters in the amelioration of arsenic toxicity in plants. Plant gasotransmitters and molecules with hormonal activity. Academic, pp 167–85

Seth T, Sejal Asija M Khan IR, Iqbal N (2023) Chapter 3—Emerging role of nitric oxide in combating abiotic stresses in crops. In: Khan MIR, Iqbal N, Poór P, Ferrante A, Pratap Singh V, Tripathi DK, Fotopoulos V (eds) Nitric oxide in developing plant stress resilience. Plant gasotransmitters and molecules with hormonal activity. Academic, pp 31–68

Sharma P, Singh SP, Tripathi RD, Tong YW (2023a) Chromium toxicity and tolerance mechanisms in plants through cross-talk of secondary messengers: an overview of pathways and mechanisms. Environ Pollut 320:121049. https://doi.org/10.1016/j.envpol.2023.121049

Sharma P, Sri Swetha Victoria V, Praneeth Kumar P, Karmakar S, Swetha M, Reddy A (2023b) Cross-talk between insulin resistance and nitrogen species in hypoxia leads to deterioration of tissue and homeostasis. Int Immunopharmacol 122:110472. https://doi.org/10.1016/j.intimp.2023.110472

Singh A, Pandey H, Pal A, Chauhan D, Pandey S, Gaikwad DJ, Sahu C, Atta K (2023a) Linking the role of melatonin in plant stress acclimatization. S Afr J Bot 159:179–190. https://doi.org/10.1016/j.sajb.2023.05.034

Singh A, Rajput VD, Sharma R, Ghazaryan K, Minkina T (2023b) Salinity stress and nanoparticles: insights into antioxidative enzymatic resistance, signaling, and defense mechanisms. Environ Res 235:116585. https://doi.org/10.1016/j.envres.2023.116585

Singh P, Saini N, Bhola L, Mishra P, Aamir M (2023c) Chapter 20—Crosstalk of biostimulants with other signaling molecules under metal stress. In: Gill SS, Tuteja N, Khan NA, Gill R (eds) Biostimulants in alleviation of metal toxicity in plants: emerging trends and opportunities. Biostimulants and protective biochemical agents. Academic, pp 469–509

Singh SK, Srikanth GS, Puranik S, Shukla L (2023d) Chapter 9—Chemical talk within plant holobiont: a fascinating conversation. In: Swapnil P, Meena M, Harish, Marwal A, Vijayalakshmi S, Zehra A (eds) Plant-microbe interaction—Recent advances in molecular and biochemical approaches. Academic, pp 165–203

Singh T, Bisht N, Ansari MM, Chauhan PS (2024) The hidden harmony: exploring ROS-Phytohormone nexus for shaping plant root architecture in response to environmental cues. Plant Physiol Biochem 206:108273. https://doi.org/10.1016/j.plaphy.2023.108273

Sodhi GK, Saxena S (2023) Plant growth promotion and abiotic stress mitigation in rice using endophytic fungi: advances made in the last decade. Environ Exp Bot 209:105312. https://doi.org/10.1016/j.envexpbot.2023.105312

Wu X, Liao W (2023) Chapter 10—Role of H2S in antioxidant regulation of plants exposed to arsenic toxicity. In: Singh VP, Tripathi DK, Fotopoulos V, Lemos Batista B, Tortella Fuentes G, Da Silva Lobato AK (eds) The role of gasotransmitters in the amelioration of arsenic toxicity in plants. Plant gasotransmitters and molecules with hormonal activity. Academic, pp 187–206

Yadav DS, Rai K, Gupta A, Agrawal SB, Agrawal M (2023) Chapter 2—Cross-talk between ROS and phytohormones signaling determines crop sensitivity against ozone. In: . In: Van Breusegem F, Mittler R (eds) Advances in botanical research. Ozone pollution and plant health: understanding the impacts and solutions for sustainable agriculture, vol 108. Academic, pp 47–81

Zhong M, Yue L, Qin H, Wang G, Xiao L, Cheng Q, Lei B, Huang R, Yang X, Kang Y (2023) TGase-induced Cd tolerance by boosting polyamine, nitric oxide, cell wall composition and phytochelatin synthesis in tomato. Ecotoxicol Environ Saf 259:115023. https://doi.org/10.1016/j.ecoenv.2023.115023

Strigolactone in Action- Signalling, Transport, and the Control of Plant Growth and Development

5

Ritu Bala, Urvasha Patyal, Khushboo, and Manpreet Kaur

Abstract

Synthesized from carotenoids, strigolactones (SLs) are plant hormones and signalling molecules that aid root-parasitic plants and symbiotic fungi in identifying their host plants. Additionally, they regulate several developmental processes that alter how roots and shoots behave to adapt to their environment. Through deliberate manipulation of plant growth, SLs also enable adjustments to resource allocation, allowing plants to respond to nutritional deficits. SL initiates signalling networks by interacting with ethylene, auxin, cytokinin, abscisic acid, and several other plant phytohormones. The present chapter will focus on the origin and evolution of strigolactones, biosynthesis, and signalling mechanism, transportation and control of plant development, interaction of strigolactones with other hormones in the growth, and development of plants and future aspects of strigolactones.

Keywords

Strigolactones · Hormones · Structural diversity · Auxin

R. Bala · Khushboo
Department of Microbiology, School of Bioengineering and Biosciences, Lovely Professional University, Phagwara, Punjab, India

U. Patyal
Department of Biosciences and Technology, Maharishi Markandeshwar University, Ambala, Haryana, India

M. Kaur (✉)
Department of Biotechnology, School of Bioengineering and Biosciences, Lovely Professional University, Phagwara, Punjab, India

© The Author(s), under exclusive license to Springer Nature Singapore Pte Ltd. 2024
M. Faizan, S. Hayat (eds.), *Plant Growth Regulators: Resilience for Sustainable Agriculture*, https://doi.org/10.1007/978-981-97-2918-0_5

5.1 Introduction

Plant growth regulators also known as phytohormones which contribute to the plant cycle are synthesised from plants. These phytohormones include responses to biotic and abiotic stress (Peleg and Blumwald 2011). There are a few examples of phytohormones which induce plant signal transmission as well as interact with each other to coordinate the biochemical, physiological, and morphological function of plants by maintaining or controlling the allocation of nutrients and implementing a source and sink transition (Iqbal et al. 2014). There are a number of compounds which have been found recently and they are as important as the phytohormones which include brassinosteroids (BRs) known or belong to the steroidal plant hormones class and jasmonic acid. Strigolactone is an organic molecule which has recently been found (Smith and Li 2014). Strigolactone was discovered in cotton plants as root exudates in 1966 as they are carotenoid derivates that occur naturally in various plants. "Strigol" was the first naturally occurring strigolactone as they are known to be found in stems as well as stems and from there, they transferred in the xylem (Pandey et al. 2016). Sorghum, cotton, maize red clover, cowpea as well as several other monocots and dicots are found as a producer of strigolactone. Around 25 strigolactones such as strigol, orobanchol, solanacol, and sorgomol have been extracted from many different types of plants (Cook et al. 1966). Out of these, the butanolide ring (D-ring), and Tricyclic ring (ABC ring) are the most common and naturally occurring strigolactone which creates a bond by coupling with enol- ether bridge in canonical strigolactone or is less constructive in non- canonical strigolactone (Kohlen et al. 2011).

In 2008, strigolactones were recognized and categorized as a new class of plant hormone with various roles, such as inhibiting the outgrowth of buds and communicating with different types of microorganisms. Strigolactones act as signalling molecules in endogenous and exogenous pathways (Koltai 2015). They also play a crucial role within the rhizosphere by encouraging the alliance between arbuscular mycorrhizal fungi (obligate heterotrophs mainly present in the roots of plants, especially in land plants) and plant roots in nutrient-deficient environments (Jia et al. 2018a, b) (Fig. 5.1).

There are some known factors and effects of strigolactone in different species on the functions of plants (Gomez-Roldan et al. 2008).

1. *Oryza sativa* is a plant species which helps in the enhancement of chlorophyll content and suppresses the tillering.
2. *Arabidopsis thaliana* which helps in the repression of lateral formation of roots increases the seed germination, increases the elongation of root hair as well as increases and maintains the resistance against the abiotic stress, decreases the drought stress and salinity.
3. *Glycine max* helps in the nodulation enhancement.
4. *Bambusoideae* increases the senescence of the leaf as this is a complex process which includes different factors such as dark treatment, deficiency of nutrients, ageing as well as flowering (Umehara et al. 2008).

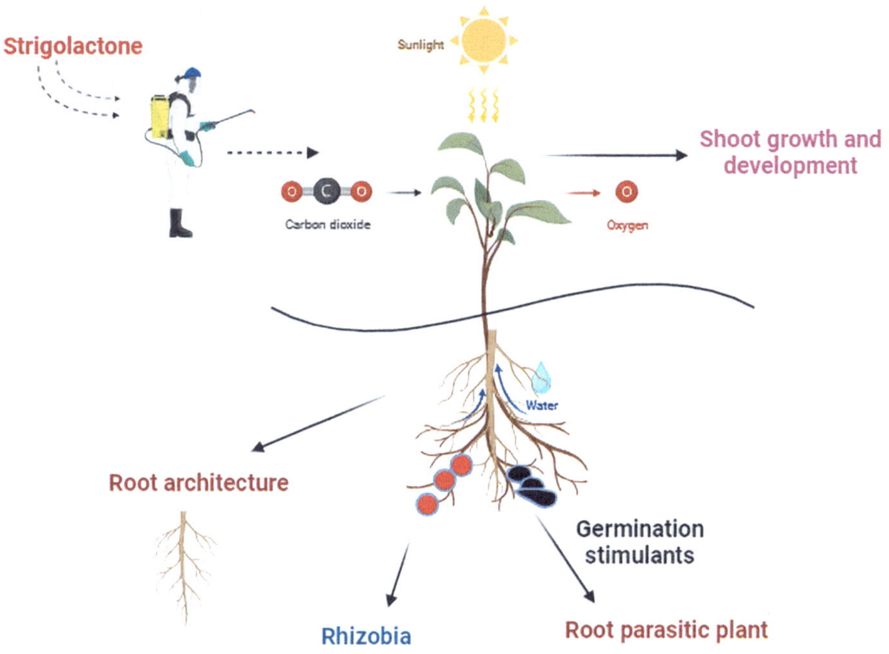

Fig. 5.1 Role of strigolactone in plants

5.2 Biosynthesis and Signalling Mechanism

Strigolactones are a new type of plant hormone that have recently been discovered in plants (Cardoso et al. 2014), but the studies showed that they play important roles in the plant, rather than just supporting the seed germination of the parasitic weed. Inside the plant, the main function of the strigolactone is as a hormone and controls many physiological processes, including shoot and root growth. Once secreted into the soil, it stimulates the growth of parasitic weeds. Although the harmful effects of inducing parasitic weed growth are well-known, they also stimulate the association of the plant to arbuscular mycelium fungi, which are very beneficial to the plant. Previous studies have focused on biosynthesis, signalling, structural diversity, crosstalk between strigolactones, regulation of strigolactone biosynthesis, and potential of strigolactones in plant improvements and human cancer treatments.

5.3 Structural Diversity of Strigolactones

A type of strigolactone, strigol, which is known to be very active as a plant hormone, was first isolated from cotton plant root exudates. The molecular formula of strigolis identical to that of the gibberellic acid, and it is well dissolved in solvents with a polar nature, but comparatively not dissolved in hexane which is non-polar solvent (Smith 2014). Strigol has later been extracted from a variety of plant species including maize, sorghum, and proso millet (Xie 2016) which are host plants of the parasitic weed Striga. Strigol, which is a type of sesquiterpene lactones, is produced not only by these host plants but also by non-host plant root exudates. Orobanchol is also a type of strigolactone isolated from red clover (Wakabayashi et al. 2022). Alectrol, another kind of strigolactone is one of the constituents of cowpea root exudates (Zwanenburg et al. 2016) and it contains a tertiary hydroxy group. Sorgolactone, yet another type of strigolactone, has been identified in sorghum root exudates and has a distinct structure compared to strigol. However, all the mentioned strigolactones are highly unstable and inactive, even in high concentrations within the plant.

5.4 Biosynthesis of Strigolactones

Production of strigolactones from cleavage of carotenoids has been identified from a variety of plants (Jia et al. 2018a, b). Production of strigolactones mainly takes place in the roots of the plants which are non-parasitic. An iron-containing protein encoded by D27 plays critical roles in the biosynthetic pathway of strigolactones in rice (Saeed et al. 2017). Three types of more axillary branching (MAX) loci including MAX3, MAX4, and MAX1 are involved in the subsequent steps of strigolactone biosynthesis in *A. thaliana* (Czarnecki et al. 2014). The respective genes have been studied in *P. sativum* (RMS5, RMS1) and *O. sativa* (D17/HTD1, D10).

In *Arabidopsis* sp., the studies have shown that the MAX3, which encodes a CCD7 protein, and MAX4 encoding a CCD8 protein act upstream of MAX1, which encodes a P450 protein in the further conversion into strigolactones. All the aforementioned MAX genes are known to act in the same pathway (Czarnecki et al. 2014). The CCD7 protein encoded by MAX3 is confined into the plastids and present in 4 reduced concentrations (Booker et al. 2004). They are involved in enzymatic cleavage of many carotenoid substrates leading to strigolactone biosynthesis (Jia et al. 2018a, b). Preliminary steps of MAX/RMS/D pathway play an essential role in strigolactone biosynthesis (Vogel et al. 2009). The perception of strigolactones is mediated by MAX4 which encodes CCD8 protein and acts downstream of auxin (Booker et al. 2004). Lack of function of MAX4 gives rise to functionless MAX3, and this finding implies that both genes are expressed in a single pathway (Hellens et al. 2023).

The genome of all the plants constitutes MAX3 and MAX4 orthologs. The biosynthetic pathway of strigolactones has found to act together with the other plant hormones such as cytokinin and auxin (Omoarelojie et al. 2019). Orthologs of MAX4 in *Arabidopsis* and RMS1 in *P. sativum* have been found (Foo et al. 2005). In *P. sativum*, both RMS5 and RMS1 are implicated in the carotenoid cleavage processes. Dad1 gene in *Petunia hybrida* is involved in the fragmentation of polyene chains, which leads to a production of the intermediate products in the strigolactone biosynthesis pathway (Lee et al. 2020). The cleavage of carotenoid compounds is also facilitated by the D10 gene, which is present in *O. sativa* (Wei et al. 2022). An enzyme that converts carlactone (CL), a kind of intermediary in the strigolactone biosynthesis pathway, into ent-2′-epi-5DS is encoded by the Os900 gene in rice. An enzyme encoded by Os1400 carries out the hydroxylation of this product into orobanchol, a kind of strigolactone (Mashiguchi et al. 2022).

Yet unidentified enzymes are needed for *Arabidopsis* to oxidise carlactone into strigolactones. Lateral branching oxidoreductase (LBO), a different gene, is involved in the latter stages of this biosynthesis pathway, which results in the production of molecules similar to strigolactones (Yang et al. 2022). NCED (9-cis-epoxycarotenoid dioxygenase) is necessary for the cleavage of carotenoids and certain following reactions lead to the production of various strigolactone types (Jia et al. 2018a, b). Additionally, the results show that two different types of regulators known as nodulation signalling pathways I and II govern the phases in the strigolactone biosynthesis pathway (McAdam et al. 2017). All the information above suggests that distinct enzymes in various plants regulate the multiple reaction stages involved in strigolactone production (Fig. 5.2).

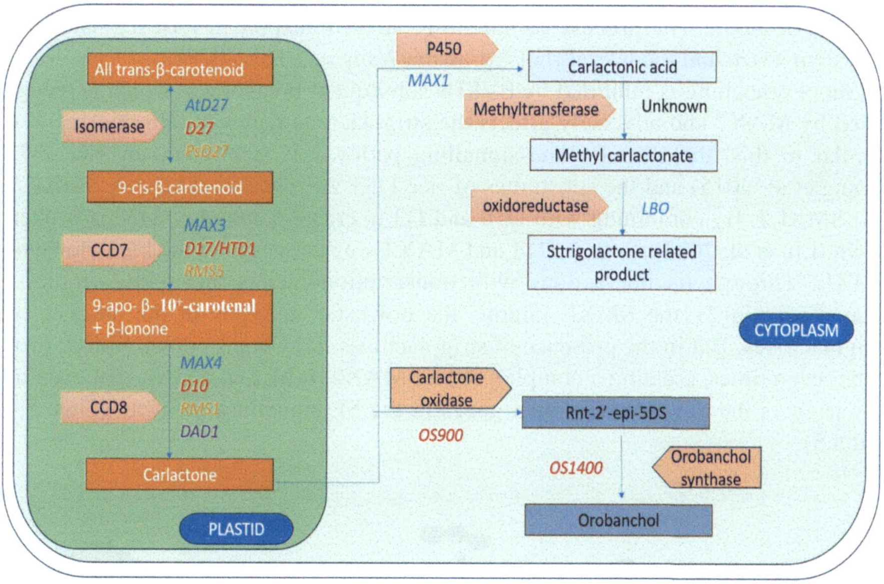

Fig. 5.2 The suggested biosynthesis pathway of strigolactones in plants. Genes involved in the strigolactone biosynthesis pathway of *Arabidopsis*, *Oryza sativa*, *Pisum sativum*, and *Petunia hybrida* are shown in blue, red, brown, and purple colors respectively. The respective encoded proteins are displayed in pink color boxes

5.5 Strigolactone Signaling Pathway

The D14, DAD2, and D3 genes in *O. sativa* (Yao et al. 2018), RMS4 in *P. sativum* (Dun et al. 2013), and the MAX2 gene in *Arabidopsis* (Shen et al. 2007) are the primary genes implicated in the strigolactone signalling system. The degrading activity of transcriptional regulators controls the strigolactone signalling pathway (Duan et al. 2019). In plants, auxin signalling and strigolactone signalling work together. By encoding the protein / hydrolase, D14 plays a crucial role in the pathway that initiates the strigolactone signal and converts it into its active form.

The primary pathway in *Arabidopsis* for strigolactone signal transduction is MAX2 (Nelson et al. 2009), which codes for the LRR. MAX2 is ubiquitous throughout the plant (Shen et al. 2007). By interacting with SCF proteins in strigolactone signalling, it modifies the ubiquitination and consequent destruction of target proteins (Tal et al. 2022) (Fig. 5.3).

In eudicots, strigolactones undergo downstream signalling through the induction of BRC1 expression by strigolactone. By inhibiting the PIN gene expression through the strigolactone signalling pathway, Shy2-short hypocotyl 2 at root tips limits the transfer of auxin. The process of inhibiting auxin transport in turn regulates the meristem's size and the lateral roots' growth (Wang and Jiao 2018).

Shoot branching is inhibited by BES1's subsequent breakdown, which is recognized by MAX2 and adversely affects the strigolactone pathway (Hu et al. 2020). Similar to this, the strigolactone signalling pathway is suppressed by rice D53 (Wang et al. 2015) and the substitutes of rice D53, *Arabidopsis* SMXL6, SMXL7, and SMXL8. By combining with D14 and D3 to create a complex, D53 is broken down (Liu et al. 2023). Both AtD14 and MAX2 are necessary for the breakdown of SMXL. Through its interactions with transcription factors and TPR2 (topless-related protein 2), the SMXL inhibits the downstream genes in the absence of strigolactones. But in the presence of strigolactones, D14 breaks down SMXL proteins seven times, creating a complex of SCFMAX2, D14, and SMXL that eventually triggers the downstream target genes in the SL signalling pathway (Figs. 5.4 and 5.5).

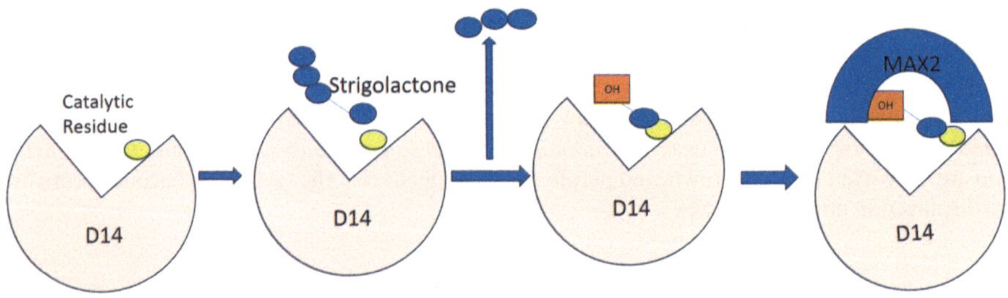

Fig. 5.3 Strigolactone signalling in plant cells of *Arabidopsis thaliana* and *Oryza sativa*

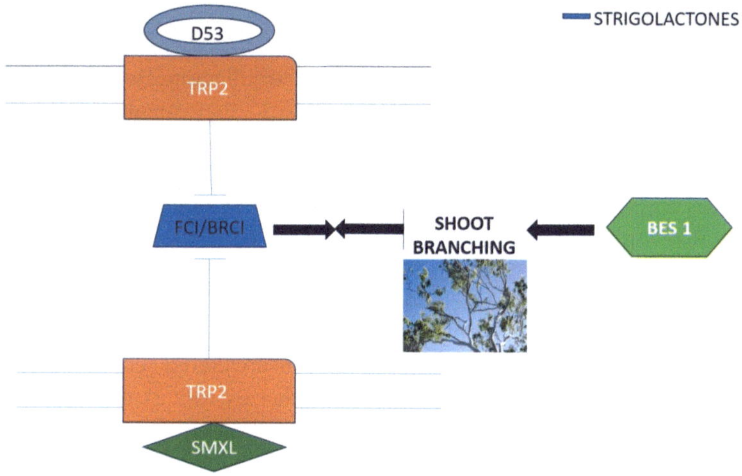

Fig. 5.4 Promotion of shoot branching in the absence of strigolactones

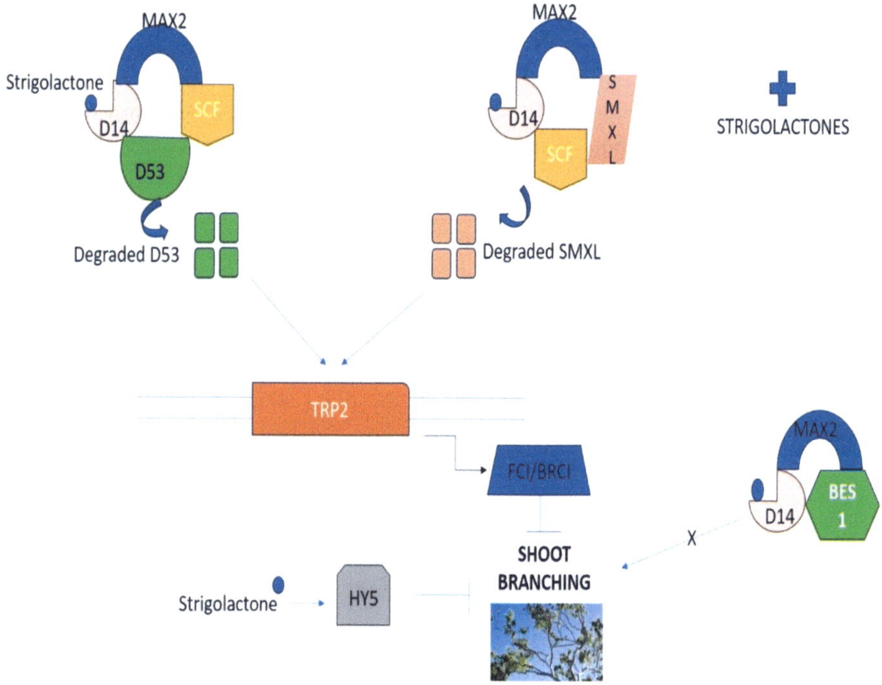

Fig. 5.5 Suppression of shoot branching in the presence of strigolactones

5.5.1 Transportation and Control of Plant Development

Carotenoids are responsible for the production of strigolactone, which is made from isoprene, also known as terpenes, and building blocks that can also be described as terpenoids and isoprenoids. Carotenoids carry a major role in photosynthesis as they work against oxidative damage by protecting the photosynthetic apparatus and light energy absorption providing the colour in fruits such as orange, and yellow as well as red pigments in tomatoes, carrots, and bananas. The control of plant response to environmental stress is responsible for the presence of abscisic acid. The formation of car-lactone involves the three chloroplast enzymes which convert the beta-carotene to lactone which is then oxidised in the cytosol to produce strigolactone. The ATP-dependent transporter protein has been discovered recently which helps in the strigolactone transportation out of the cell either by root exudation or by long-distance translocation within the plant.

5.5.2 Strigolactone Movement from Root to Shoot

To adapt to the environmental conditions or coordinate the plant's growth, the plant's transport system has been evolved by the plants in endogenous hormones. For AM fungi strigolactone functions as a symbiotic signal as they are exuded from roots into the soil when the conditions of nitrogen and phosphorous are low. From the roots to the shoots, the strigolactone provides the long-distance signals. There are some mutants such as max, rms, and dad which are suggested and recommended by strigolactone are used for grafting which also suggests the production of strigolactone around the shoot corner increases the elongation of auxiliary bud. This process takes place as the grafted WT scions into the strigolactone deficient rootstock mutants which display the normal phenotype (Fang et al. 2020). The hyper-branching phenotype can be rescued by WT rootstock which belongs to strigolactone-deficient mutants. The biosynthetic precursors of strigolactone have the ability to move from the base towards the apex of the plant shoot as they are basically synthesized in roots.

1. In the presence of phosphorous and nitrogen in roots under limited conditions, strigolactone can be actively produced.
2. The signal integration of shoot architecture in the shoot of the plant or the availability of nutrients in the soil helps in the derivation of strigolactone in roots.
3. The mobile activity of strigolactone is achieved by some possible candidates such as experiments between the deficient- mutants and strigolactone for example grafting allows the biosynthetic strigolactone intermediate movement to speculate.
4. There are some studies which demonstrated that max1 rootstocks may restore the max4 scions' shoot branching phenotype, indicating that CL may be a mobile signal as it is the only known intermediate upstream and downstream of MAX4. Furthermore, CL accumulated in the max1 at a very high level. Furthermore, it

was observed that the shoot branching phenotype of Atd27 scions could be restored by max1 rootstocks but not by max4 rootstocks, indicating that 9-cis-carotene and 9-cis-apo-10′-carotenal are immobile. Because plastids are the organelles that carry out the production of SLs from all-trans-α-carotene to CL, these physiological results make sense (Ha et al. 2014).

5. Further evidence that CLA, MeCLA, and/or their unidentified metabolites may serve as a long-range mobile signal comes from the recovery of the max1 scions' branching phenotype by lbo rootstocks. The rosette branches of the lbo mutant were likewise drastically reduced when grafting lbo scions onto WT rootstocks, indicating that 1′-OH-MeCLA, which LBO synthesizes from MeCLA, is one of the potential options for the long-distance signal from the root to the shoot.
6. The xylem sap's LC-MS/MS analysis offers an additional indicator to take into account the movement of SLs in addition to physiological research. Several SLs, including ORO, were found in the xylem sap of tomatoes and Arabidopsis. On the other hand, xylem saps taken from a variety of plant species, including tomato and Arabidopsis, did not include any known SLs, such as ORO and SL biosynthetic intermediates like CL, CLA, and MeCLA. Additionally, they verified that within 20 h of being supplied to rice roots, stable isotope-labelled ORO and 4DO were found in shoots but not in the xylem sap, indicating that rice uses

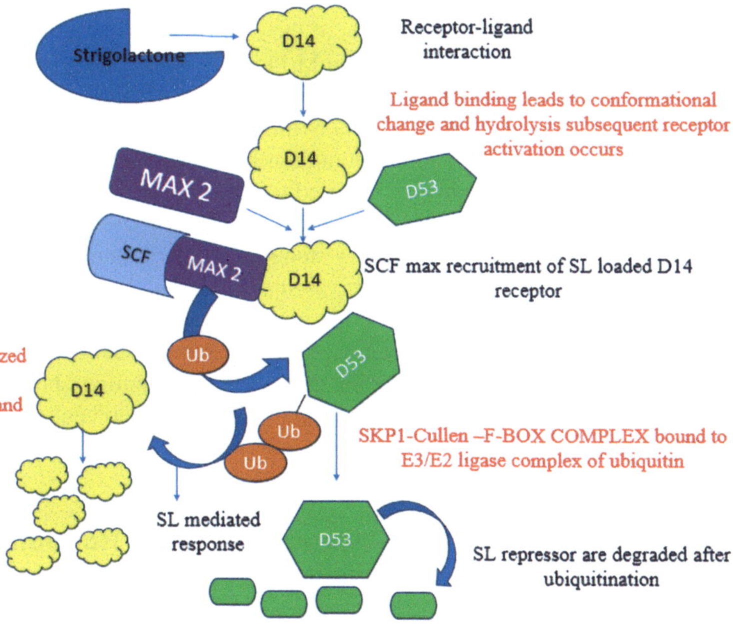

Fig. 5.6 Perception of strigolactone and its signalling pathway. Interaction between α/β hydrolase D14, the ligand-binding moiety of the SL receptor complex, and the F-box MAX2, the co-receptor moiety. This kind of interaction encourages MAX2 to attach to its target or targets, which ubiquitinates and degrades the latter via the action of proteasome machinery and subsequent signalling cascades

directed cell-to-cell movement rather than xylem-mediated transport of SLs (Haider et al. 2018). On the other hand, it's possible that unknown SL-related molecules are in motion (Fig. 5.6).

5.5.3 SL Transport Mediated by ABCG/PDR

Reverse genetic analysis in 2012 demonstrated the involvement of PDR1, a G-type ABC transporter, in the transport of SLs in petunia. At the hypodermal passage cells (HPCs) in roots, where the hyphae of AM fungus penetrate during mycorrhization, there was a strong expression of PDR1. SLs from HPCs were released into the rhizosphere by PDR1, as evidenced by the upregulation of PDR1 expression in roots in response to phosphate deficiency and AM fungal invasion. In actuality, the pdr1 mutant showed less interaction with AM fungus and dramatically lowered the levels of ORO in root exudates.

Subsequent subcellular localization study showed that PDR1 is located at the outer lateral membrane of HPCs above root tips and at the apical membrane of hypodermal cells in the root tip. Exogenously administered isotope-labelled GR24 moved through the pdr mutant far more slowly than it did in the WT. PDR1 is suggested to play two roles: SL export toward the soil from HPCs above the root tip and SL shootward transfer from root tips where SL is biosynthesized since DAD1/CCD8 is expressed in the cortex cells of the root tip. More recently, it was shown that the homolog of PDR1, MtABCG59/MtPDR23, is expressed in the cortical cells and root tip of Medicago truncatula, a plant that does not have HPCs. Comparable to that of the petunia PDR1, MtABCG59 transcript levels. In addition to SL transport in the root tissue, PDR1 and its homologs may have a role in SL transport in the aerial parts. The pdr1 mutant was reported to show the hyper-branching phenotype (Jamil et al. 2011). Moreover, a knockdown line of NtPDR6, a close homolog of PDR1 in tobacco, has more secondary stems with lateral branches compared with WT. PDR1 is shown to express at the vasculature and nodes near leaf axils but not in the dormant buds. Similarly, NtPDR6 expression was observed in nodes at lower parts of the stem. Their mutant phenotypes and expression patterns indicate that the PDR-type transporters function in SL transport for inhibiting shoot branching in Solanaceae species. However, it is still uncovered which molecule(s) is a substrate for PDR1.

5.6 Interaction of Strigolactone with Other Hormones in the Growth and Development of Plants

5.6.1 Abscisic Acid and Strigolactone

The ABA importer genes ABCG22/AT5G06530 and ABCG40/AT1G15520 were downregulated in the SL signalling-deficient mutant max2 of Arabidopsis under well-watered and dehydrated conditions, demonstrating the importance of the correlation between ABA and SLs for controlling multiple physiological mechanisms and adaptation to environmental changes in plants. By primarily triggering ABA signaling, SLs promote tolerance to salt stress and drought (Kapulnik and Koltai 2014). The SL biosynthesis-deficient mutant of *L. japonicus* showed reduced resistance to drought, which was linked to slower stomata closure and was thought to be caused by ABA insensitivity. Strigolactones and ABA work together to control branching or tillering, and in Arabidopsis, ABA functions as a downstream component of SLs and BRC1.

In order to induce HB40/OsHOX12 expression, activate AtNCED3/OsNCED1 expression, and promote ABA accumulation in the lateral buds of Arabidopsis or the shoot bases of rice, SLs mediating auxiliary bud outgrowth are involved in degrading SMXL6 and releasing BRC1 transcriptional repression. In rice, ABA supply does, in fact, prevent the growth of tiller buds and the development of unproductive upper tillers, although its impact is not as great as that of SLs.

1. **Cytokinin and Strigolactone**:
   - The transcriptional regulation of BRC1 in Arabidopsis and peas, as well as the orthologous gene FINE CULM 1 in rice, is integratively promoted by strigolactones and CK through direct interactions in buds. By serving as a crucial hub for regulatory signals governing bud outgrowth, BRC1 is known to modify the bud activation potential in a number of species.
   - The antagonistic relationship between CK and SL mediates auxin's inhibitory effect on bud outgrowth. Through the activity of cytokinin oxidase/dehydrogenase 9 (OsCKX9), SLs in rice modify the shoot architecture by activating CK catabolism. Consequently, SLs may influence the CK content through auxin crosstalk in addition to the induced activation of OsCKX9.
   - Furthermore, elevated sugar levels were discovered to block SL perception, particularly by focusing on SL signalling directly. When it comes to plant adaptation to drought, cytokinins and SLs have opposing regulatory roles (Kobae et al. 2018). It was discovered that CKs and CK-signalling elements adversely affect plant drought tolerance in Arabidopsis CK-depleted and CK-signalling mutants.
2. **Auxin and Strigolactones**:
   - Root growth and shoot branching are jointly regulated by strigolactones and auxin. PIN-dependent auxin transport, notably its canalization, is closely associated with the regulation of shoot branching mediated by SL. This is corroborated by the observations that the PIN1 auxin efflux carrier accumulates more in the basal plasma membrane of the SL biosynthesis-deficient mutants MAX4–5 and D27–1 that GR24 treatment can induce PIN1 endocytosis and decrease auxin transport during SL biosynthesis in Arabidopsis but not in response mutants.
   - Using D14- and MAX2-mediated signalling pathways, SLs prevent auxin feedback on PIN polarity and clathrin-mediated endocytosis of PIN proteins (Kretzschmar et al. 2012). However, in auxin transport inhibitor 1-N57 naphthylphthalamic acid-treated shoots, exogenous SLs are still able to restrict

bud expansion, indicating the possibility of an additional mechanism by which SLs block bud growth (Ma et al. 2017). This implies additionally that SL influences bud outgrowth directly without relying on polar auxin transfer.
- Polar auxin transport is primarily dependent on the auxin efflux protein PINs, which form local auxin maxima to form the basis for root initiation and elongation. GR24 reduced IAA distribution and modulated AR formation in rice by downregulating the levels of PIN family genes. However, in the presence of exogenous auxin, the PIN gene expression level in the PR tip of Arabidopsis was not affected by GR24 treatment (Manandhar et al. 2018).

3. **Ethylene and Strigolactones**:
   - Strigolactones have also been shown to regulate RH elongation and interact with ET signalling. SLs had no effect on RH morphogenesis in ET signalling-deficient ein2 and etr1 mutants.
   - Even in the presence of GR24, treatment with the ET biosynthesis inhibitor aminoethoxyvinylglycine did not increase RH elongation in Arabidopsis. Instead, GR24 treatment alone did. This suggests that ET is essential for SL-mediated RH elongation promotion. In response to nutrient shortages in the soil, SLs modify the balance between the auxin and ET signalling pathways to initiate distinct developmental programs, thereby regulating their biosynthesis in roots. SLs interact with ET and facilitate the transmission of auxin signalling when P is adequate.
   - When SLs and auxin pathways are combined, ET creates a crosstalk junction that controls the development of RH. Together, these hormones most likely form a purposefully coordinated network that controls how a plant grows and reacts to unfavourable growing environments (Nguyen et al. 2016).

5.7 Future Aspects of Strigolactone in Action Signalling Transport and Control of Plant Development

Strigolactones are a class of plant hormones that play crucial roles in various aspects of plant development, including shoot branching, root architecture, and interactions with symbiotic fungi. However, keep in mind that research in the field of plant biology is continually evolving, and there may have been further developments or discoveries since then. The future aspects of strigolactones in plant development are likely to involve deeper insights into their signalling pathways, transport mechanisms, and regulatory roles. Here are some potential directions for future research:

1. Signalling Pathways: Investigating the intricate signalling pathways of strigolactones involves understanding how these hormones transmit signals within the plant. Researchers aim to identify the proteins, genes, and molecular events involved in strigolactone signalling. Discovering downstream components in the signalling cascade is crucial for comprehending how strigolactones regulate various aspects of plant development.

2. Transport Mechanisms: Studying the transport mechanisms of strigolactones involves exploring how these hormones are synthesized in one part of the plant and transported to other regions. Understanding the transport dynamics helps researchers grasp how strigolactones coordinate developmental processes in different tissues and organs.
3. Interactions with Other Hormones: Investigating crosstalk with other hormones is about exploring how strigolactones interact with other signalling pathways in plants. For example, understanding how strigolactones coordinate with auxins, cytokinins, and gibberellins to modulate growth and development. Unravelling these interactions provides insights into the complex regulatory networks that control plant physiology.
4. Environmental Responses: Research on environmental responses aims to uncover how strigolactone signalling adapts to changes in the environment. This includes responses to light conditions, nutrient availability, and stress factors. Understanding the environmental sensitivity of strigolactone signalling helps elucidate how plants integrate external cues to optimize their growth and development.
5. Crop Improvement: Applying strigolactone knowledge for crop improvement involves using the information gained from research to enhance desirable traits in crops. This could include modifying strigolactone-related pathways to optimize plant architecture, increase yield, or improve stress resistance in agricultural settings.
6. Molecular and Genetic Approaches: Utilizing molecular and genetic tools involves employing techniques like CRISPR-Cas9 to manipulate specific genes related to strigolactone biosynthesis and signalling. These approaches enable researchers to create genetically modified plants with altered strigolactone pathways, providing a means to study the functions of individual genes.
7. Biotechnological Applications: Exploring biotechnological applications involves considering how the knowledge of strigolactones can be harnessed for practical purposes, such as engineering crops with improved traits. This could lead to the development of crop varieties that are more resilient, productive, or adapted to specific environmental conditions.
8. Synthetic Biology Approaches: Synthetic biology techniques could be employed to design and engineer strigolactones with specific properties. This might include creating custom compounds for targeted agricultural applications or ecological purposes, offering new tools for researchers and potentially expanding the range of applications for strigolactones.

By addressing these aspects, researchers aim to gain a comprehensive understanding of strigolactone-mediated signalling and its implications for plant development and agriculture. Ongoing advancements in these areas contribute to the broader field of plant biology and have potential applications in sustainable crop production.

References

Booker J, Auldridge M, Wills S, McCarty D, Klee H, Leyser O (2004) MAX3/CCD7 is a carotenoid cleavage dioxygenase required for the synthesis of a novel plant signaling molecule. Curr Biol 14:1232–1238. https://doi.org/10.1016/j.cub.2004.06.061

Cardoso C, Zhang Y, Jamil M, Hepworth J, Charnikhova T, Dimkpa SON, Meharg C, Wright MH, Liu J, Meng X, Wang Y, Li J, McCouch SR, Leyser O, Price AH, Bouwmeester HJ, Ruyter-Spira C (2014) Natural variation of rice strigolactone biosynthesis is associated with the deletion of two *MAX1* orthologs. Proc Natl Acad Sci U S A 111:2379–2384. https://doi.org/10.1073/pnas.1317360111

Cook CE, Whichard LP, Turner B, Wall ME, Egley GH (1966) Germination of witchweed (Striga lutea Lour.): isolation and properties of a potent stimulant. Science 154(3753):1189–1190

Czarnecki O, Yang J, Wang X, Wang S, Muchero W, Tuskan GA, Chen J-G (2014) Characterization of more axillary growth genes in populus. PLoS One 9:e102757. https://doi.org/10.1371/journal.pone.0102757

Duan J, Yu H, Yuan K, Liao Z, Meng X, Jing Y, Liu G, Chu J, Li J (2019) Strigolactone promotes cytokinin degradation through transcriptional activation of *CYTOKININ OXIDASE/DEHYDROGENASE 9* in rice. Proc Natl Acad Sci U S A 116:14319–14324. https://doi.org/10.1073/pnas.1810980116

Dun EA, De Saint Germain A, Rameau C, Beveridge CA (2013) Dynamics of Strigolactone function and shoot branching responses in *Pisum sativum*. Mol Plant 6:128–140. https://doi.org/10.1093/mp/sss131

Fang Z, Ji Y, Hu J, Guo R, Sun S, Wang X (2020) Strigolactones and brassinosteroids antagonistically regulate the stability of the D53–OsBZR1 complex to determine FC1 expression in rice tillering. Mol Plant 13(4):586–597

Foo E, Bullier E, Goussot M, Foucher F, Rameau C, Beveridge CA (2005) The branching gene *RAMOSUS1* mediates interactions among two novel signals and auxin in pea. Plant Cell 17:464–474. https://doi.org/10.1105/tpc.104.026716

Gomez-Roldan V, Fermas S, Brewer PB, Puech-Pagès V, Dun EA, Pillot JP, Letisse F, Matusova R, Danoun S, Portais JC, Bouwmeester H (2008) Strigolactone inhibition of shoot branching. Nature 455(7210):189–194

Ha CV, Leyva-González MA, Osakabe Y, Tran UT, Nishiyama R, Watanabe Y, Tanaka M, Seki M, Yamaguchi S, Dong NV, Yamaguchi-Shinozaki K (2014) Positive regulatory role of strigolactone in plant responses to drought and salt stress. Proc Natl Acad Sci U S A 111(2):851–856

Haider I, Andreo-Jimenez B, Bruno M, Bimbo A, Floková K, Abuauf H, Ntui VO, Guo X, Charnikhova T, Al-Babili S, Bouwmeester HJ (2018) The interaction of strigolactones with abscisic acid during the drought response in rice. J Exp Bot 69(9):2403–2414

Hellens AM, Chabikwa TG, Fichtner F, Brewer PB, Beveridge CA (2023) Identification of new potential downstream transcriptional targets of the strigolactone pathway including glucosinolate biosynthesis. Plant Direct 7:e486. https://doi.org/10.1002/pld3.486

Hu J, Ji Y, Hu X, Sun S, Wang X (2020) BES1 functions as the co-regulator of D53-like SMXLs to inhibit BRC1 expression in Strigolactone-regulated shoot branching in *Arabidopsis*. Plant Commun 1:100014. https://doi.org/10.1016/j.xplc.2019.100014

Iqbal N, Umar S, Khan NA, Khan MI (2014) A new perspective of phytohormones in salinity tolerance: regulation of proline metabolism. Environ Exp Bot 100:34–42

Jamil M, Charnikhova T, Cardoso C, Jamil T, Ueno K, Verstappen F, Asami T, Bouwmeester HJ (2011) Quantification of the relationship between strigolactones and Striga hermonthica infection in rice under varying levels of nitrogen and phosphorus. Weed Res 51(4):373–385

Jia K-P, Baz L, Al-Babili S (2018a) From carotenoids to strigolactones. J Exp Bot 69:2189–2204. https://doi.org/10.1093/jxb/erx476

Jia KP, Baz L, Al-Babili S (2018b) From carotenoids to strigolactones. J Exp Bot 69(9):2189–2204

Kapulnik Y, Koltai H (2014) Strigolactone involvement in root development, response to abiotic stress, and interactions with the biotic soil environment. Plant Physiol 166(2):560–569

Kobae Y, Kameoka H, Sugimura Y, Saito K, Ohtomo R, Fujiwara T, Kyozuka J (2018) Strigolactone biosynthesis genes of rice are required for the punctual entry of arbuscular mycorrhizal fungi into the roots. Plant Cell Physiol 59(3):544–553

Kohlen W, Charnikhova T, Liu Q, Bours R, Domagalska MA, Beguerie S, Verstappen F, Leyser O, Bouwmeester H, Ruyter-Spira C (2011) Strigolactones are transported through the xylem and play a key role in shoot architectural response to phosphate deficiency in nonarbuscular mycorrhizal host Arabidopsis. Plant Physiol 155(2):974–987

Koltai H (2015) Cellular events of strigolactone signalling and their crosstalk with auxin in roots. J Exp Bot 66(16):4855–4861

Kretzschmar T, Kohlen W, Sasse J, Borghi L, Schlegel M, Bachelier JB, Reinhardt D, Bours R, Bouwmeester HJ, Martinoia E (2012) A petunia ABC protein controls strigolactone-dependent symbiotic signalling and branching. Nature 483(7389):341–344

Lee HW, Sharma P, Janssen BJ, Drummond RSM, Luo Z, Hamiaux C, Collier T, Allison JR, Newcomb RD, Snowden KC (2020) Flexibility of the petunia strigolactone receptor DAD2 promotes its interaction with signaling partners. J Biol Chem 295:4181–4193. https://doi.org/10.1074/jbc.RA119.011509

Liu S, Wang J, Song B, Gong X, Liu H, Hu Q, Zhang J, Li Q, Zheng J, Wang H, Xu HE, Li J, Wang B (2023) Conformational dynamics of the D53−D3−D14 complex in Strigolactone signaling. Plant Cell Physiol 64:1046–1056. https://doi.org/10.1093/pcp/pcad067

Ma N, Hu C, Wan L, Hu Q, Xiong J, Zhang C (2017) Strigolactones improve plant growth, photosynthesis, and alleviate oxidative stress under salinity in rapeseed (Brassica napus L.) by regulating gene expression. Front Plant Sci 8:1671

Manandhar S, Funnell KA, Woolley DJ, Cooney JM (2018) Interaction between strigolactone and cytokinin on axillary and adventitious bud development in Zantedeschia. J Plant Physiol Pathol 6:1

Mashiguchi K, Seto Y, Onozuka Y, Suzuki S, Takemoto K, Wang Y, Dong L, Asami K, Noda R, Kisugi T, Kitaoka N, Akiyama K, Bouwmeester H, Yamaguchi S (2022) A carlactonoic acid methyltransferase that contributes to the inhibition of shoot branching in *Arabidopsis*. Proc Natl Acad Sci U S A 119:e2111565119. https://doi.org/10.1073/pnas.2111565119

McAdam EL, Hugill C, Fort S, Samain E, Cottaz S, Davies NW, Reid JB, Foo E (2017) Determining the site of action of Strigolactones during nodulation. Plant Physiol 175:529–542. https://doi.org/10.1104/pp.17.00741

Nelson DC, Riseborough J-A, Flematti GR, Stevens J, Ghisalberti EL, Dixon KW, Smith SM (2009) Karrikins discovered in smoke trigger *Arabidopsis* seed germination by a mechanism requiring gibberellic acid synthesis and light. Plant Physiol 149:863–873. https://doi.org/10.1104/pp.108.131516

Nguyen KH, Ha CV, Nishiyama R, Watanabe Y, Leyva-González MA, Fujita Y, Tran UT, Li W, Tanaka M, Seki M, Schaller GE (2016) Arabidopsis type B cytokinin response regulators ARR1, ARR10, and ARR12 negatively regulate plant responses to drought. Proc Natl Acad Sci U S A 113(11):3090–3095

Omoarelojie LO, Kulkarni MG, Finnie JF, Van Staden J (2019) Strigolactones and their crosstalk with other phytohormones. Ann Bot 124:749–767. https://doi.org/10.1093/aob/mcz100

Pandey A, Sharma M, Pandey GK (2016) Emerging roles of strigolactones in plant responses to stress and development. Front Plant Sci 7:434

Peleg Z, Blumwald E (2011) Hormone balance and abiotic stress tolerance in crop plants. Curr Opin Plant Biol 14(3):290–295

Saeed W, Naseem S, Ali Z (2017) Strigolactones biosynthesis and their role in abiotic stress resilience in plants: a critical review. Front Plant Sci 8:1487. https://doi.org/10.3389/fpls.2017.01487

Shen H, Luong P, Huq E (2007) The F-box protein MAX2 functions as a positive regulator of photomorphogenesis in *Arabidopsis*. Plant Physiol 145:1471–1483. https://doi.org/10.1104/pp.107.107227

Smith SM (2014) Q&A: what are strigolactones and why are they important to plants and soil microbes? BMC Biol 12:19. https://doi.org/10.1186/1741-7007-12-19

Smith SM, Li J (2014) Signalling and responses to strigolactones and karrikins. Curr Opin Plant Biol 21:23–29

Tal L, Palayam M, Ron M, Young A, Britt A, Shabek N (2022) A conformational switch in the SCF-D3/MAX2 ubiquitin ligase facilitates strigolactone signalling. Nat Plants 8:561–573. https://doi.org/10.1038/s41477-022-01145-7

Umehara M, Hanada A, Yoshida S, Akiyama K, Arite T, Takeda-Kamiya N, Magome H, Kamiya Y, Shirasu K, Yoneyama K, Kyozuka J (2008) Inhibition of shoot branching by new terpenoid plant hormones. Nature 455(7210):195–200

Vogel JT, Walter MH, Giavalisco P, Lytovchenko A, Kohlen W, Charnikhova T, Simkin AJ, Goulet C, Strack D, Bouwmeester HJ, Fernie AR, Klee HJ (2009) SlCCD7 controls strigolactone biosynthesis, shoot branching and mycorrhiza-induced apocarotenoid formation in tomato: SlCCD7 controls tomato branching. Plant J 61:300–311. https://doi.org/10.1111/j.1365-313X.2009.04056.x

Wakabayashi T, Ueno K, Sugimoto Y (2022) Structure elucidation and biosynthesis of orobanchol. Front Plant Sci 13:835160. https://doi.org/10.3389/fpls.2022.835160

Wang Y, Jiao Y (2018) Auxin and above-ground meristems. J Exp Bot 69:147–154. https://doi.org/10.1093/jxb/erx299

Wang L, Wang B, Jiang L, Liu X, Li X, Lu Z, Meng X, Wang Y, Smith SM, Li J (2015) Strigolactone signaling in *Arabidopsis* regulates shoot development by targeting D53-like SMXL repressor proteins for ubiquitination and degradation. Plant Cell 27:3128–3142. https://doi.org/10.1105/tpc.15.00605

Wei H, Movahedi A, Liu G, Li Y, Liu S, Yu C, Chen Y, Zhong F, Zhang J (2022) Comprehensive analysis of carotenoid cleavage dioxygenases gene family and its expression in response to abiotic stress in poplar. Int J Mol Sci 23:1418. https://doi.org/10.3390/ijms23031418

Xie X (2016) Structural diversity of strigolactones and their distribution in the plant kingdom. J Pestic Sci 41:175–180. https://doi.org/10.1584/jpestics.J16-02

Yang R, Liu W, Sun Y, Sun Z, Wu Z, Wang Y, Wang M, Wang H, Bai S, Fu C (2022) *LATERAL BRANCHING OXIDOREDUCTASE* , one novel target gene of squamosa promoter binding protein-like 2, regulates tillering in switchgrass. New Phytol 235:563–575. https://doi.org/10.1111/nph.18140

Yao R, Wang L, Li Y, Chen L, Li S, Du X, Wang B, Yan J, Li J, Xie D (2018) Rice DWARF14 acts as an unconventional hormone receptor for strigolactone. J Exp Bot 69:2355–2365. https://doi.org/10.1093/jxb/ery014

Zwanenburg B, Pospíšil T, Ćavar Zeljković S (2016) Strigolactones: new plant hormones in action. Planta 243:1311–1326. https://doi.org/10.1007/s00425-015-2455-5

Interpreting the Genetic Symphony: Strigolactones and Their Regulatory Effect on Plant Growth and Development

6

Savita Jangde, Saumya Shruti, and Padmanabh Dwivedi

Abstract

Strigolactones (SLs) are important plant hormones that regulate multiple physiological processes and interact with the environment to control plant growth and development. They have a vital function in the synthesis of other important plant hormones, including SLs, which are necessary for the overall health and production of plants. The biosynthesis pathway of strigolactones (SLs) is essential for the regulation of plant growth and development. This mechanism involves the conversion of carotenoids into carlactone by oxidation. Furthermore, SLs play a role in controlling nutrition by stimulating the growth of cells, increasing the branching of hyphae, and triggering the germination of spores. SLs modulate root architecture and growth by interacting with other plant hormones such as auxin, CK, gibberellin, and zaxinone. Moreover, plants react to environmental signals, assisting them in effectively distributing resources and adapting their growth patterns to varying circumstances. SLs play a role as mediators in both mutually beneficial and harmful interactions within the plant kingdom. They govern the signalling of abscisic acid (ABA), the dormancy of seeds, and the responses to stress. Structural Ligands (SLs) play a crucial role in various areas of plant development and growth, such as controlling the opening and closing of stomata, increasing resistance to environmental stress, and directing the process of ageing and survival in deciduous trees. They are a significant class of growth regulators that have immense promise in controlling diverse biological processes in plants. Their many regulatory roles in distinct plant tissues are affected by a

S. Jangde · P. Dwivedi (✉)
Department of Plant Physiology, Institute of Agricultural Sciences, Banaras Hindu University, Varanasi, India
e-mail: pdwivedi25@bhu.ac.in

S. Shruti
Department of Plant Physiology and Ecology, Anand Agricultural University, Anand, India

© The Author(s), under exclusive license to Springer Nature Singapore Pte Ltd. 2024

M. Faizan, S. Hayat (eds.), *Plant Growth Regulators: Resilience for Sustainable Agriculture*, https://doi.org/10.1007/978-981-97-2918-0_6

variety of physiological and genetic variables. This chapter seeks to investigate the substantial influence of several genes and genetic variables, facilitated by strigolactones, on the growth, development, and physiological activities of plants.

Keywords

Strigolactones · Phytohormones · Butenolide ring · SL-signalling · Stomatal regulation

6.1 Introduction

Plants frequently face challenging conditions, which can negatively impact their productivity. Phytohormones have a function, in regulating physiological processes in plants and enabling their interaction with the surrounding environment (Li et al. 2020). When Cook et al. (1966) examined the capacity of a signalling molecule generated by cotton roots to promote the germination of parasitic weed seeds, they found strigolactones (SLs). SLs are an intriguing class of plant hormones that function as miniaturized conductors, orchestrating multiple aspects of plant development and growth. The terpenoid small-compound, SLs control plant shoot branching (Umehara et al. 2008; Lopez-Raez et al. 2008), a crucial agronomic characteristic that affects crop yields. Furthermore, SL induces hyphal branching of symbiotic arbuscular mycorrhizal fungi, which promotes host plant growth by collecting important mineral nutrients from the soil and stimulates the germination of root-parasitic weeds (Cook et al. 1966, Yoneyama et al. 2012), causing devastating damage to crops. Considering the importance of SL in agriculture, the mechanisms and signaling pathways of SL have been extensively studied to provide a broader overview of SL functions. However, the mechanisms underlying SL recognition by plants are poorly understood. SLs are a type of plant hormone distinguished by a tricyclic lactone moiety linked to a butenolide ring. The tricyclic lactone, which forms a ring structure with an oxygen atom, gives the molecule stability, whereas the butenolide ring, with its double bond and carbonyl group, contributes to overall biological activity (Wang and Bouwmeester 2018). This basic structure serves as the foundation for the various SLs compounds found in various plant species. The precise arrangement of atoms and functional groups in SLs are critical to their physiological functions, which include the regulation of shoot branching, root development, and plant interactions with symbiotic fungi. While side chain variations and substitutions exist, the core structure remains essential to SLs diverse roles in plants. A three-part architecture is at the heart of their design:

1. The tricyclic lactone ring: The molecule's stable foundation is formed by this central ring system, which resembles a three-ringed circus tent. Consider it the stage on which the drama of plant development takes place.
2. The butenolide ring: The butenolide ring, which is attached to the main stage like a smaller, spotlight-stealing platform, adds an important twist. Its distinct shape

significantly contributes to the molecule's biological activity, allowing it to interact with specific receptors and cause downstream effects.

3. The final touches: The butenolide ring is adorned with hydroxyl and methylene groups, which act as designers, fine-tuning the molecule's fit and function. These minor changes further customize the molecule's interactions with its targets, ensuring that its regulatory roles are carried out precisely.

6.2 Control of Strigolactones Biosynthesis

6.2.1 Biosynthesis Overview

Based on the field of stereochemistry of the B-C ring bond, SLs can be categorised into two types: Strigol and orobanchol (Fig. 6.1). Both have a conserved R-configuration at the C-2′ position. Bioactivators involved in different SL bioactivities are located in the region connecting the D-ring to the core and vary depending on the type of SL (Zwanenburg and Pospisil 2013).

The SL biosynthesis pathway has been uncovered in recent investigations. The crucial precursor CL is produced from all-trans carotene through the activity of an isomerase (D27) and two carotenoid-cleaving dioxygenases (CCD7 and CCD8) as shown in Table 6.1. The SL synthesis pathway takes place in plastids, and the resultant CL is then moved to the cytoplasm. CL undergoes oxidation followed by further ring closure and functionalization by enzymes from the CYP711 family (MAX1), resulting in the production of SL and SL-like substances (Zhang et al. 2014). Within Arabidopsis, the enzyme MAX1 transforms CL into carlactonic acid,

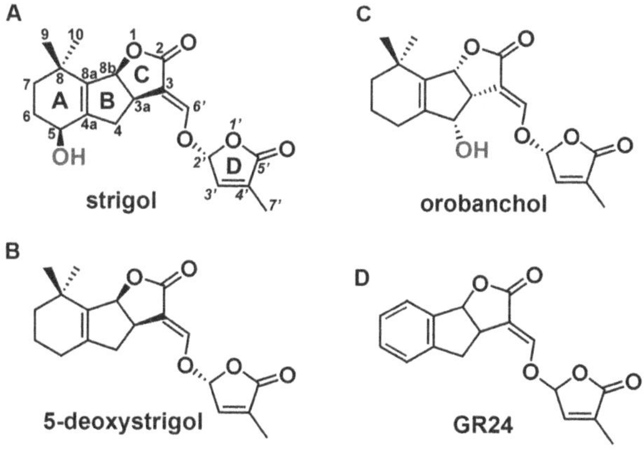

Fig. 6.1 The structure of strigolactones. (**a**) The structure of strigol, the first identified SL representative; (**b**) The structure of 5-deoxystrigol, a prototype for other β-oriented C-ring SLs (strigol-configured SLs); (**c**) The structure of orobanchol, an instance of SLs with an α-oriented C-ring (orobanchol-configured SLs); (**d**) The structure of GR24, a man-made version of SLs

Table 6.1 Gene associated with the synthesis and signalling of strigolactone in plant species

Arabidopsis thaliana	*Oryza sativa*	Function of the proteins
SL	*D27*	*Carotenoid isomerase*
Biosynthesis	*OsCCD7/D17/*	*Carotenoid cleavage dioxygenase*
AtD27,	*HTD1*	*Carotenoid cleavage dioxygenase*
MAX3	*OsCCDB/D10*	*Cytochrome P450 enzymes have varying catalytic activities*
MAX4	*OsMAX1*	*in Arabidopsis and rice. They are involved in the oxidation*
MAX1	*Os01g0935400*	*of 5-deoxystrigol to carlactonic acid and orobenchol, as*
LBO	*D14*	*well as in the downstream process of MAX1 on MeClA,*
SL signalling	*D3*	*requiring 2-oxoglutarate and Fe(II)-dependent dioxygenase*
AtD14		*2-Oxoglutarate- and Fe(II)-dependent dioxygenase and*
MAX2		*acting downstream of MAX1 on MeCLA*
		α/β-fold hydrolase and receptor for SL
		F-box protein component of SCF complex

which is further transformed into the SL-like molecule methyl carlactone (MeCLA) by an unidentified enzyme (Abe et al. 2014).

Rice, unlike *Arabidopsis*, produces orobanchol as well as 4-deoxyorobanchol (4DO) as part of its SLs biosynthetic pathway (Kobae et al. 2018). Rice most likely uses specific enzymes for each step in SLs biosynthesis. Specific enzymes, such as carotenoid cleavage dioxygenases (CCDs) and cytochrome P450s (Fig. 6.2), are involved in the conversion of carotenoids into carlactone and subsequent steps. Unlike rice, tomato plants primarily produce orobanchol and do not synthesize 4-deoxyorobanchol (4DO). There is only one MAX1 homolog in tomatoes, CYP711A21 (Zhang et al. 2018), and it is unable to convert 4DO into orobanchol. Until the discovery of CYP722C recently, the synthesis step for orobanchol in tomatoes was unknown (Wakabayashi et al. 2019). This enzyme, found in cowpea, works downstream of MAX1 to convert carlactone (CLA) straight into orobanchol without the requirement for a 4DO intermediate. Several plants include SL-like compounds that have a CL-type structure without the typical ABC-rings, showing the structural diversity of this compound class (Kim et al. 2014; Ueno et al. 2014). Once synthesised, all these chemicals can be transferred within the plant and into the rhizosphere. PhPDR1, a member of the ABC family, has been recognized as a possible SL transporter according to Kretzschmar et al. (2012) and Sasse et al. (2015).

SL biosynthesis primarily takes place in roots, with some production also happening in stems, and is well controlled (Al-Babili and Bouwmeester 2015). Phosphate deprivation significantly boosts the production of strigolactones (SL) as shown by López-Ráez et al. in 2008. Abundant levels of SL biosynthesis gene transcripts, particularly CCD7 and CCD8, in SL-deficient and SL-insensitive mutants across several species suggest that SL biosynthesis is regulated by feedback mechanisms (Mashiguchi et al. 2021). This is a summarized synopsis of SLs biosynthesis.

(a) Carotenoid Cleavage: Carotenoids are pigments that are involved in photosynthesis. Their cleavage is the first step in the biosynthesis of SLs. The first few

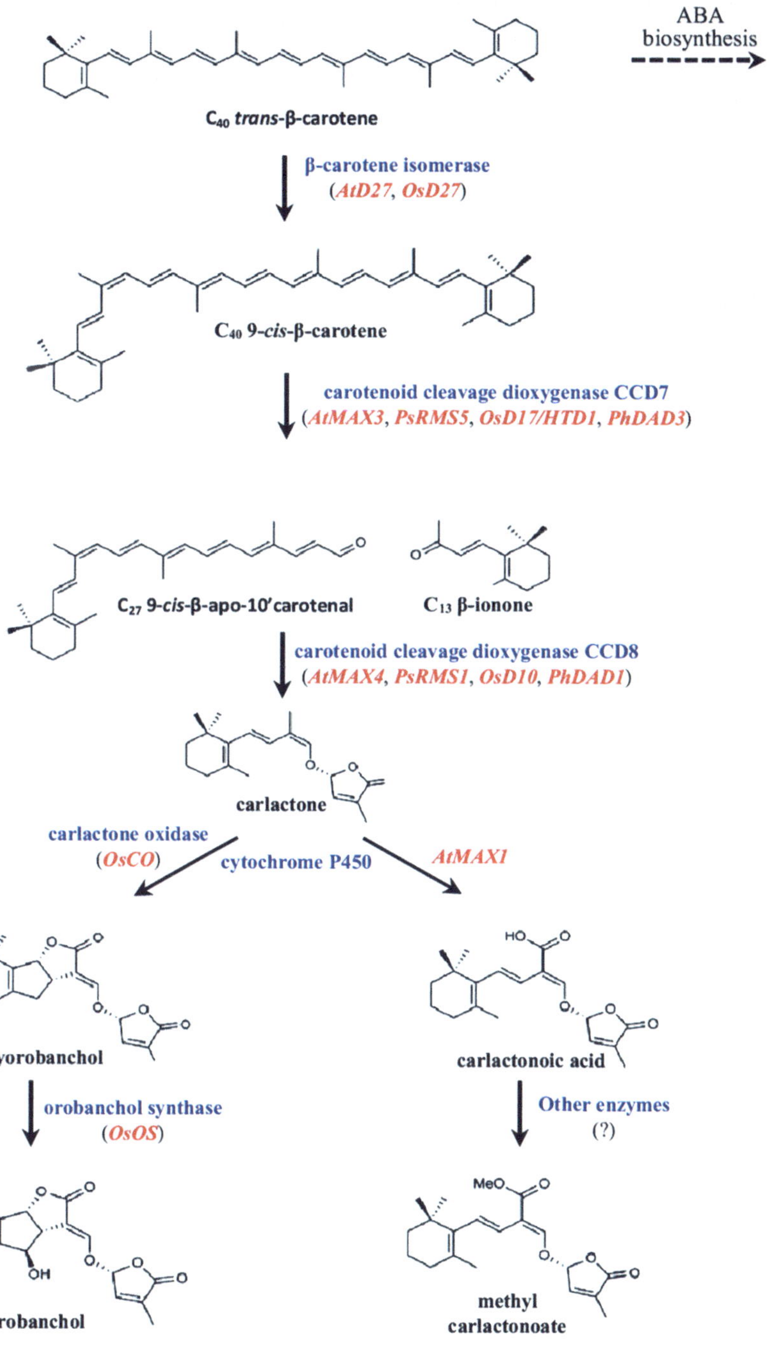

Fig. 6.2 The Strigolactone biosynthesis pathway in *Arabidopsis* (At), rice (Os), pea (Ps), and petunia (Ph) illustrates substrates in black, proteins in blue, and genes encoding the enzymes involved in red. Abscisic acid (ABA) biosynthesis and strigolactone biosynthesis compete for the same substrate.trans-β-carotene (Aliche et al. 2020)

steps of this process are catalyzed by carotenoid cleavage dioxygenases, or CCDs.

(b) Carotenoid Oxidation: Subsequent oxidation reactions take place with the products of carotenoid cleavage. This oxidation step involves specific enzymes, such as 9-cis-epoxycarotenoid dioxygenases (NCEDs).

(c) Butenolide Formation: The distinctive SLs butenolide ring is created by cycling the oxidized carotenoid derivatives. In this cyclization process, the enzyme MORE AXILLARY GROWTH 3 (MAX3) is involved.

(d) Lactonization: Additional changes result in the formation of the tricyclic lactone moiety after butenolide formation. In this lactonization step, enzymes such as MORE AXILLARY GROWTH 1 (MAX1) are involved.

6.2.2 Nutritional Control

As shown in recent studies, SL biosynthesis is regulated in several ways. Signals of phosphorus shortage are important modulators of SL biosynthesis. Studies have demonstrated that root exudates from plants low in phosphorus stimulate the growth of hyphae in arbuscular mycorrhizal fungi (Elias and Safir 1987; Nagahashi and Douds 2000). These results support the theory that colonization by AM fungi facilitates efficient uptake of mineral nutrients, in particular, phosphorus is transferred from the soil to the host plant. (Harrison 2005). Subsequently, SLs were discovered to be symbiotic signals in AM fungi that can promote cell proliferation, hyphal branching, and spore germination. According to these results, SLs are essential for the rhizosphere's AM fungus-mediated nutrient uptake (Akiyama et al. 2005; Besserer et al. 2006). Numerous plant species exhibit markedly elevated SL levels in phosphorus-deficient environments, according to analysis of root exudates (Andreo-Jimenez et al. 2015). Furthermore, it has been demonstrated that insufficiency of nitrogen raises SL levels; however, this kind of control is probably species-specific (Andreo-Jimenez et al. 2015; Yoneyama et al. 2012). These results suggest a close link between plant nutrient acquisition strategy and low nutrient-dependent SL synthesis.

The control of SL biosynthesis in plants, especially in response to phosphorus availability, has been elucidated at the molecular level. Expression of rice genes CCD7/D17, D27, CCD8/D10, CYP711A5/Os1900 and CYP711A2/Os900 associated with natural 4-deoxyorobanchol (4DO) levels in rice roots. These genes are regulated based on phosphorus levels. The transcription factors NSP1 and NSP2 help regulate MtD27 in the legume *Medicago truncatula* under phosphorus-deficient conditions. The MtD27 and CYP711A12/MtMAX1 are downregulated in nsp1 and nsp2 knockout mutants, which lowers SLs synthesis and hinders arbuscular mycorrhizal (AM) fungal infection. The rice orthologs of NSP1 and NSP2 are also connected to the biosynthesis. Furthermore, studies in tomato showed that mycorrhizal-induced Sl-IAA27 expression positively regulated the expression of CYP711A21/Sl-MAX1, Sl-NSP1, Sl-D27, and, indicating that NSP1/Sl-MAX1 in the phosphorus starvation response 2 suggests that a conserved role of mediated control is played in tomato (Guillotin et al. 2017). Moreover, the D14-LIKE pathway, associated with SL signalling and essential for arbuscular mycorrhizal (AM) symbiosis, positively controls rice SL biosynthesis by inducing NSP2 (Choi et al. 2020). This suggests that the D14-LIKE pathway and SL biosynthesis under phosphorus-deficient conditions may interact with each other and further study of this relationship is needed.

6.2.3 Regulation of Feedback

Furthermore, SL biosynthesis is controlled by unique signals. Overall, feedforward and feedback control mechanisms work together to precisely and precisely control endogenous levels of plant hormones to maintain homeostasis. CCD7 and CCD8 could serve as conserved target areas for negative feedback control in strigolactone production. Research on plants lacking or unresponsive to strigolactones (SL) in different species such as Arabidopsis, pea, petunia, and rice has shown increased expression of CCD7 and CCD8 genes (Mashiguchi et al. 2021). Applying GR24, a synthetic analogue of strigolactone, externally, totally stopped the increase of CCD7 and CCD8 in both normal and plants lacking strigolactones (Umehara et al. 2008; Mashiguchi et al. 2009). Furthermore, AtD27 is regulated by feedback regulation in Arabidopsis, since it is increased in the biggest mutant (Waters et al. 2012). Furthermore, according to LC-MS/MS analysis, rice or arabidopsis mutant plants that are not susceptible to strigolactones showed increased levels of 4-deoxyorobanchol and strigolactone biosynthetic intermediates, including carlactone and carlactonoic acid (Seto et al. 2014; Umehara et al. 2008; Arite et al. 2009; Abe et al. 2014). These data provide convincing evidence for a negative feedback control theory to control SL biosynthesis.

6.2.4 Hormonal and Other Controls

It has been suggested that SLs interact with other plant hormones. Since auxin suppresses axillary bud outgrowth and is essential for apical dominance, its impact on SL biosynthesis has drawn significant attention. In *Arabidopsis*, pea, and rice, auxin has been demonstrated to induce the expression of the CCD7 and CCD8 genes (Zou et al. 2006; Johnson et al. 2006; Arite et al. 2009; Hayward et al. 2009; Xu et al. 2015a, b). Studies have shown that auxin boosts the activity of CCD7/MAX3 an CCD8/MAX4 in the lower stem sections of *Arabidopsis* plants (Hayward et al. 2009). The auxin-resistant iaa12/bodenlos mutant exhibited reduced transcript levels of MAX3 and MAX4 due to constitutive suppression of auxin signalling. Moreover, exogenous application of GR24 to rosette axils prevented the hyper-branching phenotype of the iaa12 mutant (Hayward et al. 2009). According to Brewer et al. (2009), GR24 may also lessen the enhanced proliferation of auxin signalling mutants in Arababidopsis. These findings suggest that auxin may activate the CCD7 and CCD8 genes in order to function upstream of SL action.

Cytokinin (CK) is another hormone that has been suggested to act antagonistically with SL and can regulate SL biosynthesis. CK has been demonstrated to suppress auxin-induced MAX4 expression in Arabidopsis roots discovered that benzyladenine, a synthetic cytokinin analogue, inhibits the expression of D27, D17, and D10 genes in the tiller nodes, tiller buds, and roots of rice (Mashiguchi et al. 2021). A recent investigation indicated that externally applied natural and synthetic cytokinins (CK) reduced 4-deoxyorhormomycin (4DO) levels in phosphorus-deficient rice roots. However, the regulation of strigolactone (SL) production and phosphorus availability signalling by CK is likely to be independent (Yoneyama et al. 2020). CK has been demonstrated to decrease the expression of CYP711A2, CYP711A3, D27, D17, and D10 (Yoneyama et al. 2020). The data indicate that CK has an adverse impact on SL biosynthesis.

Gibberellin (GA) has been found to strongly suppress strigolactone (SL) production in rice and L. japonicus, in addition to auxin and CK, as demonstrated by Mori et al. (2020) and Ito et al. (2017). Bioactive GA application effectively decreased 4DO and ORO levels in rice roots, even at 100 ppm (Ito et al. 2017). GA suppressed SL production by acting on the GID1-DELLA signalling pathway and reducing the expression of SL biosynthetic genes including D27, D17, D10, CYP711A2, and CYP711A3. Additionally, GA-deficient tangyun rice, with elevated 4DO values compared to control plants, demonstrated insensitivity to GR24 regarding tiller bud length (Ito et al. 2017). Gibberellic acid signalling may play a role in strigolactone-mediated suppression of tiller bud elongation in rice. Analysis of the transcriptome showed that gibberellic acid treatment decreases the expression of strigolactone biosynthesis genes such MAX4 and LBO in *Arabidopsis* (Lantzouni et al. 2017).

Additional studies are being conducted to determine the physiological significance of the crosstalk between GA and SL. Zaxinone, a new naturally occurring compound generated from carotenoids, has been demonstrated to control strigolactone biosynthesis in Arabidopsis and rice according to studies by Wang et al. (2019) and Ablazov et al. (2020). Zaxinone hinders SL production in rice roots in both standard and phosphorus-lacking environments. Zaxinone was shown to downregulate the transcript levels of D27, D17, D10, and CYP711A2 in phosphate-starved roots independently of SL signaling (Wang et al. 2019). Zaxinone, on the other hand, induced MAX3 and MAX4 in Arabidopsis roots under normal and phosphate-depleted conditions. MeCLA levels were also found to be higher in zaxinone-treated roots (Ablazov et al. 2020).

SL biosynthesis is also influenced by a variety of developmental signals and environmental cues. In Arabidopsis, for example, transcription of MAX1, MAX3, and MAX4 was induced during leaf senescence (Ueda and Kusaba 2015). In addition, the application of GR24 triggered leaf senescence in mutants that were unable to synthesise SLs, and these mutants displayed a delayed senescence phenotype indicating that secondary metabolites produced in mature leaves have a role in the ageing process of Arabidopsis leaves (Ueda and Kusaba 2015).

Abiotic stressors have been shown to regulate the production of strigolactones in several plant species, as evidenced by studies conducted by Ha et al. (2014), Liu et al. (2015), and Visentin et al. (2016). Future research will determine the molecular mechanism that controls the regulation of SL production at the transcriptional or post-transcriptional level. SLs have been demonstrated to operate as plant hormones and allelochemicals.

6.3 Control of Physiological Functions of Strigolactones

6.3.1 Root Architecture and Growth

SLs suppress shoot branching while encouraging root growth. They influence root architecture by suppressing lateral root formation and promoting primary root elongation. This is necessary for plants to efficiently search the soil for water and nutrients. SLs influence gene expression and signaling pathways, which helps to regulate root architecture and growth (Marzec and Melzer 2018).

(a) Lateral Root Growth Inhibition: SLs have been shown to suppress lateral root growth. This process is regulated by the MAX pathway, which involves several key genes.
   1. MAX1 (More Axillary Growth 1): Encodes an enzyme involved in SL biosynthesis. Mutations in MAX1 can affect SL's inhibitory effect on lateral root growth.
   2. MAX2 (More Axillary Growth 2): This gene encodes an F-box protein that is a component of the SCF (Skp1-Cul1-F-box protein) E3 ubiquitin ligase complex. MAX2 participates in the SL response and acts downstream in the signaling pathway, regulating lateral root development (Challis et al. 2013).

(b) Promotion of Primary Root Elongation: SLs have been shown in studies to promote primary root elongation (Sun et al. 2016). This aspect is influenced by several genes:
   1. D27 (DWARF27): This gene encodes an enzyme that is required for the biosynthesis of SL precursors. D27 mutations can affect primary root elongation.
   2. D3 (DWARF3): This gene encodes an F-box protein that participates in the SL signaling pathway. By regulating downstream targets, it influences primary root elongation.

(c) Evidence has also been found that auxin may interact with SLs to affect tuberization, root development, and secondary growth organisation. SLs either enhance auxin transport or activate the transcription of the auxin receptor TIR1 to affect and regulate auxin pathways (Faizan et al. 2020). SLs influence root development by interacting with the auxin signaling pathway:
   1. AUX1 (AUXIN RESISTANT 1): Encodes a protein that transports auxin. SLs can influence root growth and branching by regulating AUX1 expression and auxin distribution.
   2. PIN Protein Regulation: PIN proteins (PIN-FORMED) are auxin efflux carriers that regulate the directional flow of auxin in plant tissues. SLs can

influence the expression and localization of PIN proteins, thereby influencing auxin distribution and, as a result, root growth and patterning.

(d) Interaction with Other Signaling Pathways: SLs may interact with other signaling pathways involving transcription factors and regulators, influencing root architecture and growth further. For example, SLs can influence root growth and development by modulating the expression of genes involved in abiotic stress responses.

6.3.2 Shoot Branching Inhibition

SLs are essential in the complex regulation of shoot branching inhibition in plants. These specialized plant hormones function as key signaling molecules, suppressing the outgrowth of axillary buds, which are dormant buds found in leaf axils. SLs inhibit shoot branching through a complex interaction with other plant hormones, particularly auxins. SLs have been shown to modulate auxin transport, influencing auxin distribution within plants. This, in turn, inhibits lateral shoot activation and growth. Furthermore, SLs help to establish and maintain apical dominance, where the terminal bud inhibits the growth of lateral buds, shaping the overall shoot architecture. SLs ' regulatory influence on shoot branching is not only intrinsic, but also responsive to environmental cues. In nutrient-deficient soils, for example, SL production increases as part of an adaptive strategy, resulting in increased shoot branching inhibition. This dynamic control provided by SLs enables plants to allocate resources wisely and adapt their growth patterns to changing environmental conditions, ultimately optimizing their overall developmental strategies. An overview of genes involve in sheet branching inhibition.

(a) **Repression of axillary bud outgrowth**
   1. D14 (DWARF14): This gene encodes a receptor protein that is involved in the perception of SLs. D14 mutations can disrupt SL signaling and interfere with axillary bud inhibition.
   2. MAX2 (More Axillary Growth 2): This gene encodes an F-box protein that is a component of the SCF (Skp1-Cul1-F-box protein) E3 ubiquitin ligase complex. MAX2 regulates axillary bud growth and branching via the SLs signaling pathway.
   3. SMXL (LIKE SMAX1) Family: MAX2-regulated SMXL proteins interact with D53 (DWARF53) to regulate axillary bud outgrowth. They contribute to the mechanism by which SLs inhibit branching.

(b) **Downregulation of Bud Growth-Promoting Genes: SLs affect the expression of genes involved in bud growth**:
   1. TB1 (TEOSINTE BRANCHED 1): TB1 is a transcription factor that inhibits the growth of axillary buds. SLs have the ability to increase TB1 expression, inhibiting bud growth and branching.

2. SPY (SPINDLY): SPY is a gibberellin signaling gene. SLs can inhibit branching by downregulating SPY expression and thus indirectly affecting the gibberellin pathway.

6.3.3 Symbiotic and Parasitic Interactions

SLs mediate both symbiotic and parasitic interactions within the plant kingdom. On the symbiotic side, SLs function as signaling compounds, establishing a critical link between plants and arbuscular mycorrhizal fungi in the soil. In exchange for plant-derived carbohydrates, these fungi form symbiotic relationships with plant roots, facilitating nutrient uptake, particularly phosphorus. SLs act as chemical signals, attracting beneficial fungi and initiating mycorrhizal associations. On the parasitic front, SLs are also important in plant interactions with root parasitic plants like Striga and Orobanche. Through the use of SLs as chemical cues, these parasitic plants are able to identify and locate the roots of their host plants, establishing a parasitic connection that causes the host to withdraw nutrients. An enzyme required for the biosynthesis of SLs precursors is encoded by the gene D27 (DWARF27). D27 mutations may have an impact on a plant's capacity to form a symbiotic relationship with mycorrhizal fungi. Parasitic plants manipulate the same SL signaling pathway that promotes symbiosis with mycorrhizal fungi to create a harmful relationship with their hosts (Akiyama and Hayashi 2006). The parasitic plant perceives SLs, which activates the genes DWARF14-LIKE (DLK) and HYPOSENSITIVE TO LIGHT (HTL). These genes produce receptors in parasitic plants that identify SLs and react to them, causing the parasitic plant to germinate and attach itself to the host root. Thus, SLs represent the diversity of plant signaling molecules by coordinating both beneficial and detrimental interactions within the complex network of plant-microbe relationships.

6.3.4 Control of Abscisic Acid (ABA) Signaling

SLs have the ability to affect the ABA signaling pathway, which is necessary for seed dormancy and stress reactions. They may improve ABA responses, which may have an effect on seed dormancy and plant stress tolerance. Through complex molecular mechanisms involving multiple genes and signaling pathways, SLs regulate ABA signaling. Plant hormone ABA is essential for stress reactions, germination of seeds, and other developmental processes. SLs have the ability to alter ABA signaling and reactions, especially under stressful circumstances (Wu et al. 2022). SLs have the ability to affect downstream ABA signaling components or ABA receptor signaling. They could alter ABA perception and signaling by modifying the expression or activity of ABA receptors, such as PYR/PYL/RCAR (Pyrabactin resistance1/PYR1-LIKE/regulatory component of ABA receptors). SLs have the potential to modify the activity of protein phosphatases (like PP2Cs) and protein kinases (like SnRK2s) that are involved in ABA signaling, thereby affecting the

ABA signaling cascade (Korek and Marzec 2023). SLs have the ability to control the expression of ABA-responsive genes that are involved in osmoregulation, stress responses, and other ABA-influenced processes. SLs have the ability to control the expression of transcription factors that respond to ABA, such as members of the ABF/AREB (ABRE BINDING FACTOR/ABRE-BINDING PROTEIN) family, which in turn controls the expression of ABA-responsive genes that are downstream. SLs decrease the expression of enzymes that biosynthesize ABA, like NCED9 (Ton et al. 2012). SLs have the ability to control ABA responses in stressful situations, which helps plants adapt to a variety of environmental stressors. Through the regulation of stress-responsive genes involved in drought tolerance, osmotic stress, and other stress-related processes, the interaction between SLs and ABA signaling aids in plants' ability to respond to stress (Fig. 6.3).

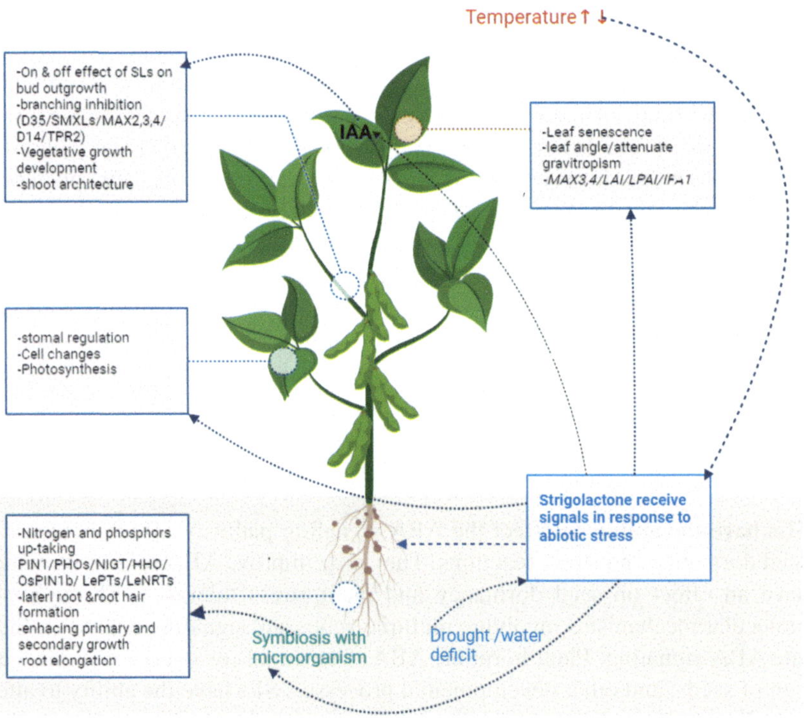

Fig. 6.3 Effects of strigolactone analogues (e.g., GR24) on plant biological properties, vegetative growth, and signal transduction of plants under abiotic stress related to soil and rhizosphere microbiome. This figure shows the role of SL in nitrogen and phosphorus uptake. Changes in root and shoot structure identify the influence of SL in modifying plant architecture to cope with abiotic stress. This figure shows that SL influences shoot branching. The plant uses SLs to develop defense mechanisms against various abiotic stresses such as drought, senescence, temperature, light, and nutrient deficiencies

6.3.5 Stomatal Regulation

A large body of research indicates that SLs are engaged in various aspects of plant development and growth. Furthermore, a growing body of research suggests that SLs are becoming integrators of different signals related to plant stress adaptation. Specifically, two recent studies (Liu et al. 2015; Piisila et al. 2015) indirectly suggested that SLs might be involved in controlling stomatal apertures due to the finding that max2 mutants had more widely opening stomata or higher stomatal conductance. According to studies, SLs are important in causing stomatal closure, which may prevent water loss and increase plant tolerance to environmental stress. According to the suggested model (Fig. 6.4), OST1 and MPK9/12 are primarily responsible for mediating ABA signaling, while SLs are transduced by unidentified components even though H_2O_2, NO, and SLAC1 are required for both ABA- and SL-mediated stomatal responses. Our research also shows that SL-triggered stomatal closure appears to be independent of endogenous ABA and certain ABA-signaling components. Further research is necessary to understand the precise molecular mechanism by which SLs mediate the stomatal response. Further research is necessary to understand the precise molecular mechanism by which SLs mediate the stomatal response. According to reports, ABA-induced stomatal closure depends on H_2O_2-activated Ca^{2+} channels (Pei et al. 2000). Consequently, it will be interesting to look into the possibility that SL-triggered stomatal closure also requires H_2O_2-activated Ca^{2+} channels. Further research is still needed to determine which downstream signaling component (or components) in guard cells transduces the SL signal. Guard cell-expressed calcium-dependent protein kinases (CDPKs) are likely to play a role in the signalling pathway of SL-induced stomatal closure as they are believed to assist in guard cell signal transduction and ion channel regulation (Kim et al. 2010).

6.3.6 Senescence and Plant Survival

As a seasonal physiological senescence process, deciduous trees lose their leaves in order to prepare for the harsh winter weather and hibernation. Many variables, including temperature, light intensity, stress, age, and phytohormone concentration, control the rate of leaf senescence (Xie et al. 2016). Since ore9/max2 displays a delayed senescence phenotype, SLs have an impact on leaf senescence. In their investigation into the function of SLs in leaf senescence, Ueda and Kusaba (2015) discovered that SLs displayed a delayed senescence phenomenon, suggesting that SLs regulate leaf senescence (Fig. 6.2). Additionally, Ueda and Kusaba (2015) investigated the connection between phosphorus deficiency, senescence, and SLs. They discovered that the application of GR24 accelerated the process of senescence in all SL-deficient mutants under conditions of phosphate deficiency and dark-induced senescence. The observation that GR24 is more prevalent in *d10* mutant genes than *d17* mutant genes implies that SLs control leaf senescence in response to phosphate deficiency (Brewer et al. 2013). The relationship between photosynthesis

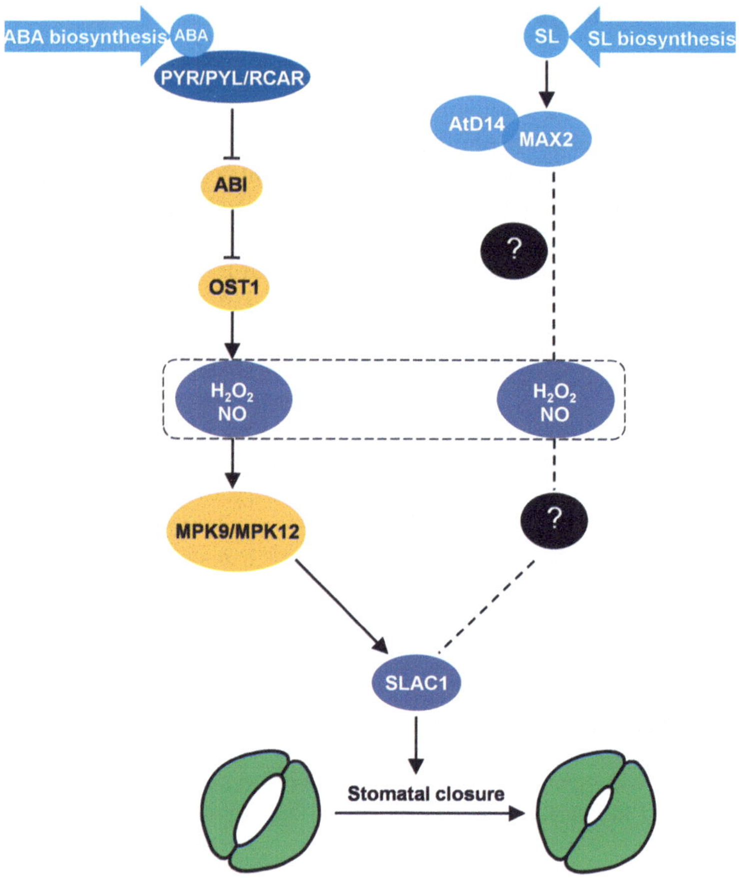

Fig. 6.4 An illustrative model demonstrating the signalling pathways for stomatal closure in guard cells of *Arabidopsis thaliana*, regulated by strigolactone (SL) and abscisic acid (ABA). By stimulating the generation of hydrogen peroxide (H_2O_2) and nitric oxide (NO) and triggering the activation of the anion channel SLOW ANION CHANNEL-ASSOCIATED 1 (SLAC1) to promote stomatal closure, SLs function similarly to ABA. Our genetic and pharmacological studies have shown that many signalling components responsible for ABA-induced stomatal closure are not required for stomatal closure driven by SL. The dashed line depicts theoretical cell responses; black arrows symbolize positive regulation; black bars symbolize negative regulation; and unidentified substances potentially involved in SL-mediated stomatal closure are denoted by a question mark. PYR stands for PYRABACTIN RESISTANCE; RCAR stands for REGULATORY COMPONENTS OF ABA RECEPTORS; PYL stands for PYR1-LIKE

and ethylene biosynthesis was mentioned in the intricate analysis. Both IAA and IAA + GR24 treatments increased the expression of many ethylene biosynthesis-related genes, indicating their role in the regulation of ethylene biosynthesis (Zhan et al. 2018).

6.4 Perspectives

Targeting the manipulation of SL biosynthesis would be a prudent strategy for agrochemicals or genetic modifications that aim to improve crop yield and quality in sustainable low-input agriculture and prevent harm caused by root parasitic weeds (Marzec et al. 2020). The application of SLs in agriculture could be accelerated by advances in sequencing, gene editing, biochemistry, and synthetic biology; however, these advances won't happen quickly unless we also figure out the many unidentified SL reaction stages and the purposes of the SL products in plants. Forward genetics has been superseded by gene expression analyses and reverse genetics in the search for SL mutants. In the event that the expression of the missing enzyme genes is not changed, alternative screening techniques will be required. Additionally, additional study is extremely difficult due to the instability of some bioactive SLs. It might be necessary to use a wider variety of cutting-edge synthetic SL mimics and more resolution detection techniques on smaller tissue samples. Furthermore, the identification of SL, its production process, and any ancestral or substitute SL receptors are critically needed. To properly comprehend how SL structure relates to function, tissue-specific gene targets and cellular processes of SLs will ultimately be needed.

6.5 Conclusion

Future growth regulators of great promise are SLs, which perform a wide range of regulatory tasks in different plant tissues under the influence of numerous physiological and genetic factors. They interact with other organisms in the ecosystem and have important network processes both inside the plant organs. SLs' primary biological partners in a variety of plant biological processes are microorganisms. Since the effects of global climate change pose a threat to plants and their products, SLs may be a helpful tool in promoting plant growth and productivity in the face of biotic and abiotic stress. Under the conditions of climate change, improved plant growth under stress will promote sustainability. Investigation into the functions of phyllosphere microorganisms is still necessary, though, because they are involved in the development of above-ground plants. SLs are necessary as a connecting factor in the interaction between the soil, rhizosphere, and plant roots which bolster the idea of "asking for help." Plant health greatly benefits from the soil and plant microbiomes. Research on the impact of SLs on different plant species, particularly fruit trees, is necessary given the significance of SLs in relation to microbes. Furthermore, research on the impact of SLs on specific plant phenotypes is still lacking. The application of lactone derivatives in organic and sustainable farming as a phytohormone warrants further research due to the need to examine the impact of naturally extracted SLs on plant crops concerning environmental factors.

References

Abe S, Sado A, Tanaka K, Kisugi T, Asami K, Ota S, Kim HI, Yoneyama K, Xie X, Ohnishi T, Seto Y (2014) Carlactone is converted to carlactonoic acid by MAX1 in Arabidopsis and its methyl ester can directly interact with AtD14 in vitro. Proc Natl Acad Sci U S A 111(50):18084–18089

Ablazov A, Mi J, Jamil M, Jia KP, Wang JY, Feng Q, Al-Babili S (2020) The Apocarotenoid zaxinone is a positive regulator of strigolactone and abscisic acid biosynthesis in Arabidopsis roots. Front Plant Sci 11:578

Akiyama K, Hayashi H (2006) Strigolactones: chemical signals for fungal symbionts and parasitic weeds in plant roots. Ann Bot 97(6):925–931

Akiyama K, Matsuzaki K, Hayashi H (2005) Plant sesquiterpenes induce hyphal branching in arbuscular mycorrhizal fungi. Nature 435:824–882

Al-Babili S, Bouwmeester HJ (2015) Strigolactones, a novel carotenoid- dervived plant hormone. Annu Rev Plant Biol 66:161–186

Aliche EB, Screpanti C, De Mesmaeker A, Munnik T, Bouwmeester HJ (2020) Science and application of strigolactones. New Phytol 227(4):1001–1011

Andreo-Jimenez B, Ruyter-Spira C, Bouwmeester HJ, Lopez-Raez JA (2015) Ecological relevance of strigolactones in nutrient uptake and other abiotic stresses, and in plant- microbe interactions below-ground. Plant Soil 394:1–19

Arite T, Umehara M, Ishikawa S, Hanada A, Maekawa M, Yamaguchi S, Kyozuka J (2009) d14, a strigolactone-insensitive mutant of rice, shows an accelerated outgrowth of tillers. Plant Cell Physiol 50(8):1416–1424

Besserer A, Puech-Pages V, Kiefer P, Gomez-Roldan V, Jauneau A, Roy S, Portais JC, Roux C, Becard G, Sejalon-Delmas N (2006) Strigolactones stimulate arbuscular mycorrhizal fungi by activating mitochondria. PLoS Biol 4(7):226

Brewer PB, Dun EA, Ferguson BJ, Rameau C, Beveridge CA (2009) Strigolactone acts downstream of auxin to regulate bud outgrowth in pea and Arabidopsis. Plant Physiol 150(1):482–493

Brewer PB, Koltai H, Beveridge CA (2013) Diverse roles of strigolactones in plant development. Mol Plant 6(1):18–28

Challis RJ, Hepworth J, Mouchel C, Waites R, Leyser O (2013) A role for more axillary growth1 (MAX1) in evolutionary diversity in strigolactone signaling upstream of MAX2. Plant Physiol 161(4):1885–1902

Choi J, Lee T, Cho J, Servante EK, Pucher B, Summers W, Bowdens W, Rahimi M, An K, An G, Bownmeester HJ (2020) The negative regulator SMAX1 controls mycorrhizal symbiosis and strigolactone biosynthesis in rice. Nat Commun 11(1):2114

Cook CE, Whichard LP, Turner B, Wall ME, Egley GH (1966) Germination of witchweed (Striga lutea Lour.): isolation and properties of a potent stimulant. Science 154(3753):1189–1190

Elias KS, Safir GR (1987) Hyphal elongation of Glomus fasciculatus in response to root exudates. Appl Environ Microbiol 53(8):1928–1933

Faizan M, Faraz A, Sami F, Siddiqui H, Yusuf M, Gruszka D, Hayat S (2020) Role of strigolactones: signalling and crosstalk with other phytohormones. Open Life Sci 15(1):217–228

Guillotin B, Etemadi M, Audran C, Bouzayen M, Becard G, Combier JP (2017) Sl-IAA27 regulates strigolactone biosynthesis and mycorrhization in tomato (var. MicroTom). New Phytol 213(3):1124–1132

Ha CV, Leyva-Gonzalez MA, Osakabe Y, Tran UT, Nishiyama R, Watanabe Y, Tanaka M, Seki M, Yamaguchi S, Dong NV, Yamaguchi-Shinozaki K (2014) Positive regulatory role of strigolactone in plant responses to drought and salt stress. Proc Natl Acad Sci U S A 111(2):851–856

Harrison MJ (2005) Signaling in the arbuscular mycorrhizal symbiosis. Ann Rev Microbiol 59:19–42

Hayward A, Stirnberg P, Beveridge C, Leyser O (2009) Interactions between auxin and strigolactone in shoot branching control. Plant Physiol 151(1):400–412

Ito S, Yamagami D, Umehara M, Hananda A, Yoshida S, Sasaki Y, Yajima S, Kyuzuka J, Ueguchi-Tanaka M, Matsuoka M, Shirasu K (2017) Regulation of Strigolactone biosynthesis by gibberellin signaling. Plant Physiol 174(2):1250–1259

Johnson X, Brcich T, Dun EA, Goussot M, Haurogne K, Beveridge CA, Rameau C (2006) Branching genes are conserved across species. Genes controlling a novel signal in pea are coregulated by other long-distance signals. Plant Physiol 142(3):1014–1026

Kim TH, Bohmer M, Hu H, Nishimura N, Schroeder JI (2010) Guard cell signal transduction network: advances in understanding abscisic acid, CO2, and Ca2+ signaling. Annu Rev Plant Biol 61:561–591

Kim HI, Kisugi T, Khetkam P, Xie X, Yoneyama K, Uchida K, Yokota T, Nomura T, McErlean CS, Yoneyama K (2014) Avenaol, a germination stimulant for root parasitic plants from Avena strigosa. Phytochemistry 103:85–88

Kobae Y, Kameoka H, Sugimura Y, Saito K, Ohtomo R, Fujiwara T, Kyozuka J (2018) Strigolactone biosynthesis genes of rice are required for the punctual entry of arbuscular mycorrhizal fungi into the roots. Plant Cell Physiol 59(3):544–553

Korek M, Marzec M (2023) Strigolactones and abscisic acid interactions affect plant development and response to abiotic stresses. BMC Plant Biol 23(1):314

Kretzschmar T, Kohlen W, Sasse J, Borghi L, Schlegel M, Bachelier JB, Reinhardt D, Bours R, Bouwmeester HJ, Martinoia E (2012) A petunia ABC protein controls strigolactone-dependent symbiotic signaling and branching. Nature 483(7389):341–344

Lantzouni O, Klermund C, Schwechheimer C (2017) Largely additive effects of gibberellin and strigolactone on gene expression in Arabidopsis thaliana seedlings. Plant J 92(5):924–938

Li G, Zhang C, Zhang G, Fu W, Feng B, Chen T, Peng S, Tao L, Fu G (2020) Abscisic acid negatively modulates heat tolerance in rolled leaf rice by increasing leaf temperature and regulating energy homeostasis. Rice 13:1–16

Liu J, He H, Vitali M, Vinsentin I, Charnikhova T, Haider I, Schubert A, Ruyter-Spira C, Bouwmeester HJ, Lovisolo C, Cardinale F (2015) Osmotic stress represses strigolactone biosynthesis in Lotus japonicus roots: exploring the interaction between strigolactones and ABA under abiotic stress. Planta 241:1435–1451

Lopez-Raez JA, Charnikhova T, Gomez-Roldan V, Matusova R, Kohlen W, De Vos R, Verstappen F, Puech-Peges V, Becard G, Mulder P, Bouwmeester H (2008) Tomato strigolactones are derived from carotenoids and their biosynthesis is promoted by phosphate starvation. New Phytol 178(4):863–874

Marzec M, Melzer M (2018) Regulation of root development and architecture by Strigolactones under optimal and nutrient deficiency conditions. Int J Mol Sci 19(7):1887

Marzec M, Situmorang A, Brewer PB, Brąszewska A (2020) Diverse roles of MAX1 homologues in rice. Genes (Basel) 11(11):1348

Mashiguchi K, Sasaki E, Shimada Y, Nagae M, Ueno K, Nakano T, Yoneyama K, Suzuki Y, Asami T (2009) Feedback-regulation of strigolactone biosynthetic genes and strigolactone-regulated genes in Arabidopsis. Biosci Biotechnol Biochem 73(11):2460–2465

Mashiguchi K, Seto Y, Yamaguchi S (2021) Strigolactone biosynthesis, transport and perception. Plant J 105(2):335–350

Mori N, Nomura T, Akiyama K (2020) Identification of two oxygenase genes involved in the respective biosynthetic pathways of canonical and non-canonical strigolactones in Lotus japonicus. Planta 251:1–6

Nagahashi G, Douds DD (2000) Partial separation of root exudate components and their effects upon the growth of germinated spores of AM fungi. Mycol Res 104(12):1453–1464

Pei ZM, Murata Y, Benning G, Thomine S, Klusener B, Allen GJ, Grill E, Schroeder JI (2000) Calcium channels activated by hydrogen peroxide mediate abscisic acid signaling in guard cells. Nature 406(6797):731–734

Piisila M, Keceli MA, Brader G, Jakobson L, Jossaar I, Sipari N, Kollist H, Palva ET, Kariola T (2015) The F-box protein MAX2 contributes to resistance to bacterial phytopathogens in Arabidopsis thaliana. BMC Plant Biol 15:1–17

Sasse J, Simon S, Gubeli C, Liu GW, Cheng X, Friml J, Bouwmeester H, Martinoia E, Borghi L (2015) Asymmetric localizations of the ABC transporter PaPDR1 trace paths of directional Strigolactone transport. Curr Biol 25(5):647–655

Seto Y, Sado A, Asami K, Hanada A, Umehara M, Akiyama K, Yamaguchi S (2014) Carlactone is an endogenous biosynthetic precursor for strigolactones. Proc Natl Acad Sci U S A 111(4):1640–1645

Sun H, Tao J, Gu P, Xu G, Zhang Y (2016) The role of strigolactones in root development. Plant Signal Behav 11(1):1110662

Ton S, Kamiya Y, Kawakami N, Nambara E, McCourt P, Tsuchiya Y (2012) Thermoinhibition uncovers a role for strigolactones in Arabidopsis seed germination. Plant Cell Physiol 53(1):107–117

Ueda H, Kusaba M (2015) Strigolactone regulates leaf senescence in concert with ethylene in Arabidopsis. Plant Physiol 169(1):138–147

Ueno K, Furumoto T, Umeda S, Mizutani M, Takikawa H, Batchvarova R, Sugimoto Y (2014) Heliolactone, a non-sesquiterpene lactone germination stimulant for root parasitic weeds from sunflower. Phytochemistry 108:122–128

Umehara M, Hanada A, Yoshida S, Akiyama K, Arite T, Takeda-Kamiya N, Magome H, Kamiya Y, Shirasu K, Yoneyama K, Kyozuka K (2008) Inhibition of shoot branching by new terpenoid plant hormones. Nature 455(7210):195–200

Visentin I, Vitali M, Ferrero M, Zhang Y, Ruyter-Spira C, Novak O, Strnad M, Lovisolo C, Schubert A, Cardinale F (2016) Low levels of strigolactones in roots as a component of the systemic signal of drought stress in tomato. New Phytol 212(4):954–963

Wakabayashi T, Hamana M, Mori A, Akiyama R, Ueno K, Osakabe K, Osakabe Y, Suzuki H, Takikawa H, Mizutani M, Sugimoto Y (2019) Direct conversion of carlactonoic acid to orobanchol by cytochrome P450 CYP722C in strigolactone biosynthesis. Sci Adv 5(12):9067

Wang Y, Bouwmeester HJ (2018) Structural diversity in the strigolactones. J Exp Bot 69(9):2219–2230

Wang JY, Haider I, Jamil M, Fiorilli Y, Saito V, Mi J, Baz L, Kountche BA, Jia KP, Guo X, Balakrishna A (2019) The apocarotenoid metabolite zaxinone regulates growth and strigolactone biosynthesis in rice. Nat Commun 10(1):810

Waters MT, Brewer PB, Bussell JD, Smith SM, Beveridge CA (2012) The Arabidopsis ortholog of rice DWARF27 acts upstream of MAX1 in the control of plant development by Strigolactones. Plant Physiol 159:1073–1085

Wu F, Gao Y, Wang W, Sui N, Zhu J (2022) Biological functions of strigolactones and their crosstalk with other phytohormones. Front Plant Sci 13:821563

Xie X, Yoneyama K, Kisugi T, Nomura T, Akiyama K, Asami T, Yoneyama K (2016) Structure- and stereospecific transport of strigolactones from roots to shoots. J Pestic Sci 41(2):55–58

Xu JX, Zha MR, Li Y, Ding YF, Chen L, Ding CQ, Wang SH (2015a) The interaction between nitrogen availability and auxin, cytokinin, and strigolactone in the control of shoot branching in rice (Oryza sativa L.). Plant Cell Rep 34:1647–1662

Yoneyama K, Xie X, Kim HI, Kisugi T, Nomura T, Sekimoto H, Yokota T, Yoneyama K (2012) How do nitrogen and phosphorus deficiencies affect strigolactone production and exudation? Planta 235:1197–1207

Yoneyama K, Xie X, Nomura T, Yoneyama K (2020) Do phosphate and cytokinin interact to regulate strigolactone biosynthesis or act independently? Front Plant Sci 11:438

Zhan Y, Qu Y, Zhu L, Shen C, Feng X, Yu C (2018) Transcriptome analysis of tomato (Solanum lycopersicum L.) shoots reveals a crosstalk between auxin and strigolactone. PLoS One 13(7):0201124

Zhang Y, van Dijk AD, Scaffidi A, Flematti GR, Hofman M, Charnikhova T, Verstappen F, Hepworth JO, Van Der Krol S, Leyser O, Smith SM (2014) Rice cytochrome P450 MAX1 homologs catalyze distinct steps in strigolactone biosynthesis. Nat Chem Biol 10(12):1028–1033

Zhang Y, Cheng X, Wang Y, Dıez-Simon C, Flokova K, Bimbo A, Bouwmeester HJ, Ruyter-Spira C (2018) The tomato MAX1 homolog, SlMAX1, is involved in the biosynthesis of tomato strigolactones from carlactone. New Phytol 219:297–309
Zou J, Zhang S, Zhang W, Li G, Chen Z, Zhai W, Zhao X, Pan X, Xie Q, Zhu L (2006) The rice HIGH-TILLERING DWARF1 encoding an ortholog of Arabidopsis MAX3 is required for negative regulation of the outgrowth of axillary buds. Plant J 48:687–698
Zwanenburg B, Pospisil T (2013) Structure and activity of strigolactones: new plant hormones with a rich future. Mol Plant 6:38–62

Karrikins and Their Role in Mitigating Nutrient Deficiency and Stress in Plants

7

Aina Inam, Lalarukh Azhar, Sadia Afzal, Sumera Javad, and Ozair Choudhary

Abstract

Karrikins have recently gained recognition as essential players in plant growth and stress management, acting as vital signaling molecules akin to plant hormones. Their significance lies in their ability to facilitate seed germination and bolster plant development, even in the face of diverse environmental adversities. Current scientific investigations have delved into the intricate molecular pathways through which karrikins operate, shedding light on their profound impact on plant physiology and their role in helping plants adapt to changing environmental dynamics. This chapter provides an in-depth overview of karrikins, emphasizing their multifaceted contributions to plant physiology, with a focus on nutrient uptake, seed germination, and stress response mechanisms. Notably, karrikins exert a wide range of effects in the rhizosphere, influencing the development of both roots and shoots, regulating nutrient distribution, and modulating seed dormancy. Furthermore, parallels between karrikins and strigolactones are highlighted, showcasing their shared characteristics and potential synergistic applications in bolstering plant growth and resilience against environmental stressors. Looking ahead, future research endeavors aim to elucidate the intricate downstream signaling pathways of karrikins, to explore the role of post-translational modifications, and to unravel the genetic underpinnings of karrikin-mediated responses to environmental challenges. These prospective investigations hold promise in deepening our understanding of karrikin function and expanding their applications in sustainable agriculture. Ultimately, karrikins emerge as promising agents capable of addressing nutrient deficiencies and enhancing crop

A. Inam · L. Azhar · S. Afzal · S. Javad (✉)
Department of Botany, Lahore College for Women University, Lahore, Pakistan

O. Choudhary
Ontario Institute of Agrologist, Harriston, ON, Canada

© The Author(s), under exclusive license to Springer Nature Singapore Pte Ltd. 2024

M. Faizan, S. Hayat (eds.), *Plant Growth Regulators: Resilience for Sustainable Agriculture*, https://doi.org/10.1007/978-981-97-2918-0_7

productivity in the context of shifting climatic conditions, paving the way for resilient and environmentally sustainable agricultural practices.

Keywords

Karrikins · Stress management · Plant development · Environmental adversities · Molecular pathways · Plant physiology · Nutrient uptake · Sustainable agriculture

7.1 Nutrient Deficiencies in Plants

Agriculture (crop and livestock) is the backbone of developing countries and is prime source of income for rural inhabitants. Cash crops and medicinal plants contribute a major share to global economy in terms of GDP. About 98% inhabitants rely upon soil for plant food (Asule et al. 2023); as soil dispense nutrients to plants (Kopittke et al. 2019).

Plants need 14 elements for their development and procreation. These elements are categorized as micronutrients and macronutrients depending upon plant needs. Micronutrients are needed in smaller amounts (less than 1% of plant dry mass) but play crucial roles in plant metabolism. These include Boron (B), Zinc (Zn), Manganese (Mn), Iron (Fe), Copper (Cu), Molybdenum (Mo), and Chlorine (Cl). Plants need larger amounts (over 95% of plant dry mass) of macronutrients to fulfill their energy needs. These include Nitrogen (N), Phosphorus (P), Potassium (K), Magnesium (Mg), Sulphur (S), and Calcium (Ca). Soil fertility acts as a contributing factor to supply these nutrients to plants, thereby impact plant growth parameters (Stevens et al. 2018). Soil edifice, porosity, pH, and water content play a pivotal role in soil fertility. Elevated levels of pH lead to nutrient scarcity, while acidic pH may cause metal toxicity (Stevens et al. 2018). Acidic soil with deficit nutrients cause remarkable decline in plant biomass (Kopittke et al. 2019; Asule et al. 2023).

7.1.1 Symptoms and Effects on Growth and Development

Nutrient deficiency in plants is usually caused due to their deficiency in the growing media like soil, but it may also be caused due to some problems and issues in the plant metabolism. Deficiency of one nutrient may cause some complex reactions to start and to cause some more problems with the absorption of other nutrients. These nutrient deficiencies can lead to various symptoms in plants along with growth and yield malfunctioning. Deficiency of different nutrients varied widely among various plant species. Plants respond differently against deficiencies of certain nutrients by showing clear symptoms including stunted growth, necrosis, chlorosis, senescence, and premature fruit drop. Among plant nutrients, Boron is constituent of cell wall, enzymes, and chlorophyll. It maintains ionic balance and membrane integrity. It affects plant growth by impacting photosynthesis (Shireen et al. 2018). Zinc regulates enzymatic activities and various abiotic stresses in plants. It impacts sprouting, water balance, stomatal opening, and membrane potential (Umair et al. 2020). Zinc deficiency creates leaf chlorosis and purplish foliage (Mattiello et al. 2015). Manganese acts as cofactor in photosynthesis. It is involved in glycosylation and ROS mitigation in plants. Scarcity of manganese causes reduction in biomass and plant yield (Alejandro et al. 2020). Iron is crucial for DNA formulation, respiration, and photosynthesis. It is constituent of cytochromes and chlorophyll. It is part of various enzymes and helps to maintain cellular structures. Iron deficiency causes chlorosis of leaves (Rout and Sahoo 2015). Copper acts as a cofactor to drive various cellular processes. It helps to regulate structural and metabolic pathways in plants. Deficiency and excess amount of copper may prove to be lethal for plants (Yruela 2005). Molybdenum facilitates N_2 fixation in plants (Evans 1956). Along with pterin, molybdenum acts as cofactor to catalyze metabolic processes (Mendel 2011). Chlorine is pivotal to regulate osmosis, stomatal conductance, O_2 release, and pathogenic infection mitigation (Chen et al. 2010). Chlorine scarcity causes chlorosis, necrosis, and wilting of leaves (Broyer et al. 1954). Plants take N_2 as nitrates that is incorporated to constitute proteins, nucleic acids, growth regulators, and chlorophyll molecules. N_2 scarcity reduces leaf surface area as well as plant resistance against biotic and abiotic stresses (Torres-Olivar et al. 2014). Phosphorus is important for replication, ATP biosynthesis and photosynthesis and plants uptake phosphorus as orthophosphate. Inadequate amount of phosphorus causes reduction in plant biomass. Potassium has some other vital roles in plant physiology. Insufficient amount of potassium, magnesium, and calcium impacts size of leaves along with causing chlorosis, necrosis, and decline in sink functioning (Cakmak and Kirkby 2008). Scarcity of nutrients causes remarkable decline in leaf, shoot, and root growth parameters of Spathiphyllum (Yeh et al. 2000).

7.1.2 Need of Problem Solving

Currently agricultural sector is experiencing threats of human ventures. Deforestation, habitat loss, extensive use of resources, and heavy tillage practices have altered climate and landscape for subsistence of plants (Pablo-Rodríguez et al. 2023). Population explosion, various biotic and abiotic components are contributors of nutrient deficit soil (Stevens et al. 2018). Human population is increasing enormously; will be 10 billion upto 2050 (Asule et al. 2023) putting pressure on natural resources. Feeding of this over populous globe is a major future concern. Food quality and limited arable acreage are challenging to meet global food needs. Repeated tillage practices on arable land lead to nutrient deficit soil, highly unsuitable for cultivation. Under such circumstances, use of chemical fertilizers is most quick option to cope nutrient deficiency. Due to lack of modern technology and limited resources, developing countries still rely upon conventional methods for tillage practices to get high yield.

7.1.3 Drawbacks of Conventional Methods

Plants depend on soil to meet their nutrient requirements. On average nutrient deficit soil is a leading cause of plant biomass decline. Several conventional methods (i.e., crop rotation, cover cropping, organic compost, manure, and chemical fertilizers) are used as common practice to enhance soil fertility for better crop yield to meet human standards. Among all of the conventional methods, synthetic fertilizers (NPK) are most efficient and frequently used method to enhance nutrient availability for plants. Despite their benefit to enhance crop yield instantly, the synthetic fertilizers impact environment, cause soil degradation and potential long-term dependency on external inputs.

- **Environmental Impact:** Synthetic fertilizers need water for their efficacy. They readily dissolve in water to release nutrients for plant use but meanwhile also percolate to deeper soil layers, thereby contaminating underground water table. Nitrate leads to eutrophication of above ground water (Lerner and Harris 2009), thereby disrupts food chain (Wezel et al. 2018). An example is eutrophication of Baltic Sea due to heavy tillage activities (Lerner and Harris 2009). Intensive utilization of animal manure (Romero et al. 2021) and synthetic fertilizers leads to release of nitrous oxide and CO_2, contributors of acid rain and greenhouse effect respectively (Francaviglia et al. 2023).
- **Soil Degradation:** Normally farmers remain unaware about nature and nutrient needs of both soil and plants. They do not try to know which and to what extent a specific nutrient is deficient in soil. Farmers use synthetic fertilizers in unchecked and bulk amount. Thereby repeated applications and excessive amount of certain nutrients, including zinc, arsenic, lead, copper, and cadmium (Thompson-Morrison et al. 2023), in synthetic fertilizers for tillage can cause nutrient toxicity and soil degradation by contributing to elevate salinity, loss of organic matter, and deterioration of soil texture (Francaviglia et al. 2023) and food chain (Thompson-Morrison et al. 2023).
- **Dependency on External Inputs:** Synthetic fertilizers do not remain in soil for longer durations. They usually washed out by rain or irrigation water, leaving soil deficient of nutrients. So plants depend on external inputs of synthetic fertilizers for better yield, which can make agriculture more vulnerable to price fluctuations and food chain deterioration (Van der Ploeg 2017).
- **Microbial Impact:** Extensive applications of synthetic fertilizers also impact soil microbiota (Francaviglia et al. 2023). Soil microbes are essential contributors of boosting soil aeration, fertility, texture, and soil health. Soil decomposers are best drivers of biogeochemical cycles for sustainable environment. Some microbes form mycorrhizal association with angiosperms rhizosphere thereby play positive role in growth parameters (Fang et al. 2023). Heavy practices of nitrogen fertilizers contribute to acidic soil pH, which retards microbial activities (Fang et al. 2023). Chemicals in synthetic fertilizers may cause toxicity for soil microbes leading to deterioration of biodiversity (Wezel et al. 2018) and ecosystem balance. Fang et al. (2023) revealed NPK in proportion of only

600:120:300 mg/plant is most suitable for cultivation and micriobial population (Fang et al. 2023).
- **Health Concerns:** Utilization of fungicides and fertilizers is consistently elevating through last decade in Europe, in similar way to rest of agricultural world. Foliar applications of pesticides to boost crop yield could be fatal to humans. Contaminated water could cause diarrhea, dysentery, typhoid, bacterial infections, hepatitis, cholera, salmonellosis etc. with severe health concerns (Wezel et al. 2018).
- **Socio Economic Status:** Synthetic fertilizers are produced in larger industrial units at the cost of huge energy consumption by combustion of fossil fuels. Which ultimately elevates price of fertilizers? Hence fertilizers prove to be expensive for small-scale farmers of developing world leading to economic unsustainability and socio economic differences (Van der Ploeg 2017).
- **Short-Term Solutions:** Although synthetic fertilizers provide nutrients quickly but have short-term advantages to crops. Heavy tillage practices on limited acreage to feed over populous globe of 7.9 billion inhabitants (Asule et al. 2023) need repeated applications of synthetic fertilizers which ultimately have drastic effects on global sustainability (Nájera et al. 2015; Asule et al. 2023).

The twentieth century was the era of "green revolution technology" (Jain 2010). Synthetic fertilizers and herbicides were immensely utilized to enhance tillage productivity and crop biomass (Jain 2010). Extensive utility of natural resources and synthetic fertilizers by anthropogenic activities during green revolution (Jain 2010) posed serious threats to the environment. Now current era of twenty-first century is experiencing the upshots of preceding era as pollution, smog, greenhouse effect, and global warming. Global warming is the leading cause of altered climate mainly due to elevation of CO_2 (approx. 3–6 °C) in mid Miocene (Tripati et al. 2009). Agricultural sector is at verge of destruction due to changing global climatic patterns. Climatic changes boosted the hazards of wildfires in fire-prone regions of globe. WHO estimates envisaged alleviated level of wildfire through the planet, which in turn is directly or indirectly linked with human ventures (https://www.who.int/health-topics/wildfires) (assessed on 12 December 2023). Abrupt climatic changes (i.e., temperature fluctuations, drought, saline and waterlogged conditions) are major constraint in successful cultivation practices contributing to yield reduction (Kępczyński and Kępczyńska 2023; Garcia et al. 2021).

7.1.4 Introduction to Karrikins

In recent era, Karrikins grabs immense attention in agri-sciences due to their plant friendly role in sprouting under various stresses. Karrikins also play pivotal roles in plant life span after seed germination, so it can be used as growth regulator (Van Staden et al. 2006) to enhance crop yield under emerging environmental challenges. Karrikins (KAR1) is found to activate signal transduction pathways of antioxidant enzymes to generate stress responses in plants (Kępczyński and Kępczyńska 2023) to boost crop yield.

7.2 Karrikins

A class of phytohormones, Karrikins are found to contribute to plant development by coping with stressors. Karrikins are signaling molecules that have roles in promoting seed germination of various plants. Karrikins impart their roles in nutrient deficit, saline, dearth, and frosty conditions, either single or in combination with other hormones (Yang et al. 2019). Karrikin polypeptide KAI2 is found to cope with drought. Preliminary phases of seed germination are most critical in plant life cycle, because seeds are more susceptible to surrounding climatic conditions like cold, frost, warmth, topsoil edifice, and water availability (Ghebrehiwot et al. 2008; Orzolek 1991). Karrikins break seed dormancy in vast range of cultivated, ornamental as well as certain weeds, both in fire-susceptible and fire-exempted regions (Kępczyński and Kępczyńska 2023). Even in minute quantity of about 1 nm, Karrikins sufficiently improve seed vigor (Chiwocha et al. 2009). Upon a shot of red wavelength, KAR1 exposure to Arabidopsis seeds readily initiates sprouting. Karrikins break seed quiescence state and promote cotyledon enlargement and hypocotyl reduction (Nelson et al. 2011). Karrikins exhibit species specific responses. KAR1 break quiescence state in Arabidopsis seeds but meanwhile nonresponsive in certain weeds and fungal hyphae (Nelson et al. 2011). In the field of biotechnology, KAR1 has pivotal role in somatic embryogenesis and growth (Kępczyński and Kępczyńska 2023). Applications of karrikins also contribute to ex situ seed sprouting in orchids (Kępczyński and Kępczyńska 2023). Six hours KAR1 treatment to seeds of *Avena fatua* is found sufficient for sprouting (Kępczyński et al. 2010). One hour pre-sowing treatment to maize seeds by KAR1 significantly increased longevity and plant biomass (Van Staden et al. 2006). KAR1 promotes pollen tube development and elongation among various plants including *Aloe maculate* (Kumari et al. 2015). Karrikinolides promote germination in *Ergostis tef* during heat and water stress (Ghebrehiwot et al. 2008). KAR1 share common characteristics with gibberellins and other hormones with wide spread significance in cultivation for plant growth and weed control (Light et al. 2009). Plant derived fire has dual (direct and indirect) impact on ecosystem. Amongst direct impacts, smoke derived KAR1 initiates flowering in fire-lilies (Keeley 1993). Nano scale amount of KAR1 contributes to enhance drought resistance in bent grass (Tan et al. 2023). RT-PCR results of Tan et al. (2023) envisaged activation of shock resistant genes to enhance leaf vigor during drought conditions (Tan et al. 2023). Cytometry assay of KAR1 revealed their role for rapid cell division in tomato plants (Jain and Van Staden 2006). Karrikinolides enhanced stomatal efficiency and cell vigor of *Brassica oleracea* (Shah et al. 2020a, b). KAR1 upregulates SOD, CAT, POD, and APX pathways in *Sapium sebiferum* to cope with water scarcity and salinity (Shah et al. 2020a, b). Effective amount of Karrikins differs among different plant genera (Bączek-Kwinta 2022). Researches revealed metabolic role of Karrikinolide in *Salvia miltiorrhiza* for tanshionone-I synthesis (Zhou et al. 2019).

7.2.1 History

Although discovered in recent era, history of Karrikins is traced back to smoke of wildfires. Smoke holds thousands of chemicals in it (Maga 1988). Despite the fact that smoke has deleterious chemicals, it breaks seed dormancy, initiates seedling growth and flowering. These observations led to the idea that smoke may contain certain active agents to break seed dormancy in fire susceptible regions. These ideas put forth the discovery of 71 active agents from aqueous smoke (Baldwin et al. 1994). Among various active constituents of smoke, Karrikinolides (Flematti et al. 2004; Elsadek and Yousef 2019), Cyanohydrins (Flematti et al. 2011), Butenolides (Light et al. 2010), and Hydroquinones (Kamran et al. 2017; Elsadek and Yousef 2019) are pivotal with agronomic perspectives.

Karrikinolides and Cyanohydrins are more thermostable, water soluble, and long lasting having a half-life of decades. Research on smoke based chemicals spans over past three decades. Investigations on smoke begin in fire susceptible Mediterranean zones and surpassed to Europe; where illegitimate ignition is often practiced (Zahed and Bączek-Kwinta 2023). De Lange and Boucher first envisaged impact of smoke in breaking seed dormancy (De Lange and Boucher 1990).

The presence of Karrikin (KAR1) was determined in 2003 from smoke of plants by bioactivity guided fraction (Van Staden et al. 2004; Chiwocha et al. 2009). Uptill now six types of Karrikins are described from plant burning activities (Nelson et al. 2012). VITOTOX® and Ames assays demonstrated non-genotoxic effects of Karrikins at a larger concentrations range (Light et al. 2009; Verschaeve et al. 2006). KAR1 is strongly agile, thermostable, and persistent component of plant smoke (Van Staden et al. 2004). Chromatography based assays of carbs (agar and cellulose) and smoke water led the identification of Karrikins (Bączek-Kwinta 2022). Karrikins is named after "Karrik" (Australian aboriginal term) meaning smoke (Dixon et al. 2009; Flematti et al. 2009). Later the term "Karrikinolide" was designated to KAR1 based on specific lactone group (Flematti et al. 2009).

7.2.2 Composition

Karrikins belongs to butenolide signaling molecules and 150 Da in size (Yao and Waters 2020). Pyrolysis of certain carbs like cellulose $(C_6H_{10}O_5)_n$, xylose$(C_5H_{10}O_5)$ and glucose $(C_6H_{12}O_6)$ is responsible for production of Karrikins during combustion of plant material (Yang et al. 2019; Nelson et al. 2009). Six forms of Karrikins are distinguished so far, ranging from KAR1-KAR6, which have variations in positioning of CH_3 groups (Yang et al. 2019; Nelson et al. 2012). KAR1 (3-methyl-2H-furo[2,3-c]pyran-2-one) is major butenolide also called Karrikinolide (Dixon et al. 2009; Flematti et al. 2004). Assembly of Karrikins resembles to strigolactones (another class of butenolide phytohormones found in root exudes). Karrikins possess D-loop in butenolide moeity (Nelson et al. 2011). Distillation of 3-Hydroxy-4H-pyran-4-one and CH_3CH_2COOH resulted in production of propionyl ester, which undergoes cycloaddition and dehydration to manufacture KAR1 (Flematti et al. 2009) (Fig. 7.1).

Fig. 7.1 Types of karrikins

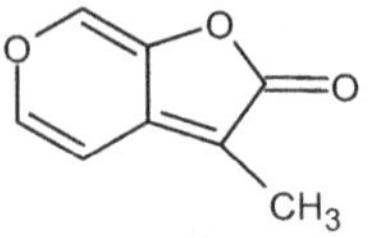

3-methyl-2*H*-furo[2,3-*c*]pyran-2-one (KAR1)

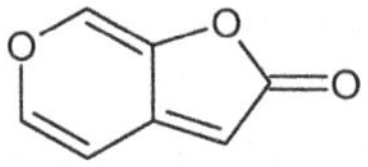

4a,5-dihydrocyclopenta[*c*]pyran-6(7*H*)-one (KAR2)

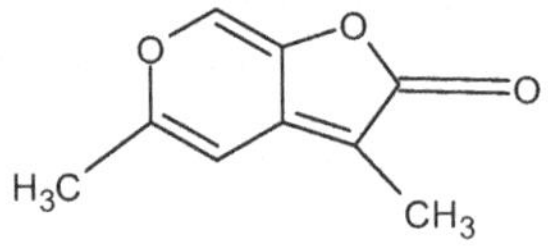

3,5-dimethyl-2*H*-furo[2,3-*c*]pyran-2-one (KAR3)

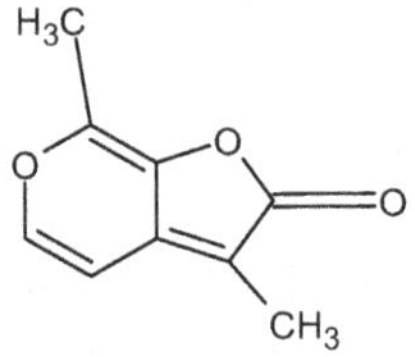

3,7-dimethyl-2*H*-furo[2,3-*c*]pyran-2-one (KAR4)

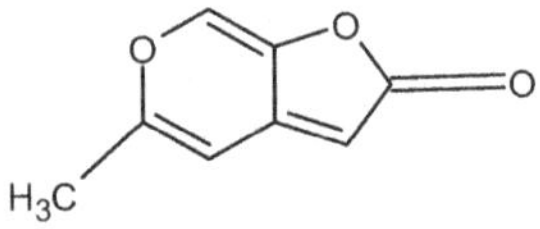

5-methyl-2*H*-furo[2,3-*c*]pyran-2-one (KAR5)

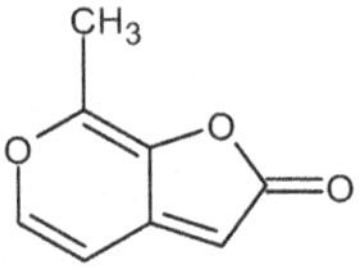

7-methyl-2*H*-furo[2,3-*c*]pyran-2-one (KAR6)

7.2.3 Sources

Naturally Karrikins occur scarcely in surrounding world (Yao and Waters 2020) but could be derived from combustion of plant materials and wildfires. Major sources of Karrikins are plant derived smoke and smoke water. Combustion process of plants releases smoke, containing volatile compounds from which Karrikins can be obtained. Smoke water is produced by passing smoke of plant fire through water and has longer lifetime (Brown and van Staden 1997). Smoke water promotes seedling development of about 1300 floral species (Soós et al. 2019). Smoke water contained six types of Karrikins, among which KAR1 and KAR2 are widely studied. KAR1 is a universal component of wildfires, formed by burning of intact cellulose (Flematti et al. 2009). Proportion of KAR1 is about 6–38 times greater when compared with other family members (Nelson et al. 2012). Smoke composition depends upon composition of plant species, as proportion of active constituents and secondary metabolites varied among plant groups leading to varied range of smoke composition with different effects on growth parameters.

Aqueous smoke of white-willow and lemon-eucalyptus propounds most positive impact on plant propagation (Elsadek and Yousef 2019). Karrikins are also found in cellulose paper and filter paper smoke. Karrikins can also be synthesized chemically from pyromeconic acid (Flematti et al. 2005), D-xylose (Xiao et al. 2022), reaction of d-xylose, and glycine (Light et al. 2005). KAR1 and KAR2 can be synthesized by reaction of Ti-crossed aldol with dihydraopyrans and methylpyruvate (Nagase et al. 2008). Sun et al. (2008) produced some analogs of KAR1 synthetically (Sun et al. 2008). Light et al. (2005) produced synthetic growth initiator KAR1 by burning of polypeptides with carbohydrates. Chromatography and spectroscopy assays confirmed consistency of synthetic KAR1 with wildfire-derived KAR1 from plant matter (Light et al. 2005). Trimethylbutenolide (TMB) antagonizes the effect of KAR1 (Pošta et al. 2017). Gupta et al. (2020) produced aqueous smoke from grass to elucidate impact of Karrikins. The UHPLC-ESI(+)-MS/MS assays revealed stimulatory role of Karrikins 1 and 2 in Lettuce embryogenesis (Gupta et al. 2020). Elsadek and Yousef Elsadek and Yousef (2019) produced aqueous smoke (source of Karrikins) using plant litter to explore embryogenesis and post embryogenesis induction in fire-free ecosystems (Elsadek and Yousef 2019). According to research, aqueous smoke caused decline in ABA (growth inhibitor) concentration, while enhanced amylase concentration among certain members of Cucurbitaceae and Asteraceae (Elsadek and Yousef 2019).

7.2.4 Mechanism of Action

Mechanism of action of Karrikins involves signaling pathway that stimulates seed germination of various plant species. Karrikins interact specific receptors of plant seeds triggering a cascade of molecular events that leads to activation of germination. Three functionally conserved protein groups are critical for Karrikin signaling in an array of plants including *Arabidopsis* and peas (Yao and Waters 2020). These protein groups are KAI2, MAX2, and SMXL, among which KAI2 (an active hydrolase containing binding region) recognizes Karrikins and is responsible for making configurationally modifications at their binding regions. MAX2 acts as ligase to catabolize proteasome. Signaling protein SMXL is correlated with HSPs and incorporates in silencing of transcription.

7.2.4.1 Karrikins Signaling and Changes at Transcription Level

Recent molecular studies on plants treated with smoke solution have shed light on the actions of karrikins (KARs) and their impact at the transcriptional level. Smoke-treated lettuce exhibited differential expression of genes related to germination, cell wall expansion, translational regulation, cell division cycle, carbohydrate metabolism, and abscisic acid regulation. This response was validated through real-time polymerase chain reaction, showcasing upregulated transcript abundance of specific genes. Similar investigations in maize revealed smoke-induced abscisic acid and stress-related responses, suggesting better adaptation to abiotic stresses. In Arabidopsis, KAR1 treatment resulted in increased expression of light-induced genes, highlighting the role of karrikins in priming light responses. The complexity of plant responses to smoke and karrikins suggests that karrikins are just one component of the intricate chemistry of smoke (Machin et al. 2020).

7.2.4.2 Karrikins Signaling Pathway

The molecular mechanism of karrikin action has been elucidated in *Arabidopsis,* revealing a signaling pathway involving receptors, conformational changes, and downstream protein interactions. KAR1 is perceived by the α/β-hydrolase KAI2, inducing a conformational change in KAI2. Activated KAI2 then enhances interaction with MAX2 (MORE AUXILLARY GROWTH2), an F-box protein, leading to the formation of a Skp-Cullin-F-box (SCF) ubiquitin ligase complex involving SMAX1 (SUPPRESSOR OF MAX2 1). SMAX1, a putative substrate, undergoes polyubiquitination and degradation in the 26S proteasome. Crystallographic techniques and ligand-binding experiments have provided insights into KAI2's recognition of karrikins and its interaction with downstream components in the signaling pathway. Studies with KAI2 mutants underscore the essential role of KAI2 in perceiving KAR1 (Khatoon et al. 2020) (Fig. 7.2).

7.3 Karrikins and Nutrient Uptake

Karrikins are the new member in the world of phytohormones and have ability to do so many talented artworks, which include multifunctional roles in plant's development and show multidirectional pathways in rhizosphere (Rehman et al. 2021). Karrikins play critical role in rhizosphere for nutrient uptake and its reallocation, about 80–90% plants suffered due to nutrient deficiency (Foo and Reid 2013).

Seed germination is one of the common and important physiological behavior performed by Karrikins. Karrikins were first isolated from germination based bioassays by using *Lactuca sativa* cv. Grand Rapids (Waters 2017). To understand these phenomena, the model plant *Arabidopsis thaliana* is one of the best examples. According to Nelson et al. (2010), the seedlings development in imbibed *Arabidopsis* seeds ensures that karrikins are not only an initiative of seed germination in plants after fires but they are also involved in light responses for shoot elongation. When we talk about root development, karrikins behave differently. In dicots like *Arabidopsis*, the karrikins promote adventitious roots from nodal shoot areas,

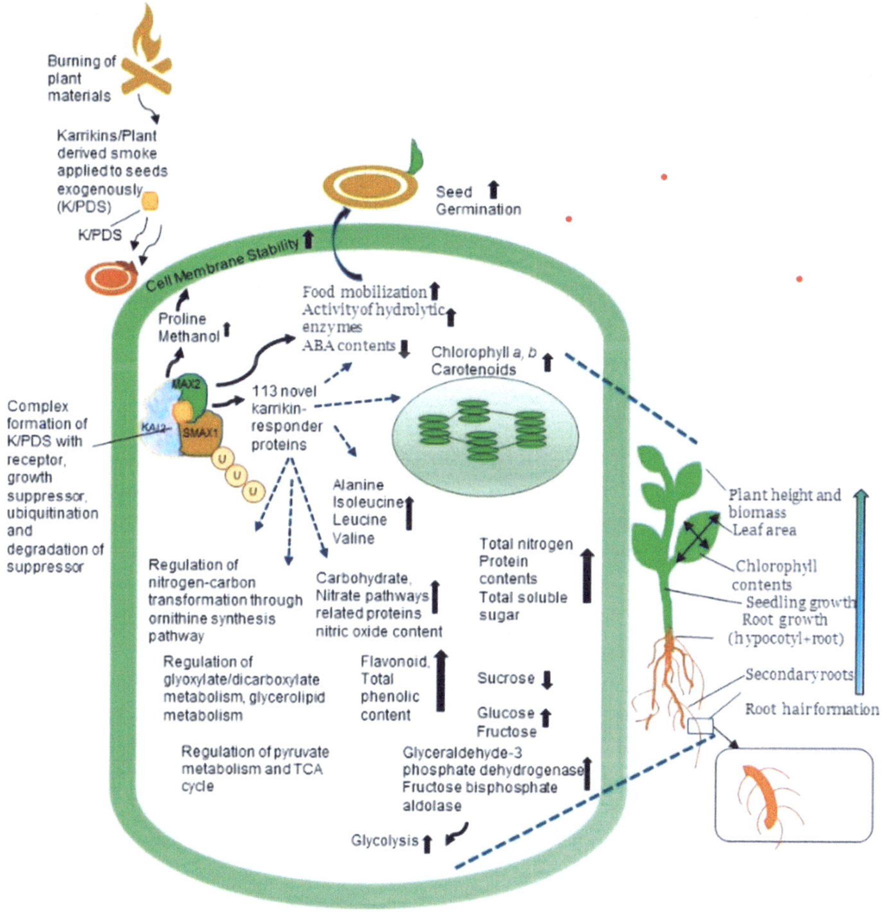

Fig. 7.2 Plant-derived smoke induces diverse cellular responses, accelerating seed germination and various metabolic processes. The diagram illustrates increased and decreased activities, with dashed lines indicating potential molecular-level relationships. Key components include KAI2 receptors, MAX2 in the SCF ligase complex, and the growth suppressor protein SMAX1 (Khatoon et al. 2020)

indicate its positive nature as primary root growth promotor. Recent studies revealed that lowering in auxin flux may occur due to exogenous supply of karrikins in Arabidopsis (Varshney and Gutjahr 2023; Rehman et al. 2021; Foo and Reid 2013). In overall review, karrikins play a role in seed germination via applying in medium or in case of seed priming.

The presence of KAR_1 and smoke-water in the growing medium shows an enhancement effect on the development of shoots and roots in rice, maize, okra, and tomato (Waters 2017). It was also demonstrated that exogenous application of strigolactone (mimic signaling hormone like karrikins) in SL-deficient rice mutants indicates the enhancement in internodal growth. This type of internodal elongation is also noticed in pea plant without disturbing the GA signaling. Therefore, it was proved that SLs and GAs work independently in enhancement of intermodal elongation (Rehman et al. 2021).

In biotic and abiotic stresses, karrikins behave diversely. In case of biotic stress, KIA*2* seeds and wild type seeds of *Arabidopsis* were experienced with elevated temperature about 30 °C and back to its normal conditions (20 °C), but wild type seeds experienced reduced germination as compared to KIA2-treated seeds. This condition is known as thermo-morphogenesis because wild type seeds recover timely after heat shock, which show positive behavior of KAI2. Similarly in case of drought stress and osmotic stress, behavior of karrikins is multidirectional. The kia*2* seeds under osmotic stress may reduce seed germination when GA treatment is applied. Arabidopsis dual mutant species probably reduce ROS production, increase root hair length and root to shoot ratio show stronger drought tolerance response (Varshney and Gutjahr 2023).

There is an interesting point still here surrounding, the karrikins behavior is quite similar to strigolactone (SL). Strigolactone is a phytohormone that helps in the development and growth of symbiotic AMF fungi in the soil and in returns it increases plant growth and its branching. Karrikins and SLs regulate many physiological processes which include photomorphogenesis, seed germination, rooting, root hair and root elongation, nodulation, and protonemal development (Foo and Reid 2013). Some phylogenetic studies also revealed that KAR receptors are present in some taxa of plants which include green algae, liverworts, or mosses (Antala et al. 2019). Recently, biochar also confirms the presence of karrikinolide, a chemical that is formed as result of pyrolysis, highly rich carbon, fine grained residue have many ecological and physiological properties related to plant's growth (Kochanek et al. 2016).

There are few examples of karrikins and its physiological roles in plants which are summarized in Table 7.1.

7.4 Nutrient Deficiencies and Karrikins

In the development stages of plant, they require involvement of many macronutrients and micronutrients. Karrikins perform multiple function in the developmental stages as an alternative source. Nutrients like nitrogen play a role in development of root to flower formation. Similarly, it is involved in the regulation of phytohormones (auxins and cytokinins) which play critical role in elongation and differential growth (Leghari et al. 2016). Phosphorus is the backbone of the genetic material and power bank of plant cell. It helps the plant in cell division and shoot growth also enhance grains crops (Malhotra et al. 2018). Calcium is also responsible for functioning of Indole acetic acid (IAA) which help in cell elongation (Tuteja and Mahajan 2007). When plants are grown in phosphorous and nitrogen deficient soils or are not provided with sufficient amounts of these nutrients, then phytohormones like strigolactones are activated that helps plant to fight against nutrient deficiency. They do so by promotion of arbuscular mycorrhiza or nitrogen fixing rhizobial bacteria or by

Table 7.1 Role of karrikins in plants' development and nutrient uptake (Antala et al. 2019)

S. no.	Plant	Conc. (M)	Means of application	Examined features	References
1.	Rice (*Oryza sativa* L.)	10^{-10}–10^{-8}	Grown in petri dishes with KAR1 solution	Seedling weight, vigor index root and shoot length, no. of lateral roots	Kulkarni et al. (2006)
2.	Tomato (*Lycopersicon esculentum* Mill.)	10^{-7}	Grown in petri dishes with KAR1 solution	Germination [80] % of abnormal seedlings vigour index, seedling weight hypocotyl and radicle length weight of 10 embryonic axis weight of 10 cotyledons	Jain and Van Staden (2006)
3.	Onion (*Allium cepa* L.)	10^{-10}	Grown in pots drenched by KAR1 solution twice a week	No. of leaves, leaf length fresh and dry leaf weight fresh bulb diameter fresh bulb diameter absolute growth harvest index genotoxicity and mutagenicity	Kulkarni et al. (2010)
4.	Pepper (*Capsicum annuum* L.), Salvia (*Salvia* sp.)	10^{-7}	Grown in pots irrigated by KAR1 solution	Seedling emergence of pepper seedling emergence of salvia seedling fresh and dry weight mean emergence time (days) catalase activity of pepper catalase activity of salvia	Demir et al. (2012)
5.	Pepper (*Capsicum annuum* L.)	10^{-7}	Seeds presoaked in KAR1 solution for 40 h	Germination, seedling emergence seedling fresh and dry weight catalase activity superoxid dismutase activity ascorbate peroxidase activity	Demir et al. (2018)

increasing lateral roots, but these actions of strigolactones are found to be karrikins dependent (Marzec and Melzer 2018). They also play their role in modulating the plant responses under magnesium deficiency. Karrikins are known to promote root growth to increase the surface area of roots in magnesium deficient soils. They control all this deficient nutrient metabolism by activating ethylene, auxins, and other growth promoters. They act as an upstream signal for these phytohormones in nutrient stress (Shah et al. 2022).

Karrikins in the fire places or in abiotic/biotic stress behave like an alternative source of nutrients. In Arabidopsis, karrikins serve a protection against drought and salinity stress by developing seed dormancy, thick cuticle formation, and stomatal closure (Varshney and Gutjahr 2023). Shah et al. (2020a, b) also mentioned that Karrikins significantly alleviate in the drought and salt stress of plant *Sapium*

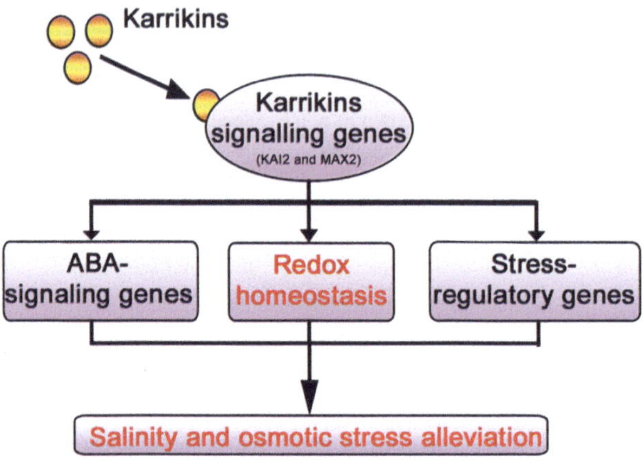

Fig. 7.3 Role of karrikins in abiotic stresses (Shah et al. 2020a, b)

sebiferum. He concluded that one nanomoler of karrikin solution improved taproot length, seedling biomass, and no. of lateral roots branches under abiotic stress. It also crosstalk with ABA, GA, CK, and ET hormones under abiotic stresses (Yang et al. 2019). Therefore, it could be conclude that karrikins are alternative source of nutrients in nutrient deficiencies and help the plants in avoidance of abiotic and biotic stresses (Fig. 7.3).

7.5 Future Prospectives

Scientific literature extensively covers the impact of Karrikin Analog Receptors (KARs) on seed germination, in vitro organogenesis, and tissue anatomy in horticultural and agricultural plants. The versatility of KARs in enhancing plant propagation for commercial production, addressing cross-breeding barriers, and promoting pollen grain germination emphasizes their potential applications in breeding and urban gardening. Furthermore, exploring KARs' role in secondary metabolite biosynthesis presents opportunities for advancements in the pharmaceutical industry, highlighting the untapped potential of this phytohormone group. Here are some future prospects of karrakins:

7.5.1 Role of KARs in Post-germination Stress Response

Despite the substantial progress in understanding the Karrikin Analog Receptors (KARs), there is a notable gap in knowledge regarding their role in plant stress response after germination. Future research should focus on unraveling the specific mechanisms and targets of KAR signaling pathways under adverse environmental conditions beyond the germination stage.

7.5.2 Downstream Targets of KAR Signaling Pathway

While the KAR pathway is well-documented up to the formation of the SCFMAX2-KAI2-SMAX1/SMXL2 complex, the downstream targets of KAR signaling pathways remain unclear, particularly under stress conditions. Investigating these targets will provide crucial insights into how KARs modulate plant responses to abiotic stresses.

7.5.3 Post-translational Modifications (PTMs) Caused by ROS/RNS

Although studies have demonstrated the involvement of Reactive Oxygen Species/Reactive Nitrogen Species (ROS/RNS) in KAR-regulated processes, the identity of the PTMs caused by ROS/RNS remains unexplored. Future research should delve into understanding the specific PTMs induced by these reactive species to enhance our comprehension of KAR-regulated resistance in plants.

7.5.4 Redox Regulation Crosstalk

The crosstalk between KAR signaling and redox signals, including those from other redox regulators such as Abscisic Acid (ABA), Strigolactones (SLs), and auxins, is not well-understood. Investigating the interplay between KARs and these redox regulators under stress conditions will provide a comprehensive understanding of how plants respond to multiple environmental challenges.

7.5.5 Genetic and Molecular Basis of KAR Crosstalk

Unraveling the genetic and molecular basis of the crosstalk between KARs and redox signals is essential for a deeper understanding of KAR-modulated abiotic stress resistance. This knowledge will be pivotal in developing strategies to enhance stress tolerance in crops and contribute to improved agricultural production and ecological sustainability.

In conclusion, future research efforts should focus on addressing these knowledge gaps to fully realize the potential of KARs as protective agents against various abiotic stresses, ultimately leading to the development of stress-tolerant crops and sustainable agricultural practices.

7.6 Conclusion

Nutrient deficiencies in plants pose a significant challenge to global agriculture, affecting crop yields and the livelihoods of millions. The imbalance of micronutrients and macronutrients, crucial for plant growth, results in symptoms like stunted growth, chlorosis, and reduced biomass. Traditional methods, such as synthetic fertilizers, have been widely used to address nutrient deficiencies, but they come with drawbacks like environmental pollution, soil degradation, and health concerns. The increasing global population and the pressure on natural resources further exacerbate these challenges. Amidst these issues, the role of karrikins in mitigating nutrient deficiencies has gained attention. Karrikins, signaling molecules found in plant smoke, have demonstrated remarkable potential in enhancing seed germination, promoting plant growth, and providing resilience against various environmental stresses. The discovery of karrikins traces back to wildfires, where the smoke was found to contain active agents capable of breaking seed dormancy. Karrikins, particularly KAR1, have been identified as key contributors to these beneficial effects. In comparison to conventional methods, such as synthetic fertilizers, karrikins offer a more sustainable and eco-friendly solution. They exhibit multifunctional roles, not only breaking seed dormancy but also influencing root and shoot development, enhancing drought resistance, and participating in nutrient uptake. The diverse applications of karrikins, from seed germination to plant growth regulation, make them a promising avenue for addressing nutrient deficiencies in a changing agricultural landscape. As global agriculture faces increasing threats from climate change and population growth, exploring innovative solutions like karrikins becomes imperative for sustainable and resilient food production.

References

Alejandro S, Höller S, Meier B, Peiter E (2020) Manganese in plants: from acquisition to subcellular allocation. Front Plant Sci 11:300

Antala M, Sytar O, Rastogi A, Brestic M (2019) Potential of karrikins as novel plant growth regulators in agriculture. Plants 9(1):43

Asule PA, Musafiri C, Nyabuga G, Kiai W, Ngetich FK, Spurk C (2023) Determinants of simultaneous use of soil fertility information sources among smallholder farmers in the central highlands of Kenya. Agriculture 13(9):1729

Bączek-Kwinta R (2022) An interplay of light and smoke compounds in photoblastic seeds. Plants 11(13):1773

Baldwin IT, Staszak-Kozinski L, Davidson R (1994) Up in smoke: I. Smoke-derived germination cues for postfire annual, *Nicotiana attenuata* Torr. Ex. Watson. J Chem Ecol 20:2345–2371

Brown NA, van Staden J (1997) Smoke as a germination cue: a review. Plant Growth Regul 22:115–124

Broyer TC, Carlton AB, Johnson CM, Stout PR (1954) Chlorine—a micronutrient element for higher plants. Plant Physiol 29(6):526

Cakmak I, Kirkby EA (2008) Role of magnesium in carbon partitioning and alleviating photooxidative damage. Physiol Plant 133(4):692–704

Chen W, He ZL, Yang XE, Mishra S, Stoffella PJ (2010) Chlorine nutrition of higher plants: progress and perspectives. J Plant Nutr 33(7):943–952

Chiwocha SD, Dixon KW, Flematti GR, Ghisalberti EL, Merritt DJ, Nelson DC et al (2009) Karrikins: a new family of plant growth regulators in smoke. Plant Sci 177(4):252–256

De Lange JH, Boucher C (1990) Autecological studies on *Audouinia capitata* (Bruniaceae). I. Plant-derived smoke as a seed germination cue. S Afr J Bot 56(6):700–703

Demir I, Ozuaydın I, Yasar F, Van Staden J (2012) Effect of smoke-derived butenolide priming treatment on pepper and salvia seeds in relation to transplant quality and catalase activity. S Afr J Bot 78:83–87

Demir I, Ozden E, Yıldırım KC, Sahin O, Van Staden J (2018) Priming with smoke-derived karrikinolide enhances germination and transplant quality of immature and mature pepper seed lots. S Afr J Bot 115:264–268

Dixon KW, Merritt DJ, Flematti GR, Ghisalberti EL (2009) Karrikinolide—a phytoreactive compound derived from smoke with applications in horticulture, ecological restoration and agriculture. Acta Hortic 813(2009):155–170

Elsadek MA, Yousef EA (2019) Smoke-water enhances germination and seedling growth of four horticultural crops. Plants 8(4):104

Evans HJ (1956) Role of molybdenum in plant nutrition. Soil Sci 81(3):199–208

Fang X, Yang Y, Zhao Z, Zhou Y, Liao Y, Guan Z, Zhao S (2023) Optimum nitrogen, phosphorous, and potassium fertilizer application increased *Chrysanthemum* growth and quality by reinforcing the soil microbial community and nutrient cycling function. Plants 12(23):4062

Flematti GR, Ghisalberti EL, Dixon KW, Trengove RD (2004) A compound from smoke that promotes seed germination. Science 305(5686):977–977

Flematti GR, Ghisalberti EL, Dixon KW, Trengove RD (2005) Synthesis of the seed germination stimulant 3-methyl-2H-furo [2, 3-c] pyran-2-one. Tetrahedron Lett 46(34):5719–5721

Flematti GR, Ghisalberti EL, Dixon KW, Trengove RD (2009) Identification of alkyl substituted 2 H-furo [2, 3-c] pyran-2-ones as germination stimulants present in smoke. J Agric Food Chem 57(20):9475–9480

Flematti GR, Merritt DJ, Piggott MJ, Trengove RD, Smith SM, Dixon KW, Ghisalberti EL (2011) Burning vegetation produces cyanohydrins that liberate cyanide and stimulate seed germination. Nat Commun 2(1):360

Foo E, Reid JB (2013) Strigolactones: new physiological roles for an ancient signal. J Plant Growth Regul 32:429–442

Francaviglia R, Almagro M, Vicente-Vicente JL (2023) Conservation agriculture and soil organic carbon: principles, processes, practices and policy options. Soil Syst 7(1):17

Garcia D, Arif S, Zhao Y, Zhao S, Ming LC, Huang D (2021) Seed priming technology as a key strategy to increase crop plant production under adverse environmental conditions

Ghebrehiwot HM, Kulkarni MG, Kirkman KP, Van Staden J (2008) Smoke-water and a smoke-isolated butenolide improve germination and seedling vigour of *Eragrostis tef* (Zucc.) Trotter under high temperature and low osmotic potential. J Agron Crop Sci 194(4):270–277

Gupta S, Hrdlička J, Ngoroyemoto N, Nemahunguni NK, Gucký T, Novák O, Van Staden J (2020) Preparation and standardisation of smoke-water for seed germination and plant growth stimulation. J Plant Growth Regul 39:338–345

Jain HK (2010) Green revolution: history, impact and future. The Green Revolution, Histor

Jain N, Van Staden J (2006) A smoke-derived butenolide improves early growth of tomato seedlings. Plant Growth Regul 50:139–148

Kamran M, Khan AL, Ali L, Hussain J, Waqas M, Al-Harrasi A, Lee IJ (2017) Hydroquinone; a novel bioactive compound from plant-derived smoke can cue seed germination of lettuce. Front Chem 5:30

Keeley JE (1993) Smoke-induced flowering in the fire-lily *Cyrtanthus ventricosus*. S Afr J Bot 59(6):638

Kępczyński J, Kępczyńska E (2023) Plant-derived smoke and karrikin 1 in seed priming and seed biotechnology. Plants 12(12):2378

Kępczyński J, Cembrowska D, Van Staden J (2010) Releasing primary dormancy in *Avena fatua* L. caryopses by smoke-derived butenolide. Plant Growth Regul 62:85–91

Khatoon A, Rehman SU, Aslam MM, Jamil M, Komatsu S (2020) Plant-derived smoke affects biochemical mechanism on plant growth and seed germination. Int J Mol Sci 21(20):7760

Kochanek J, Long RL, Lisle AT, Flematti GR (2016) Karrikins identified in biochars indicate post-fire chemical cues can influence community diversity and plant development. PLoS One 11(8):e0161234

Kopittke PM, Menzies NW, Wang P, McKenna BA, Lombi E (2019) Soil and the intensification of agriculture for global food security. Environ Int 132:105078

Kulkarni MG, Sparg SG, Light ME, Van Staden J (2006) Stimulation of rice (*Oryza sativa* L.) seedling vigour by smoke-water and butenolide. J Agron Crop Sci 192(5):395–398

Kulkarni MG, Ascough GD, Verschaeve L, Baeten K, Arruda MP, Van Staden J (2010) Effect of smoke-water and a smoke-isolated butenolide on the growth and genotoxicity of commercial onion. Sci Hortic 124(4):434–439

Kumari A, Papenfus HB, Kulkarni MG, Pošta M, Van Staden J (2015) Effect of smoke derivatives on in vitro pollen germination and pollen tube elongation of species from different plant families. Plant Biol 17(4):825–830

Leghari SJ, Wahocho NA, Laghari GM, HafeezLaghari A, MustafaBhabhan G, HussainTalpur K, Lashari AA (2016) Role of nitrogen for plant growth and development: a review. Adv Environ Biol 10(9):209–219

Lerner DN, Harris B (2009) The relationship between land use and groundwater resources and quality. Land Use Policy 26:S265–S273

Light ME, Burger BV, Van Staden J (2005) Formation of a seed germination promoter from carbohydrates and amino acids. J Agric Food Chem 53(15):5936–5942

Light ME, Daws MI, Van Staden J (2009) Smoke-derived butenolide: towards understanding its biological effects. S Afr J Bot 75(1):1–7

Light ME, Burger BV, Staerk D, Kohout L, Van Staden J (2010) Butenolides from plant-derived smoke: natural plant-growth regulators with antagonistic actions on seed germination. J Nat Prod 73(2):267–269

Machin DC, Hamon-Josse M, Bennett T (2020) Fellowship of the rings: a saga of strigolactones and other small signals. New Phytol 225(2):621–636

Maga JA (1988) Smoke in food processing. CRC Press, Boca Raton

Malhotra H, Vandana SS, Pandey R (2018) Phosphorus nutrition: plant growth in response to deficiency and excess. In: Plant nutrients and abiotic stress tolerance, pp 171–190

Marzec M, Melzer M (2018) Regulation of root development and architecture by strigolactones under optimal and nutrient deficiency conditions. Int J Mol Sci 19(7):1887

Mattiello EM, Ruiz HA, Neves JC, Ventrella MC, Araújo WL (2015) Zinc deficiency affects physiological and anatomical characteristics in maize leaves. J Plant Physiol 183:138–143

Mendel RR (2011) Cell biology of molybdenum in plants. Plant Cell Rep 30:1787–1797

Nagase R, Katayama M, Mura H, Matsuo N, Tanabe Y (2008) Synthesis of the seed germination stimulant 3-methyl-2H-furo [2, 3-c] pyran-2-ones utilizing direct and regioselective Ti-crossed aldol addition. Tetrahedron Lett 49(29–30):4509–4512

Nájera F, Tapia Y, Baginsky C, Figueroa V, Cabeza R, Salazar O (2015) Evaluation of soil fertility and fertilisation practices for irrigated maize (*Zea mays* L.) under Mediterranean conditions in Central Chile. J Soil Sci Plant Nutr 15(1):84–97

Nelson DC, Riseborough JA, Flematti GR, Stevens J, Ghisalberti EL, Dixon KW, Smith SM (2009) Karrikins discovered in smoke trigger Arabidopsis seed germination by a mechanism requiring gibberellic acid synthesis and light. Plant Physiol 149(2):863–873

Nelson DC, Flematti GR, Riseborough JA, Ghisalberti EL, Dixon KW, Smith SM (2010) Karrikins enhance light responses during germination and seedling development in *Arabidopsis thaliana*. Proc Natl Acad Sci U S A 107(15):7095–7100

Nelson DC, Scaffidi A, Dun EA, Waters MT, Flematti GR, Dixon KW et al (2011) F-box protein MAX2 has dual roles in karrikin and strigolactone signaling in Arabidopsis thaliana. Proc Natl Acad Sci U S A 108(21):8897–8902

Nelson DC, Flematti GR, Ghisalberti EL, Dixon KW, Smith SM (2012) Regulation of seed germination and seedling growth by chemical signals from burning vegetation. Annu Rev Plant Biol 63:107–130

Orzolek MD (1991) Establishment of vegetables in the field. HortTechnology 1(1):78–81

Pablo-Rodríguez JL, Bravo-Monzón ÁE, Montiel-González C, Benítez-Malvido J, Álvarez-Betancourt S, Ramírez-Sánchez O et al (2023) Linking anthropogenic landscape perturbation to herbivory and pathogen leaf damage in tropical tree communities. Plants 12(22):3839

Pošta M, Papenfus HB, Light ME, Beier P, Van Staden J (2017) Structure–activity relationships of N-and S-analogs of the seed germination inhibitor (3, 4, 5-trimethylfuran-2 (5 H)-one) for mode of action elucidation. Plant Growth Regul 82:47–53

Rehman NU, Li X, Zeng P, Guo S, Jan S, Liu Y, Xie Q (2021) Harmony but not uniformity: role of strigolactone in plants. Biomolecules 11(11):1616

Romero CM, Chunli LI, Owens J, Ribeiro GO, Mcallister TA, Okine E, Xiying HAO (2021) Nutrient cycling and greenhouse gas emissions from soil amended with biochar-manure mixtures. Pedosphere 31(2):289–302

Rout GR, Sahoo S (2015) Role of iron in plant growth and metabolism. Rev Agric Sci 3:1–24

Shah AA, Khan WU, Yasin NA, Akram W, Ahmad A, Abbas M et al (2020a) Butanolide alleviated cadmium stress by improving plant growth, photosynthetic parameters and antioxidant defense system of Brassica oleracea. Chemosphere 261:127728

Shah FA, Wei X, Wang Q, Liu W, Wang D, Yao Y, Wu L (2020b) Karrikin improves osmotic and salt stress tolerance via the regulation of the redox homeostasis in the oil plant *Sapium sebiferum*. Front Plant Sci 11:216

Shah FA, Ni J, Chen X, Tang C, Wu L (2022) KAI2-KL signalling regulates root hair elongation under magnesium deficiency by activating auxin, ethylene, and nitric oxide biosynthesis and signalling genes. BioRxiv

Shireen F, Nawaz MA, Chen C, Zhang Q, Zheng Z, Sohail H, Bie Z (2018) Boron: functions and approaches to enhance its availability in plants for sustainable agriculture. Int J Mol Sci 19(7):1856

Soós V, Badics E, Incze N, Balázs E (2019) Fire-borne life: a brief review of smoke-induced germination. Nat Prod Commun 14(9):1934578X19872925

Stevens G, Motavalli P, Scharf P, Nathan M, Dunn D (2018) Crop nutrient deficiencies and toxicities

Sun K, Chen Y, Wagerle T, Linnstaedt D, Currie M, Chmura P, Xu M (2008) Synthesis of butenolides as seed germination stimulants. Tetrahedron Lett 49(18):2922–2925

Tan ZZ, Wang YT, Zhang XX, Jiang HY, Li Y, Zhuang LL et al (2023) Karrikin1 enhances drought tolerance in creeping bentgrass in association with antioxidative protection and regulation of stress-responsive gene expression. Agronomy 13(3):675

Thompson-Morrison H, Ariantiningsih F, Arief SM, Gaw S, Robinson B (2023) Nutrients and contaminants in soils of current and former oil palm production systems from Indonesia. Land 12(12):2144

Torres-Olivar V, Villegas-Torres OG, Domínguez-Patiño ML, Sotelo-Nava H, Rodríguez-Martínez A, Melgoza-Alemán RM, Alia-Tejacal I (2014) Role of nitrogen and nutrients in crop nutrition. J Agric Sci Technol 4(1B):29

Tripati AK, Roberts CD, Eagle RA (2009) Coupling of CO_2 and ice sheet stability over major climate transitions of the last 20 million years. Science 326(5958):1394–1397

Tuteja N, Mahajan S (2007) Calcium signaling network in plants: an overview. Plant Signal Behav 2(2):79–85

Umair H, Aamer M, Umer Chattha M, Haiying T, Shahzad B, Barbanti L, Guoqin H (2020) The critical role of zinc in plants facing the drought stress. Agriculture 10(9):396

Van der Ploeg JD (2017) The importance of peasant agriculture: a neglected truth. Wageningen University and Research, Wageningen, p 27

Van Staden J, Jager AK, Light ME, Burger BV (2004) Isolation of the major germination cue from plant-derived smoke

Van Staden J, Sparg SG, Kulkarni MG, Light ME (2006) Post-germination effects of the smoke-derived compound 3-methyl-2H-furo [2, 3-c] pyran-2-one, and its potential as a preconditioning agent. Field Crop Res 98(2–3):98–105

Varshney K, Gutjahr C (2023) KAI_2 can do: karrikin receptor function in plant development and response to abiotic and biotic factors. Plant Cell Physiol 64(9):984–995

Verschaeve L, Maes J, Light ME, van Staden J (2006) Genetic toxicity testing of 3-methyl-2H-furo [2, 3-c] pyran-2-one, an important biologically active compound from plant-derived smoke. Mutat Res 611(1–2):89–95

Waters MT (2017) From little things big things grow: karrikins and new directions in plant development. Funct Plant Biol 44(4):373–385

Wezel A, Goris M, Bruil J, Félix GF, Peeters A, Bàrberi P, Migliorini P (2018) Challenges and action points to amplify agroecology in Europe. Sustainability 10(5):1598

Xiao G, Su G, Slawin AM, Westwood N (2022) From biomass to the karrikins via selective catalytic oxidation of hemicellulose-derived butyl xylosides and glucosides. Eur J Org Chem 2022(15):7–16

Yang T, Lian Y, Wang C (2019) Comparing and contrasting the multiple roles of butenolide plant growth regulators: strigolactones and karrikins in plant development and adaptation to abiotic stresses. Int J Mol Sci 20(24):6270

Yao J, Waters MT (2020) Perception of karrikins by plants: a continuing enigma. J Exp Bot 71(6):1774–1781

Yeh DM, Lin L, Wright CJ (2000) Effects of mineral nutrient deficiencies on leaf development, visual symptoms and shoot–root ratio of Spathiphyllum. Sci Hortic 86(3):223–233

Yruela I (2005) Copper in plants. Braz J Plant Physiol 17:145–156

Zahed M, Bączek-Kwinta R (2023) The impact of post-fire smoke on plant communities: a global approach. Plants 12(22):3835

Zhou J, Xu ZX, Sun H, Guo LP (2019) Smoke-isolated karrikins stimulated tanshinones biosynthesis in *Salvia miltiorrhiza* through endogenous nitric oxide and jasmonic acid. Molecules 24(7):1229

Hemin in Plants: Biosynthesis and Role in ROS Detoxification During Oxidative Stress

8

Anita, Nihar Mathur, and Gyan Singh Shekhawat

Abstract

Abiotic stress is now a global environmental issue and it seriously harms plants by causing chlorosis, growth retardation, and release of reactive oxygen species. Hemin has been essential for plants to protect against abiotic stress. Hemin is a heme-derived molecule, has been demonstrated to perform several adaptive physiological roles in plants' ability to endure an assortment of abiotic stress. Hemin reduced the oxidative damage and tolerance mechanisms have been conducted to evaluate growth parameter, stress parameter as well as non-enzymatic and enzymatic parameters. In plants, hemin-elevated antioxidant levels and antioxidative enzyme activity may give resistance againstabioticstress. The purpose of this chapter is to elucidate the role of hemin in decreasing abiotic stress and elevating tolerance in plants.

Keywords

Hemin · Reactive oxygen species · Oxidative stress · Heme oxygenase · Antioxidant

Abbreviations

ROS	Reactive oxygen species
HO 1	Heme oxygenase 1
APX	Ascorbate peroxidase
GPX	Guaiacol peroxidase

Anita · N. Mathur · G. S. Shekhawat (✉)
Plant Biotechnology and Molecular Biology Lab, Department of Botany (UGC-CAS), Jai Narain Vyas University, Jodhpur, India

© The Author(s), under exclusive license to Springer Nature Singapore Pte Ltd. 2024

M. Faizan, S. Hayat (eds.), *Plant Growth Regulators: Resilience for Sustainable Agriculture*, https://doi.org/10.1007/978-981-97-2918-0_8

SOD	Superoxide dismutase
GR	Glutathione reductase
H_2O_2	Hydrogen peroxide
Zn	Zinc
Pb	Lead
Cr	Chromium
ABAP	2,2′-Azobis-2-amidino-propane-hydrochloride

8.1 Introduction

Hemin is an analogue of protoporphyrin IX that has a coordinated chloride ligand and a ferric (Fe^{3+}) ion. It obtained through the in vitro purification of natural heme (Andersson et al. 2002). Normal heme is a widespread molecular compound consisting of ferrous and the protoporphyrin IX tetrapyrrole (Wagener et al. 2003). Hemin has been widely used in a variety of businesses during the last few years, whether as an iron supply (Sander et al. 2000), an essential food ingredient (Becker et al. 2011), potential therapy for the treatment of HIV-1 infection (Davidge et al. 2009), and an anticarcinogen (Katoh et al. 1983). Recent research in botany has revealed that hemin is an effective supplement that is essential for protecting plants against many kinds of abiotic stresses. For instance, hemin has been shown to induce the detoxification of Zn, Pb, and Cr in *Oryza sativa* (Chen et al. 2017). Additionally, hemin is linked to the mitigation of Hg-induced damage in *Medicago sativa* (Han et al. 2007) and it has demonstrated benefits in facilitating wheat acclimatization to salinity (Xie et al. 2011). So many researches have been undertaken to elucidate the fundamental mechanisms of the detoxification process enhanced by hemin. There is extensive documentation indicating that the application of hemin significantly upregulates the transcription and activity of HO1 (Hyvelin et al. 2010; Xuan et al. 2008). This is attributed to the inducible nature of HO1, which can be stimulated by its own substrate, heme. Jung et al. (1997) observed that hemin effectively inhibited lipid peroxidation triggered by either the ascorbate/FeSO4 system, a metal-catalyzed oxidation process of ABAP that generates free radicals at equal rate. Nevertheless, the biological activity of heme in the aquatic environment decreases due to dimerization of heme molecules (Fu et al. 2011). Furthermore, the application of hemin is frequently constrained by its low solubility in neutral aqueous solutions and organic solvents. In contrast, β-cyclodextrins (β-CDs) are recognized as exceptional embedding media due to their suitably sized hydrophobic cavities that exhibit both rigidity and flexibility (Fu et al. 2011). Various polymers and receptor compounds consequent from β-cyclodextrin have found extensive use in investigative and biomimetic chemistry. As an illustration, a complex involving the combination of β-CD with hemin (referred to as CDH) was produced for the purpose of determining the concentration of H_2O_2 (Liu et al. 1999; Fu et al. 2011). The aim of this chapter is to describe the most recent data about hemin biochemistry, biosynthesis, interaction with ROS, and role in plants against abiotic stress.

8.2 Biochemistry and Biosynthesis of Hemin

Hemin is known as ferric chloride heme that means iron containing porphyrin with chlorin. In 1853, Ludwik Karol Teichmann accomplished the initial crystallization of hemin from blood and observed that pigments in the blood have the capability to form microscopic crystals. Hence, hemins' crystals are sometimes denoted as "Teichmann crystals." Hans Fischer successfully made hemin andwon Nobel Prize in Chemistry in 1930. In Fischer's method (1941), defibrinated blood is treated with a solution of NaCl in acetic acid. Hemin exists as a dark purple crystalline powder or elongated needle-shaped crystals. It is insoluble in water, yet it can dissolve in acidic acetone and alkaline water. The molecular formula is $C_{34}H_{32}ClFeN_4O_4$, and weight is 651.96 g/mol. Due to lipophilic nature of hemin, it intercalates into cell membranes and interfere with its fluidity and function (Balla et al. 1991; Wagener et al. 2003).

Hemin is synthesized in plants through a number of enzymatic processes in many cellular compartments like mitochondria and chloroplasts (Medlock and Dailey 2022). In living tissues, the Krebs tricarboxylic acid cycle's intermediates are used in the production of hemin (Richmond and Salomon 1955). Similar intermediates might be involved in the formation of the tetrapyrrole nucleus in plant and animal systems, as it is now known that the Krebs cycle and oxidative phosphorylation occur in plants (Richmond and Salomon 1955). Although hemin is the oxidized form of heme, plants initially synthesize heme, which is subsequently transformed into hemin by removing one electron (Dutt et al. 2022). The creation of 5-aminolevulinic acid from succinyl-CoA and glycine by 5-aminolevulinic acid synthase (ALAS) is the initial reaction in the heme pathway (Ogun et al. 2019). After that, ALA is transported to the mitochondria. Two ALA molecules condense in the mitochondria to create porphobilinogen (PBG) (Kiening and Lange 2022). The ALA dehydratase enzyme is the catalyst for this reaction (Kiening and Lange 2022). The production of uroporphyrinogen III is the result of additional enzymatic processes in the mitochondria (Kiening and Lange 2022). This involves converting the intermediate molecules and adding four PBG molecules one following the other (Kumari 2018). Uroporphyrinogen III is transported back to the chloroplasts and decarboxylates to generate coproporphyrinogen III (Espinas et al. 2016). Coproporphyrinogen III is then transformed to protoporphyrinogen III in the chloroplasts. Protoporphyrinogen oxidase catalyzes this step then ferrous ion ($Fe^{2+)}$ is combined with protoporphyrin IX to produce heme (Espinas et al. 2016). Ferrochelatase is the catalyst for this reaction (Falb et al. 2023). Hemin is the oxidized form of heme that results from the iron in the heme molecule losing an electron (Falb et al. 2023). The synthesis of heme, which is necessary for many biological processes such as the oxygen transport carried out by haemoglobin and the catalytic activity of heme-containing enzymes, depends on this pathway (Falb et al. 2023). The pathway's intermediates might also perform other roles in cellular functioning (Fig. 8.1).

Hemin, an oxidative state of heme, is chemically equivalent to natural heme, an extensive molecular compound of ferric and tetrapyrrole protoporphyrin IX (Zhu

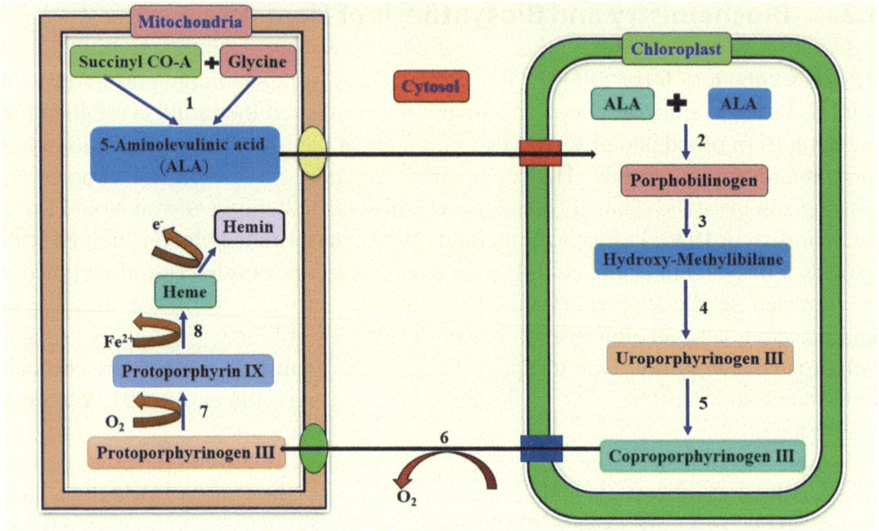

Fig. 8.1 Biosynthesis of hemin in plants: Enzymes involved in the pathway are (1) 5-aminolevulinic acid synthase, (2) ALA dehydratase, (3) Uroporphyrinogen I synthase, (4) Uroporphyrinogen III Synthase, (5) Uroporphyrinogen III decarboxylase, (6) Coporphyrinogen Oxidase, (7) Protoporphyrinogen oxidase, and (8) Ferro chelatase. Hemin is the oxidized form of heme that results from the iron in the heme molecule losing an electron

et al. 2019). Its mechanism of action involves the activation of caspases and oxidative stress, which damages microglia, neurons, and astrocytes (Robinson et al. 2009). Nevertheless, microglia that clear hemin have defensive mechanisms to avoid cell death (Robinson et al. 2009). Davidge et al. (2009) noticed that hemin can avoid oxidative damage in animal tissue through increasing the expression of HO-1. This has been well reported in the animal system, but less evidence is available in the case of the plant system about the involvement of hemin as an antioxidant defence system (Davidge et al. 2009). Xuan et al. 2008 and Xie et al. 2011 discussed the role of hemin to use defensive effects in an HO 1 dependent manner. There is various evidence that hemin has a potent cytoprotective effect on a variety of abiotic stresses in plants. For instance, hemin is linked to oxidative stress caused by Al and Cd in *Medicago sativa* (Cui et al. 2012, 2013). Furthermore, hemin is associated with ammonium tolerance in *Oryza sativa* (Xie et al. 2015), and wheat salinity adaptation by controlling ROS homeostasis (Xie et al. 2011).

8.3 Interactions Between ROS Specifically Hydrogen Peroxide and Hemin

Plant cells are produced ROS against heavy metal stress through cellular redox reaction. Then ROS generate oxidative stress through various metabolic processes (Chmielowska-Bąk et al. 2018). Oxidative stress produces many oxygen species of free radicals and non-radical substrates (Halliwell and Gutteridge 2015). The primary free radicals of notable significance include hydroxyl, alkoxyl, peroxyl, hydroperoxyl, and superoxide, while the major non-radical substrates comprise singlet oxygen, hydrogen peroxide (H_2O_2), ozone (O_3), hypochlorous acid ($HClO^-$), or peroxynitrite (Halliwell and Gutteridge 2015). The processes that produce and scavenge reactive oxygen species interact to maintain steady-state levels of ROS. Enzymatic antioxidant free oxygen radicals such as SOD are known to eliminate H_2O_2, catalase, and peroxidase as part of the ROS-scavenging pathway (Halliwell and Gutteridge 2015; Gallego et al. 2012). The accumulation of reactive oxygen species (ROS) poses a hazard due to its interactions with lipids, proteins, and nucleic acids, leading to severe damage to plant cells (Chmielowska-Bąk et al. 2018).

Hemin catalyzes the breakdown of H_2O_2 with kinetics remarkably similar to those of catalase, suggesting that free hemin has the potential to effectively degrade H_2O_2 (Kremer 1989). While the catalytic activity is expected, considering the presence of heme groups in the four active sites of catalase, the amount of H_2O_2 decomposition by hemin is significantly lower than that observed with catalase (Kremer 1989). The mechanism by which hemin functions as a catalase mimic is not thoroughly understood. It remains uncertain whether water and oxygen are the exclusive end-products of the hemin-catalyzed response (Fig. 8.2) and the specific valence situations of iron engaged in the catalytic process are not definitively identified. Apart from demonstrating catalase-like activity, hemin is capable of catalyzing a peroxidase reaction. This reaction involves the conversion of $H_2O_2$to water and occurs in the occurrence of various reducing agents (Portsmouth and Beal 1971). A possible byproduct of the cellular reaction is the production of superoxide or singlet oxygen, with GSH serving as a likely electron donor (Medeiros et al. 1987). It is important to highlight that undamaged hemin can also function as a Fenton reagent when combined with H_2O_2 and superoxide, generating harmful hydroxyl radicals at a rate even surpassing that of free iron (Huffman et al. 2000). In this recurring process, superoxide reduces hemin to heme; subsequently, in the presence of H_2O_2, the heme undergoes oxidation to regenerate hemin along with a hydroxide anion and a hydroxyl radical (Fig. 8.2).

When exposed to H_2O_2, heme can undergo oxidation, leading to the formation of the reactive ferryl-heme. It is believed that this functions as the trigger for initiating the oxidative deprivation of unsaturated fatty acids in plasma membrane (Gutteridge and Smith 1988). Although iron chelators cannot stop this process, hemopexin and albumin, which both bind hemin with a high affinity, can stop it partially and fully, respectively (Gutteridge and Smith 1988). The peroxidation is likely attributed to intact hemin rather than free iron released during breakdown by HO. The insertion of hemin into the plasma membrane heightens the sensitivity of cells to the toxicity of external H_2O_2. This is due to a significant acceleration in the rate of lipid peroxidation effectedby H_2O_2 (Balla et al. 1991). Once triggered, the immediate peroxidation of membrane lipids evolves into an autocatalytic process, relying on the occurrence of hemin but operating independently of the need for H_2O_2(Balla et al. 1991). While the specific mechanisms underlying this process remain uncertain,

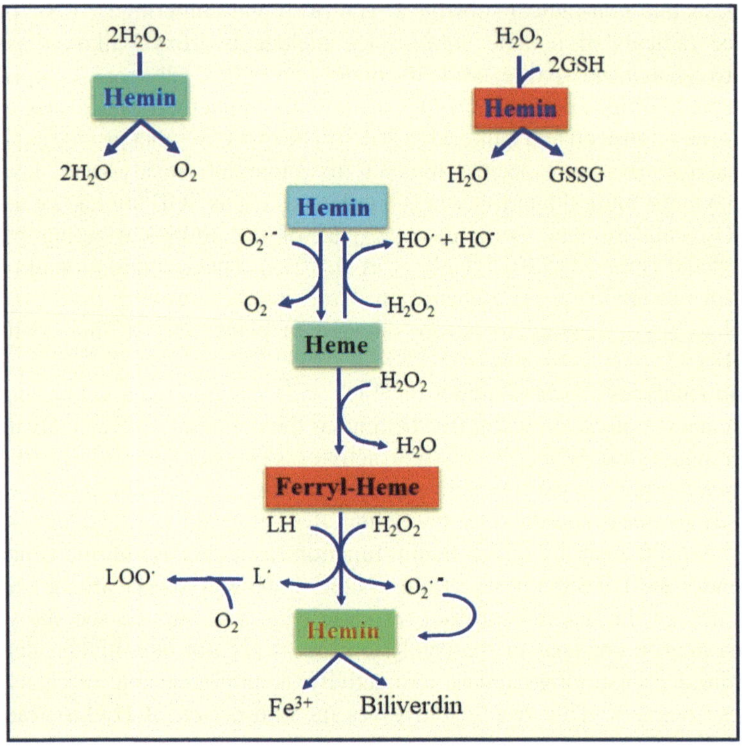

Fig. 8.2 Hemin interaction with H_2O_2 and ROS

there is a consensus that the oxidation of unsaturated lipids by hemin can be effectively inhibited by radical scavenging antioxidants, like vitamin E (Balla et al. 1991; Gutteridge and Smith 1988; Klouche et al. 2004). These antioxidants are recognized for their ability to interrupt the radical chain reaction responsible for lipid peroxidation (Fig. 8.2).

Hemin is involved not only in the decomposition of H_2O_2, but conversely, H_2O_2 also contributes to the enzyme-independent degradation of hemin (Fig. 8.2). In these processes, hemin must initially undergo reduction to heme and the reactions are likely to include superoxide, GSH or ascorbate (Nagababu and Rifkind 2004). When exposed to H_2O_2, the heme produced from hemin undergoes oxidation, forming a reactive ferryl-heme species (Nagababu and Rifkind 2004). This unstable compound is subsequently reduced back to hemin while simultaneously oxidizing H_2O_2 to superoxide. The presence of this superoxide is believed to be essential for the subsequent cleavage of the methane bridges in the tetrapyrrole rings of hemin, resulting in the release of free iron and biliverdin (Nagababu and Rifkind 2004). The greater generation of biliverdin due to the induction of HO 1 under Cd stress in soybean plants provides an antioxidant response, mitigating the effects of the initial oxidative stress (Balestrasse et al. 2005, 2008). The research suggests that the simultaneous application of biliverdin (10 μM) and cadmium (200 μM) prevents soybean plants from oxidative damage by reducing the content of TBARS (thiobarbituric acid-reactive substances). Additionally, it leads to an elevation in the activity of antioxidant enzymes includingSOD, GR, and GPX, confirming the antioxidant properties of biliverdin (Balestrasse et al. 2005).

8.4 Role of Hemin Against Abiotic Stress

Plants face a variety of biotic and abiotic challenges during their life cycle including salt, drought, heat, and heavy metals exposure (Anita and Shekhawat 2022, 2023; Kesawat et al. 2023; Shekhawat et al. 2009, 2010). Numerous plant adaptations are triggered by stress, including changed gene expression, modifications to cellular metabolism, growth rate alterations, and changes in crop yields (Verma et al. 2013; Mahawar and Shekhawat 2022). Plants must adapt their development in order to respond efficiently to stressful situations. Simultaneously, plants have developed an advanced antioxidative system to stimulate enzymes like SOD, APX, catalase (CAT), and peroxidase (POD) as well as enhance the accumulation of non-enzymatic antioxidants including glutathione (GSH) and ascorbic acid (AsA) (Zhu et al. 2019; Mahawar et al. 2018; Khator et al. 2020, 2021, Khator and Shekhawat 2020). Plant possesses the capacity for self-detoxification against abiotic stress. Hence, significant efforts have been dedicated to exploring effective strategies to mitigate the adverse effects of excessive heavy metals toxicity in plants (Zhu et al. 2019). Among these, the use of chemical additives may prove to be one of the most well-organized and active process to mitigate heavy metals toxicity in plants. As an illustration, the application of foliar sulfide has been shown to alleviate cadmium toxicity in Arabidopsis by deactivating and appropriating Cd into the cell wall (Guan et al. 2018). Additionally, the exogenous application of glycine betaine and proline enhances cadmium tolerance in mung beans by upregulating antioxidant and glyoxalase systems (Hossain et al. 2010).

Hemin application significantly reduced the cytotoxicity produced by abiotic stress, as evidenced by the longer shoot, root lengths, and the greater fresh, dry weight of the plant (Zheng et al. 2021; Mahawar et al. 2021). In response to osmotic stress and cold stress, hemin treatment enhanced plant growth, as well as the activity of antioxidant enzymes and saponin content of *Conyza blini* (Zheng et al. 2021). Hemin reduced the impact of salt stress on plant height, fresh, dry weight, leaf area, and the reduction of stress increased as hemin concentrations increased in barley (Abd El-Monem et al. 2013). Fu et al. (2011) and Mahawar et al. (2021) also reported that hemin promotes plant development. Hemin supplementation significantly mitigated the detrimental effects of excess metal accumulation on growth parameters and pigment accumulation. The defensive function of hemin in chlorophyll production may be associated with the role of HO 1 in phytochrome chromophore synthesis (Muramoto et al. 2002). In maize under cadmium stress, hemin improved the accumulation and dispersion of photosynthetic products, increased the amount of chlorophyll and the net photosynthetic rate of the leaves, and postponed leaf senescence (Sun et al. 2021). Furthermore, hemin's ability to reduce metal toxicity was linked to shielding in plants from oxidative damage brought on by ROS explosion to several heavy metal stresses. Hemin has been linked to a reduction in cadmium toxicity by decreasing the cadmium in *Medicago sativa* (Hossain et al. 2010). On the other hand, hemin was involved to improve the Cd tolerance in *Vigna radiata* (Mahawar et al. 2018, 2021), *Chinese cabbage* (Zhu et al. 2019), *Medicago sativa* (Fu et al. 2011), and maize (Piao et al. 2022; Sun et al. 2021). Beyond that, hemin improved zinc, lead, and chromium tolerance by regulating antioxidative defence in rice (Chen et al. 2017), Oryza sativa (Xie et al. 2015) and hemin improved the tolerance to zinc stress in *Oryza sativa* by regulating ROS homeostasis (Chen et al. 2018). According to Sun et al. (2020), hemin can effectively mitigate the damage caused by Cr^{6+} stress on wheat and oxidative damage of Alfalfa due to mercury stress (Han et al. 2007). Moreover, non-enzymatic as well as enzymatic antioxidants are essential in reducing oxidative stress carried out by ROS (Gill and Tuteja 2010). Cd stress markedly enhanced the concentration of AsA and GSH (Zhu et al. 2019). The disparity could be explained by the various plant species, levels of Cd, and durations of exposure. Remarkably, when hemin and Cd were co-treated, the AsA and GSH content rose more than when Cd was used alone (Zhu et al. 2019). It was in line with earlier findings (Fu et al. 2011), which suggested that hemin-induced AsA and GSH content may mitigate Cd enhanced oxidative damage by scavenging ROS generated by certain heavy metals (Zhu et al. 2019). Hemin applied exogenously improved the morphophysiological characteristics of rapeseed and considerably reduced the inhibition of NaCl stress (Zhao et al. 2023). Furthermore, hemin can increase plant stress tolerance in a variety of plant species under varied abiotic stress situations since it is a heme oxygenase-1 (HO-1) inducer (Su et al. 2022; Shekhawat et al. 2019, 2011; Shekhawat and Verma 2010).

8.5 Conclusion

Hemin is a useful and efficient supplement that has the potential to greatly decrease toxicity to plant caused by abiotic stressand appear to be accomplished by the following approaches: (1) upregulating of SOD, APX, and GR activities in plants; (2) promoting accumulation of antioxidants in plants; (3) downregulating productions of ROS (H_2O_2) in plants; (4) higher length and biomass of plants; and (5) finally resulting in plant tolerance. These results proposed that effect of hemin might be reduced by preventing roots from accumulating stress, which would lessen the biotoxicity that plants face. We deduced that the low activity of the corresponding transporters or the difference in cell wall composition could be the cause of the limited absorption of metaland other types of stress; however, further research work in this field is needed to finalize the results. Hemin could lead to resilient and sustainable agricultural production systems as agriculture continues to struggle with the effects of climate change and other environmental stresses.

Acknowledgments The author wishes to thank the University Grants Commission, New Delhi, for providing financial assistance through the Center for Advanced Study programme.

References

Abd El-Monem AA, El-Habbasha SF, Hozayn M (2013) Mitigation salinity stress effects on barley (Hordeum vulgare L.) growth, yield and some physiological aspects by hemin. J Appl Sci Res 9(3):2411–2421

Andersson JA, Egesten A, Cardell LO (2002) Hemin, a heme oxygenase substrate analog, inhibits the cell surface expression of CD11b and CD66b on human neutrophils. Allergy 57(8):718–722

Anita A, Parihar S, Shekhawat GS (2022) Phytoremediation: a sustainable approach for heavy metals uptake and accumulation. In: Choudhary M, Sharma DK, Mohil P (eds) Innovative biotechnological approaches in plant science, vol 1. Vital Biotech Publication, Kota, Rajasthan, pp 53–69

Anita A, Parihar S, Shekhawat GS (2023) Cadmium toxicity in plants: uptake, translocation and phytoremediation strategy. In: Singh RP, Singh P, Srivastava A (eds) Heavy metal toxicity: environmental concerns, remediation and opportunities. Springer, Singapore, pp 1–18. https://doi.org/10.1007/978-981-99-0397-9_1

Balestrasse KB, Noriega GO, Batlle A, Tomaro ML (2005) Involvement of heme oxygenase as antioxidant defense in soybean nodules. Free Radic Res 39:145–151

Balestrasse KB, Yannarelli GG, Noriega GO, Batlle A, Tomaro ML (2008) Heme oxygenase and catalase gene expression in nodules and roots of soybean plants subjected to cadmium stress. Biometals 21:433–441

Balla G, Jacob HS, Eaton JW, Belcher JD, Vercellotti GM (1991) Hemin: a possible physiological mediator of low-density lipoprotein oxidation and endothelial injury. Arterioscler Thromb 11(6):1700–1711

Becker EM, Cardoso DR, Skibsted LH (2011) Quenching of excited states of red-pigment zinc protoporphyrin IX by hemin and natural reductors in dry-cured hams. Eur Food Res Technol 232:343–349

Chen Q, Zhang X, Liu Y, Wei J, Shen W, Shen Z, Cui J (2017) Hemin-mediated alleviation of zinc, lead and chromium toxicity is associated with elevated photosynthesis, antioxidative capacity; suppressed metal uptake and oxidative stress in rice seedlings. Plant Growth Reg 81:253–264

Chen Q, Gong C, Ju X, Zhu Z, Shen W, Shen Z, Cui J (2018) Hemin through the Heme oxygenase 1/ferrous iron, carbon monoxide system involved in zinc tolerance in *Oryza sativa* L. J Plant Growth Reg 37:947–957

Chmielowska-Bąk J, Izbiańska K, Ekner-Grzyb A, Bayar M, Deckert J (2018) Cadmium stress leads to rapid increase in RNA oxidative modifications in soybean seedlings. Front Plant Sci 8:2219

Cui W, Li L, Gao Z, Wu H, Xie Y, Shen W (2012) Haem oxygenase-1 is involved in salicylic acid-induced alleviation of oxidative stress due to cadmium stress in *Medicago sativa*. J Exp Bot 63(15):5521–5534

Cui W, Zhang J, Xuan W, Xie Y (2013) Up-regulation of heme oxygenase-1 contributes to the amelioration of aluminum-induced oxidative stress in *Medicago sativa*. J Plant Physiol 170(15):1328–1336

Davidge KS, Motterlini R, Mann BE, Wilson JL, Poole RK (2009) Carbon monoxide in biology and microbiology: surprising roles for the "*Detroit perfume*". Adv Microbial Physiol 56:85–167

Dutt S, Hamza I, Bartnikas TB (2022) Molecular mechanisms of iron and heme metabolism. Annu Rev Nutr 42:311–335

Espinas NA, Kobayashi K, Sato Y, Mochizuki N, Takahashi K, Tanaka R, Masuda T (2016) Allocation of heme is differentially regulated by ferro chelatase isoforms in Arabidopsis cells. Front Plant Sci 7:1326

Falb N, Pati G, Furtmüller PG, Gabler T, Hofbauer S (2023) Structural aspects of enzymes involved in prokaryotic gram-positive heme biosynthesis. CSBJ

Fischer H (1941) Hemin. Org Synth 21:53. https://doi.org/10.15227/orgsyn.021.0053

Fu G, Zhang L, Cui W, Wang Y, Shen W, Ren Y, Zheng T (2011) Induction of heme oxygenase-1 with β-CD-hemin complex mitigates cadmium-induced oxidative damage in the roots of *Medicago sativa*. Plant Soil 345:271–285

Gallego SM, Pena LB, Barcia RA, Azpilicueta CE, Iannone MF, Rosales EP, Benavides MP (2012) Unravelling cadmium toxicity and tolerance in plants: insight into regulatory mechanisms. Environ Expe Bot 83:33–46

Gill SS, Tuteja N (2010) Reactive oxygen species and antioxidant machinery in abiotic stress tolerance in crop plants. Plant Physiol Biochem 48:909–930

Guan MY, Zhang HH, Pan W, Jin CW, Lin XY (2018) Sulfide alleviates cadmium toxicity in Arabidopsis plants by altering the chemical form and the subcellular distribution of cadmium. Sci Total Environ 627:663–670

Gutteridge JM, Smith A (1988) Antioxidant protection by haemopexin of haem-stimulated lipid peroxidation. Biochem J 256(3):861–865

Halliwell B, Gutteridge JMC (2015) Free radicals in biology and medicine, 5th edn. Oxford University Press, New York. https://doi.org/10.1093/acprof:oso/9780198717478.001.0001

Han Y, Xuan W, Yu T, Fang WB, Lou TL, Gao Y, Shen WB (2007) Exogenous hematin alleviates mercury-induced oxidative damage in the roots of *Medicago sativa*. J Integrat Plant Biol 49(12):1703–1713

Hossain MA, Hasanuzzaman M, Fujita M (2010) Up-regulation of antioxidant and glyoxalase systems by exogenous glycinebetaine and proline in *mung bean* confer tolerance to cadmium stress. Physiol Mol Biol Plant 16:259–272

Huffman LJ, Miles PR, Shi X, Bowman L (2000) Hemoglobin potentiates the production of reactive oxygen species by alveolar macrophages. Exp Lung Res 26(3):203–217

Hyvelin JM, Maurel B, Uzbekov R, Motterlini R, Lermusiaux P (2010) Hemin prevents in-stent stenosis in rat and rabbit models by inducing heme-oxygenase-1. J Vascul Surg 51(2):417–428

Jung YD, Chay KO, Song DU, Moon JS, Yang SY, Lee MW, Ahn BW (1997) Hemin inhibits lipid peroxidation induced by ascorbate/FeSO4 and 2, 2′-azobis-2-amidino-propane hydrochloride (ABAP). Exp Mol Med 29(3):171–175

Katoh Y, Nemoto N, Tanaka M, Takayama S (1983) Inhibition of benzo [a] pyrene-induced mutagenesis in *Chinese hamster* V79 cells by hemin and related compounds. Mutat Res Lett 121(2):153–157

Kesawat MS, Satheesh N, Kherawat BS, Kumar A, Kim HU, Chung SM, Kumar M (2023) Regulation of reactive oxygen species during salt stress in plants and their crosstalk with other signaling molecules—current perspectives and future directions. Plants J 12(4):864

Khator K, Shekhawat GS (2020) Cd-and cu-induced phytotoxicity on 2–3 leaf stage of *Cyamopsistetragonoloba* and its regulation by nitrate reductase and ROS quenching enzyme. Acta Physiol Plant 42(7):1–14

Khator K, Mahawar L, Shekhawat GS (2020) NaCl induced oxidative stress in legume crops of Indian Thar Desert: an insight in the cytoprotective role of HO1, NO and antioxidants. Physiol Mol Biol Plant 26:51–62

Khator K, Saxena I, Shekhawat GS (2021) Nitric oxide induced cd tolerance and phytoremediation potential of *B. Juncea* by the modulation of antioxidant defense system and ROS detoxification. Biometals 34(1):15–32

Kiening M, Lange N (2022) A recap of heme metabolism towards understanding protoporphyrin IX selectivity in cancer cells. Int J Mol Sci 23(14):7974

Klouche K, Morena M, Canaud B, Descomps B, Beraud JJ, Cristol JP (2004) Mechanism of in vitro heme-induced LDL oxidation: effects of antioxidants. Eur J Clin Invest 34(9):619–625

Kremer ML (1989) The reaction of hemin with H_2O_2. Eur J Biochem 185(3):651–658

Kumari A (2018) Chapter 8-heme synthesis. In: Sweet Biochemistry, pp 33–36

Liu Z, Cai R, Mao L, Huang H, Ma W (1999) Highly sensitive spectrofluorimetric determination of hydrogen peroxide with β-cyclodextrin–hemin as catalyst. Analyst 124(2):173–176

Mahawar L, Shekhawat GS (2022) Understanding the physiological mechanism of Heme oxygenase for enhanced tolerance and phytoremediation of Cd2+ in *Eruca sativa*: co-ordinated function of antioxidant defense system. J plant growth Regul 42:6307–6318

Mahawar L, Kumar R, Shekhawat GS (2018) Evaluation of heme oxygenase 1 (HO 1) in cd and Ni induced cytotoxicity and crosstalk with ROS quenching enzymes in two to four leaf stage seedlings of *Vigna radiata*. Protoplasma 255(2):527–545

Mahawar L, Popek R, Shekhawat GS, Alyemeni MN, Ahmad P (2021) Exogenous hemin improves Cd^{2+} tolerance and remediation potential in *Vigna radiata* by intensifying the HO-1 mediated antioxidant defence system. Nat Sci Rep 11:2811–2822

Medeiros MH, Wefers H, Sies H (1987) Generation of excited species catalyzed by horseradish peroxidase or hemin in the presence of reduced glutathione and H2O2. Free Rad Biol Med 3(2):107–110

Medlock AE, Dailey HA (2022) New avenues of Heme synthesis regulation. Int J Mol Sci 23(13):7467

Muramoto T, Tsurui N, Terry MJ, Yokota A, Kohchi T (2002) Expression and biochemical properties of a ferredoxin-dependent heme oxygenase required for phytochrome chromophore synthesis. Plant Physiol 130(4):1958–1966

Nagababu E, Rifkind JM (2004) Heme degradation by reactive oxygen species. Antioxid Redox Signal 6(6):967–978

Ogun AS, Joy NV, Valentine M (2019) Biochem heme synthesis

Piao L, Wang Y, Liu X, Sun G, Zhang S, Yan J, Gu W (2022) Exogenous hemin alleviated cadmium stress in maize (*Zea mays* L.) by enhancing leaf photosynthesis, AsA-GSH cycle and polyamine metabolism. Front. Plant Sci 13:993675

Portsmouth D, Beal EA (1971) The peroxidase activity of deuterohemin. Eur J Biochem 19(4):479–487

Richmond JE, Salomon K (1955) Studies on the biosynthesis of hemin in soybean nodules. Biochim Biophys Acta 17:48–55

Robinson SR, Dang TN, Dringen R, Bishop GM (2009) Hemin toxicity: a preventable source of brain damage following hemorrhagic stroke. Redox Rep 14(6):228–235

Sander A, Kretzer S, Bredt W, Oberle K, Bereswill S (2000) Hemin-dependent growth and hemin binding of *Bartonella henselae*. FEMS Microbiol Lett 189(1):55–59

Shekhawat GS, Verma K (2010) Haem oxygenase (HO): an overlooked enzyme of plant metabolism and defence. J Exp Bot 61(9):2255–2270

Shekhawat GS, Kumar D, Verma K, Singh K, Jana S, Teotia P, Yadav S (2009) Copper as a biometal: an insight on transport, toxicity and tolerance in photosynthetic organisms. In: Blanc G, Moreau D (eds) Biometals: molecular structures, binding properties and applications. Nova Science Publishers, New York., USA, pp 91–118

Shekhawat GS, Verma K, Jana S, Singh K, Teotia P, Prasad A (2010) In vitro biochemical evaluation of cadmium tolerance mechanism in callus and seedlings of *Brassica juncea*. Protoplasma 239(1):31–38

Shekhawat GS, Dixit S, Verma K, Nasybullina EI, Kosmachevskaya OV, Topunov AF (2011) Heme oxygenase: enzyme with functional diversity. J Stress Physiol Biochem 7(1):88–94

Shekhawat GS, Parihar S, Mahawar L, Khator K, Bulchandani N (2019) Bilin metabolism in plants: structure, function and haem oxygenase regulation of bilin biosynthesis. eLS, pp 1–13

Su N, Liu Z, Wang L, Liu Y, Niu M, Chen X, Cui J (2022) Improving the anthocyanin accumulation of hypocotyls in radish sprouts by hemin-induced NO BMC. BMC Plant Biol 22(1):1–13

Sun SZ, Tian ZW, Zheng QM, Sun CJ, Sun JY (2020) Alleviation effect of exogenous regulators on seed germination of wheat under Cr^{6+} stress. J Triticeae Crop 40:1104–1110. https://doi.org/10.7606/j.issn.1009-1041.2020.09.11

Sun G, Meng Y, Wang Y, Zhao M, Wei S, Gu W (2021) Exogenous hemin optimized maize leaf photosynthesis, root development, grain filling, and resource utilization on alleviating cadmium stress under field condition. J Soil Sci Plant Nutr 22:631–646

Verma S, Nizam S, Verma PK (2013) Biotic and abiotic stress signaling in plants. Genomics Proteomics Prespect 1:25–49

Wagener FA, Volk HD, Willis D, Abraham NG, Soares MP, Adema GJ, Figdor CG (2003) Different faces of the heme-heme oxygenase system in inflammation. Pharmacol Rev 55(3):551–571

Xie Y, Cui W, Yuan X, Shen W, Yang Q (2011) Heme Oxygenase-1 is associated with wheat salinity acclimation by modulating reactive oxygen species homeostasis F. J Integr Plant Biol 53(8):653–670

Xie Y, Mao Y, Xu S, Zhou H, Duan X, Cui W, Xu G (2015) Heme-heme oxygenase 1 system is involved in ammonium tolerance by regulating antioxidant defence in Oryza sativa. Plant Cell Environ 38(1):129–143

Xuan W, Zhu FY, Xu S, Huang BK, Ling TF, Qi JY, Shen WB (2008) The heme oxygenase/carbon monoxide system is involved in the auxin-induced cucumber adventitious rooting process. Plant Physiol 148(2):881–893

Zhao HM, Zheng DF, Feng NJ, Zhou GS, Khan A, Lu XT, Du YW (2023) Regulatory effects of hemin on prevention and rescue of salt stress in rapeseed (*Brassica napus* L.) seedlings. BMC Plant Biol 23(1):558

Zheng T, Zhan J, Yang M, Wang M, Sun W, Shan Z, Che H (2021) Hemin-induced increase in saponin content contributes to the alleviation of osmotic and cold stress damage to *Conyza blinii* in a heme oxygenase 1-dependent manner. J Zhejiang Univ Sci B 22(8):682–694

Zhu Z, Huang Y, Wu X, Liu Z, Zou J, Chen Y, Cui J (2019) Increased antioxidative capacity and decreased cadmium uptake contribute to hemin-induced alleviation of cadmium toxicity in *Chinesecabbage* seedlings. Ecotoxicol Environ Saf 177:47–57

Phytomelatonin: History, Biosynthesis, and Response

9

Shafia Zauq, Sumera Javad, Kiran Shahzadi, Anis Ali Shah, and Ajit Singh

Abstract

Melatonin was first discovered in animals and then in plants. In plants, it is named as phytomelatonin and considered as growth regulator. Melatonin is derived from serotonin, and it is an indole compound. It is a natural molecule and believed to be involved in various physiological roles in plants including root formation, circadian rhythms, defensive responses, fruit ripening, leaf senescence delay, photosynthesis, and seed germination. It has direct effect on response of plants to biotic and abiotic factors. It is also involved in cross talk with other signaling molecules like cytokinins, auxin, ethylene, gibberellins, abscisic acid, jasmonic acid, salicylic acid, and brassinosteroids for controlling the plants' growth and development responses. In this chapter, a comprehensive review of history, biosynthesis, and physiological roles of melatonin in plants/phytomelatonin has been done and described in detail. Various enzymes are involved in biosynthesis of melatonin. There are many other exogenous factors that can affect the endogenous level of melatonin. How phytomelatonin is involved in ameliorating pathogenic stress, drought stress, salt stress, heavy metal stress, etc., is discussed in this chapter. Endogenous level of melatonin is also studied under the effect of over expression of other genes.

S. Zauq · S. Javad (✉) · K. Shahzadi
Department of Botany, Lahore College for Women University, Lahore, Pakistan

A. A. Shah
Department of Botany, University of Education, Lahore, Pakistan

A. Singh
School of Biosciences, University of Nottingham, Selangor, Malaysia

© The Author(s), under exclusive license to Springer Nature Singapore Pte Ltd. 2024

M. Faizan, S. Hayat (eds.), *Plant Growth Regulators: Resilience for Sustainable Agriculture*, https://doi.org/10.1007/978-981-97-2918-0_9

Keywords

Phytomelatonin · Fruit maturation · Postharvest quality · Stress tolerance · Gene expression

9.1 Introduction

There are various plant growth regulators and other chemicals (jasmonic acid, salicylic acid, indole acetic acid, ethylene, hydrogen sulfide, polyamines, jasmonic acid, nitric oxides, etc.) that are involved in a strict control of plant's immune system. Their direct effect and cross talk determine the plant response to the various biotic and abiotic stresses (Gupta 2023). Melatonin is one of such growth regulators or signaling molecules that directly or indirectly help plants to develop resistance against changes caused by stress factors both in animals and plants. Melatonin is derived from two words, namely, "melas" and "tonos," meaning dark and hormone showing that this is produced in response to light and dark cycles of day and night in living beings (Zhang et al. 2022a, b; Ahn et al. 2021).

Melatonin has been found in all life forms, from bacteria to plants and animals. Chemically it is a biogenic amine (Hardeland 2016). For the first time, it was studied from the pineal gland of vertebrates in 1958. In plants, it was reported in 1995 (Fan et al. 2018). Research on melatonin expanded slowly and after 1980 melatonin was reported in most of the insects, planarians, and crustaceans. Then it was also discovered in a photoautotroph (in 1991) that was a unicellular algae named as *Gonyaulax polyedra*. Then it was also discovered in green algae and golden algae (dinoflagellates) (Fuhrberg et al. 1996; Hardeland and Fuhrberg 1996). Since then, research is going on and melatonin has been discovered in bacteria, fungi, and vascular plants.

Melatonin in vascular plants was described by Dubbels et al. (1995) and Van et al. (1995), where they worked on mono- and dicotyledons plant families. Now it has been proved that it is present in seeds, flowers, fruits, shoots, leaves, and roots of higher plants in variable amounts. Melatonin content varies considerably in various parts of plant where this is higher in sources and lower in sinks. Its quantity may be as low as in pictograms or as high as micrograms per gram of tissue (Burkhardt et al. 2001). In plants, it has been named as phytomelatonin and this term was coined by Blask et al. (2004).

It has been studied as a prominent signaling molecule and its characteristics have been extensively studied. During their life cycle, plant cells have to communicate in response to the environmental changes; climatic changes; stresses for their growth, yield, and flowering and periodic cycles; etc. Signaling molecules include cytokinins, auxins, ethylene, gibberellins, etc. These are small molecules that are made by plant cells and are transported to other cells with multiple effects. They have the capacity to initiate multiple essential reactions in plant cells particularly to cope with environmental stresses. These are called signaling molecules and melatonin is one of them.

Melatonin is, therefore, another signaling molecule, previously known in animals as neurotransmitter and discovered in plants in 1995 (Arnao 2014). It is a natural compound with antioxidant properties that augment the growth of plants by strengthening developmental metabolism and by upregulating gene expression to deal with various stresses (Wang et al. 2018). These properties characterize it as a bio-stimulator and growth regulator for sustainable crop production. It acts as a powerful scavenger and neutralizes the reactive oxygen species (ROS) in plants (Kołodziejczyk and Posmyk 2016). Phytomelatonin is known to play a significant role in various life processes including seed germination, plant growth, cell division, photoprotection, photosynthesis, apoptosis, and senescence. Work has been done to establish a relationship between phytomelatonin and other signaling molecules including cytokinins, auxin, ethylene, gibberellins, abscisic acid, jasmonic acid, salicylic acid, and brassinosteroids for controlling the plants' growth and development responses (Sun et al. 2020a).

Melatonin is derived from serotonin, and it is an indole compound. It is a natural molecule with multiple functions reported and is studied indirectly mostly. It is believed to be involved in various physiological roles in plants including root formation, circadian rhythms, defensive responses, fruit ripening, leaf senescence delay, photosynthesis, and seed germination (Ahammed et al. 2020; Nazir et al. 2020).

In vertebrates, melatonin is a controlling factor for the rhythmic body reactions like seasonal rhythms and sleep-wake cycle, where photo-stimulation of pineal glands causes its secretion. But in plants, phytomelatonin is not directly related to circadian rhythms or photoperiodism except very rare reports like in seaweed (*Lingulodinium polyedrum*) and dicotyledon (*Chenopodium rubrum*) (Arnao and Hernández-Ruiz 2018). Melatonin plays its role in modifying the root structure, process of flowering, fruit ripening, leaf senescence, pigment content of leaves, proline content and sugar content of fruits, etc. (Van et al. 1995; Back et al. 2016; McHill et al. 2017; Wang et al. 2017).

Its contribution in plants' life cycle is very important as a signaling molecule particularly when plants are facing any abiotic or biotic stress. It influences and modifies the plant defense responses against several pathogen attacks. Phytomelatonin also helps plants to fight against various abiotic stresses like radiations (UV), low temperature, heavy metal stress, drought, or salt stress (Zheng et al. 2017; Arnao and Hernández-Ruiz 2019).

9.2 Chemistry of Phytomelatonin

Phytomelatonin in pure form is a powder with an off-white coloration. Its molecular weight is 232.28 g/mol whereas it has a density of 1.175 g/cm^3. It has high melting and boiling points, that is, 118 °C and 512.8 °C, respectively (Mannino et al. 2019). It has a chemical formula of $C_{13}H_{16}N_2O_2$. The main indole chemical structure is functionalized with a 3-amide group and a 5-alkoxygroup (Fig. 9.1). Both in animals and plants, its precursor molecule is tryptophan; therefore, it is an indolamine (Omer et al. 2021). Phytomelatonin has a highly stable structure due to resonance it has.

Fig. 9.1 Chemical structure of phytomelatonin (Mannino et al. 2021)

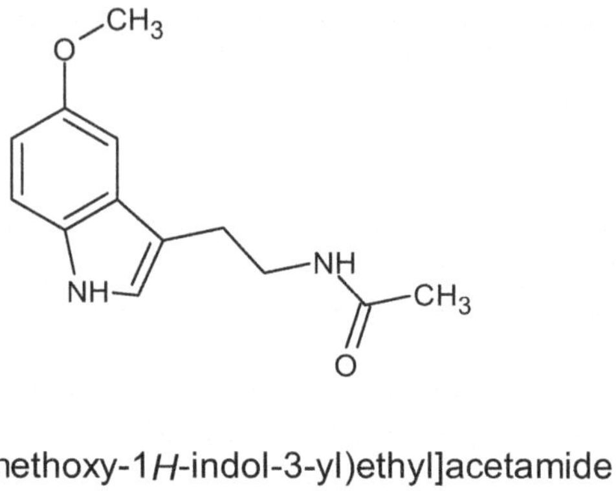

Moreover, the presence of two side groups, namely, 3-amide and 5-alkoxygroup, gives the character of amphiphilicity to the phytomelatonin. Due to this reason, phytomelatonin has the capacity to cross various biological membranes. Therefore, it can pass through any cellular or subcellular compartment (Omer et al. 2021). That is why, once it is produced inside the plant's body, it is easily distributed in various cells and their compartments (Simopoulos et al. 2005). Phytomelatonin has strong antioxidant potential that is due to the metabolites that are produced during its metabolism. It is known as phytomelatonin cascade, in which a series of new compounds with significant antioxidant potential are produced due to oxidation of melatonin (Reina and Martínez 2018; Back 2021). Among these metabolites, the most known are the following:

1. 2-Hydroxymelatonin
2. 6-Hydroxymelatonin
3. Cyclic 3-hydroxymelatonin.
4. N1-Acetyl-5-methoxykynuramine.
5. N1-Acetyl-N2-formyl-5-methoxykynuramine (Fig. 9.2).

9.2.1 N1-Acetyl-5-Methoxykynuramine (AMK)

AMK is an intermediate metabolite of phytomelatonin. It is produced by deformylation of AFMK by enzymatic or nonenzymatic methods (Hardeland et al. 2009; Gitto et al. 2010). This compound is highly active against ROS species and protects the protein from oxidative damage (Ressmeyer et al. 2003). Antioxidant potential of this metabolite is also influenced by the environmental factors (Galano et al. 2013). These factors include temperature and polarity and pH of the cell environment (Leon et al. 2006).

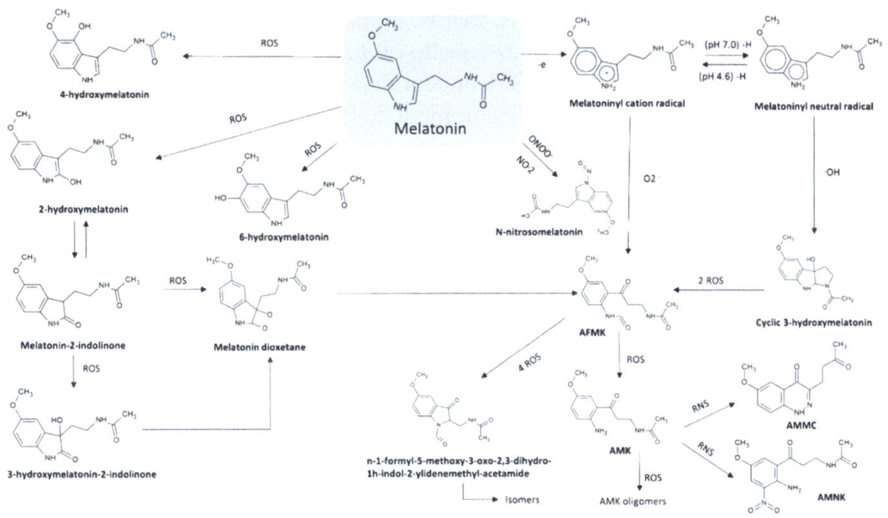

Fig. 9.2 Melatonin metabolism and its related metabolites. $ONOO^-$, peroxynitrite; O_2^-, superoxide anion; ROS, reactive oxygen species; RNS, reactive nitrogen species; OH^-, hydroxyl radical; AFMK, N1-acetyl-N2-formyl-5-methoxykynuramine; AMK, N1-acetyl-5-methoxykynuramine (Mannino et al. 2021)

9.2.2 3-Hydroxymelatonin (C3-OHM)

C3-OHM is another intermediate that is produced by the oxidation of melatonin initiated by reactive oxygen species. AFMK together with C3-OHM provide antioxidant protection to the molecules and can prevent DNA oxidation brought by Fenton reaction (Hacışevki and Baba 2018). In both in vivo and in vitro experimentations, the formation of C3-OHM is always found to be coupled with AFMK formation (Tan et al. 2007). The amount of products formed during melatonin oxidation is affected by the ratio of melatonin and oxidants. More production of AFMK was reported at higher ROS levels (Tan et al. 2003) as this condition also facilitates the oxidation of C3-OHM to AFMK.

9.2.3 Hydroxymelatonin (6-OHM)

In animal urine, initially 6-hydroxymelatonin (6-OHM) was discovered that is also one of the intermediates of metabolism of melatonin, and it is secreted as sulfate in animals. It is very evident from the experimental data that showed 6-OHM can not only prevent lipid peroxidation (Hara et al. 1997) but also avoid DNA damage induced by environmental pollutants, that is, chromium (Qi et al. 2000) and $OH^{\bullet-}$ made during Fenton reaction (López-Burillo et al. 2003). Duan et al. (2006) also described the involvement of 6-OHM in neuronal protection in an experimentally designed model involving the reperfusion-mediated injury/ischemia. It was observed that 6-OHM caused an inhibition of apoptosis by inhibiting cytochrome C and activity of caspase 3 and stabilizing the mitochondrial membrane potential. But there are also reports that 6-OHM can exhibit prooxidant activity as it was found that it can sometimes cause an oxidative DNA damage by making breaks in double strand via redox cycling (Sakano et al. 2004).

9.2.4 2-Hydroxymelatonin (2-OHM)

2-OHM is also produced during melatonin oxidation especially after scavenging of HClO (Aversa et al. 2012), oxoferryl hemoglobin (Agozzino et al. 2003) and $OH^{\bullet-}$ (Horstman et al. 2002). On the contrary to 3-OHM, in plants, main metabolite is 2-OHM after hydroxylation of melatonin. Additionally, AFMK synthesis via oxidation of 2-OHM was observed in the in vitro models of cytochrome C (Semak et al. 2005).

9.3 Biosynthesis of Phytomelatonin

Biosynthesis of melatonin in plants is different from that in animals. Its synthesis is regulated by many factors in plants; one of the major factors is light. In plants, the major sites for melatonin biosynthesis are mitochondria and chloroplasts. Different groups of enzymes are present in these organelles to synthesize melatonin via synthetic procedures. The production of melatonin is a two-way method, so if the melatonin production is blocked in mitochondria, it will start in chloroplasts (Nawaz et al. 2020). Studies have shown that l-tryptophan can be converted into melatonin via two pathways in plants, one in healthy plants and the other under senescence or biological stress (Sun et al. 2021). Under normal plant growth, l-tryptophan is fist carboxylated by l-tryptophan decarboxylase (TDC) into tryptamine in cytoplasm. The second committed enzyme trptamine-5-hydroxylase (T5H), catalyzes tryptamine into serotonin and requires acetyl-coenzyme A (Ac-CoA) on the endoplasmic reticulum. Usually in fresh plants, the third step is the acetylation of serotonin into N-acetylserotonin by serotonin N-acetyltransferase (SNAT) in the chloroplast. N-Acetylserotonin methyltransferase (ASMT) is believed to be the last enzyme; it catalyzes N-acetylserotonin into melatonin by methylation reaction in the cytoplasm. In plants under senescence and cadmium stress, the serotonin may be accumulated so that the melatonin synthesis would favorably use ASMT which catalyzes serotonin into 5-methexytryptamine and then synthesize melatonin by SNAT (Ye et al. 2019). During the last step, the conversion of acetyl-5-hydroxytryptamine to melatonin is accompanied by the co-conversion of S-adenosyl-l-methionine (SAM) to S-adenosyl-l-homocysteine (SAH).

The pathway that is followed by microorganisms and plants for biosynthesis of phytomelatonin is unique and different from that of vertebrates. As in vertebrates, tryptophan is a precursor molecule for its synthesis. The initial step, that is, tryptamine synthesis from tryptophan decarboxylation, is significantly different in plants from vertebrates (Tan et al. 2005). According to Tan and Reiter (2020), this significant change is not only found in initial step, but there are also certain differences found in final steps that were earlier thought to be similar among all organisms. In the microorganisms and plants to initiate the metabolism, tryptophan is made available from shikimic acid pathway, but on the other hand in animals, the availability of all aromatic amino acids including tryptophan is received through food.

In prokaryotes, microalgae, and plants, the synthesis of the tryptophan is quite a complex process. During this process, tryptophan combined with D-erythrose-4-phosphate and phosphoenolpyruvate and finally with carbon dioxide in phototrophs (Bandurski et al. 1995). But which pathway will be opted by the plants and in what situation needs more research?

The final subcellular sites vary from either chloroplast or cytoplasm depending on the pathways. In plants, this may have impact on and connection with the mode of action and future working of phytomelatonin (Altaf et al. 2021). Plants use both pathways of synthesis alternatively. For example, when chloroplast-based synthesis of phytomelatonin is blocked, mitochondria-based biosynthesis is started. The research is needed to find the specific place of phytomelatonin synthesis due to unavailability of adequate literature. Mitochondria in the cell is also considered as the site of origin of melatonin (Nawaz et al. 2020; Tan and Reiter 2020). SNAT (serotonin N-acetyltransferase) gene is found in mitochondrial DNA, and higher concentration of reactive oxygen species (ROS) is produced in this organelle due to which it is thought that phytomelatonin may also be produced inside the mitochondria. To manage these ROS, certain melatonin factors are synthesized and reduce their impact (Tan and Reiter 2020; Ye et al. 2019). Other than mitochondria, chloroplast is another potential candidate for production of phytomelatonin. The study of evolutionary theory also reinforces this suggestion. According to this theory, cyanobacteria were transformed to chloroplast, and non-sulfur purple bacteria were converted into mitochondria (Hardeland 2015).

The ability of plants to produce phytomelatonin was studied in the seedlings of *Hypericum perforatum* L. cv. Anthos. In its seedlings, an isotope tracer was used (Murch et al. 1997). All these discussions are still controversial about the similarity and differences of biosynthesis of melatonin in animals and phytomelatonin in plants. There are six main enzymes that are considered as key factor in the biosynthesis of phytomelatonin including caffeic acid O-methyltransferase (COMT), N-acetylserotonin O-methyltransferase (ASMT), serotonin N acetyltransferase (SNAT), tryptamine 5-hydroxylase (T5H), tryptophan decarboxylase (TDC), and tryptophan hydroxylase (TPH) as depicted in Fig. 9.3 (Arnao and Hernández-Ruiz 2018). Tryptophan is the precursor of phytomelatonin that is believed to be using above enzymes in six steps.

Tryptophan is a common compound in the biosynthetic pathway of IAA and phytomelatonin. In plants, two main pathways are suggested for IAA synthesis

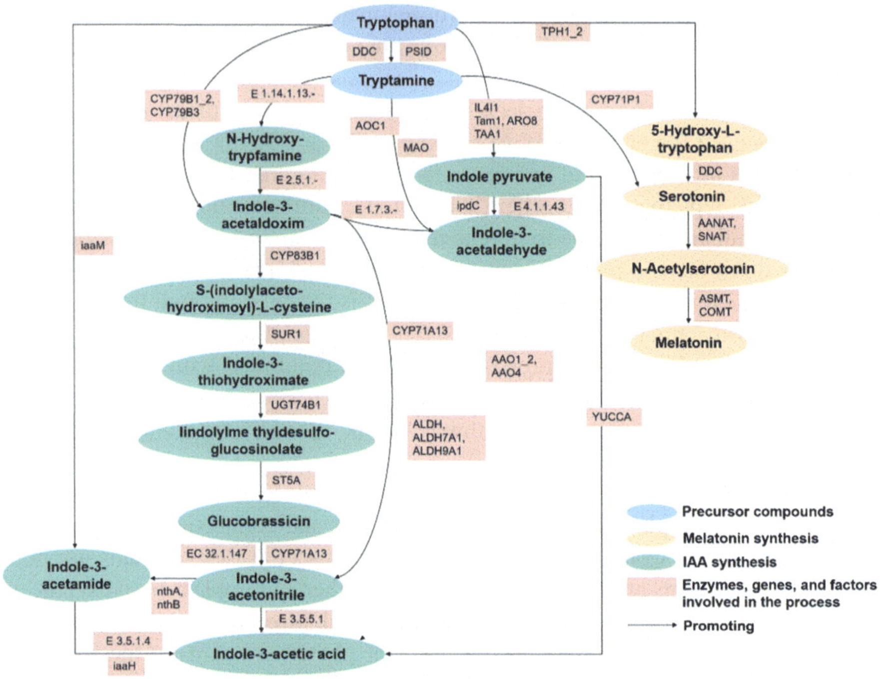

Fig. 9.3 Relationship of biosynthesis of phytomelatonin and indole acetic acid from tryptophan (taken from Zhang et al. (2022a))

where one is tryptophan-dependent and the other is tryptophan-independent pathway and have been extensively studied for the biosynthesis of IAA (Bandurski et al. 1995; Normanly et al. 2004). Figure 9.3 illustrates a diagrammatic representation of the IAA biosynthesis by tryptophan-dependent pathway. On the other hand, in other pathways present in bacteria, fungi, and plants, firstly chorismate produces indole-3-glycerol-phosphate which is then used to synthesize tryptophan and IAA.

Many biotic and abiotic factors (wounding, pathogen attack, temperature, and light) are found to be involved during IAA biosynthesis by tryptophan-dependent pathway. In plants, three possible routes for the conversion of tryptophan to IAA are (a) the indole-3-pyruvic acid, (b) the indole-3-acetaldoxime, and (c) the tryptamine pathways. The latter one appears as circular by taking in indole-3-acetaldehyde as a common metabolite with the indole-3-pyruvic acid pathway. The intermediate, that is, indole-3-acetonitrile, can be converted to IAA through the action of the enzyme nitrilase in the indole-3-acetaldoxime pathway. Family *Brassicaceae* shows that this characteristic pathway is also considered to be fed from the indolic glucosinolate pool through the action of an enzyme, namely, myrosinases. In bacteria, the indole-3-acetamide pathway has mainly been described.

Melatonin biosynthesis through serotonin (5-hydroxy tryptamine) is thought to be controlled by an acetylation reaction. This reaction is catalyzed by aryl-alkyl amine N-acetyl transferase also known as serotonin N-acetyltransferase (NAT) that acts as a regulatory enzyme. HIOMT (hydroxyl indole O-methyl transferase) is used for the catalysis of the last step. 5-Hydroxytryptophan gives rise to serotonin in a reaction which is catalyzed by an enzyme called as amino acid decarboxylase. This enzyme is also found to act on tryptophan, hence facilitating tryptamine formation. This tryptamine can be converted into serotonin by an enzyme, that is, tryptophan 5-hydroxylase, which on the other hand acts on tryptophan thus producing 5-hydroxytryptophan. NAT uses tryptamine as a substrate to form N-acetyl tryptamine which is further undergoes hydroxylation to synthesize N-acetylserotonin. All the enzymes except tryptophan decarboxylase described have been isolated not from plant tissues, but these are mostly characterized in animal tissues (Fig. 9.3). According to different databases like IntEnz, UniProt, InterPro, and Brenda databases, in many plant species, tryptophan decarboxylase has been studied. Tryptophan decarboxylase is a pyridoxal-dependent enzyme that is related to IAA biosynthesis but not to the biosynthesis of serotonin. This enzyme has been reported in all living organisms from prokaryotes to eukaryotes including animals and plants. This enzyme has been characterized in *Arabidopsis thaliana*, *Catharanthus roseus*, *Oryza sativa*, *Papaver somniferum*, and a few other plants as well (Ljung et al. 2002). There are many other enzymes involved in the biosynthesis of phytomelatonin such as tryptophan 5-hydroxylase, hydroxyl indole O-methyl transferase, and serotonin N-acetyltransferase that have not been yet studied in plants (Fig. 4). Evidently, the general activities of various enzymes such as hydroxylases, methyl transferases, and acetyl transferases have been comprehensively described in many metabolic pathways of different plants but not specifically explaining their role in the biosynthesis of phytomelatonin.

Briefly, carboxylation and hydroxylation of tryptophan are carried out in plant cells in presence of stress or senescence, and serotonin is produced. Overall, serotonin synthesis occur when plants are subjected to biological stress and senescence. Afterward methylation and acetylation of serotonin are done to give rise to phytomelatonin (Back et al. 2016). Further studies are needed to explain the deep regulatory mechanism of biosynthesis of phytomelatonin in plants in different compartments under different circumstances.

9.4 Function Performed by Phytomelatonin

9.4.1 Seed Germination

Seed germination is the first step in the life cycle of any plants and in stress conditions for plants like heavy metal stress, salt stress, drought stress, temperature stress, pathogenic stress, etc.; this is the stage that is badly affected and hinders the further growth of crop. This may also cause a poor yield even (Sharif et al. 2018a, b).

Phytomelatonin is known to affect this significant stage of plant growth as well. It is reported that if in stress, melatonin is applied to plants exogenously, it can maintain and enhance seed germination. Melatonin, due to its strong antioxidant capacity, releases the stress pressure from seeds and let them germinate (cfández

2015). Melatonin/phytomelatonin is considered to release the effect of hydrogen peroxide by inhibiting its accumulation in the cells directly or indirectly. Melatonin can do this by regulating the antioxidant enzyme production like peroxidases or can regulate the expression of antioxidant genes (Zhang et al. 2014; Balabusta et al. 2016).

Melatonin is known to increase the gibberellin content in plants by upregulating the expression of key genes of biosynthesis of gibberellin (GA). Melatonin is also known to affect the expression of genes involved in abscisic acid by downregulating it. Balance of these growth regulators affects the overall seed germination. In heavy metal stress, melatonin is known to increase seed germination via increased sugar mobilization. Melatonin enhances the production of antioxidant enzymes, thus decreasing the content of ROS species and promoting plant germination and growth in temperature stress (Li et al. 2021; Raza et al. 2022). Effect of melatonin on germination of seeds has also been studied in salt stress. It is reported that during salt stress, priming of seeds with melatonin can improve seed germination via production of GA. Same positive results of melatonin application have also been observed in field experiments in which enhanced seed morphology, vigor, and germination have been reported in the presence of melatonin (Simlat et al. 2018; Chen et al. 2020). Priming combination of melatonin has proven to be an effective tool for enhancing and speeding up the embryo physiology by activating the enzyme morphology (Rajora et al. 2022; Farooq et al. 2020).

9.4.1.1 Root Development

There are multiple reports on positive role of melatonin in root development. All of the root characters such as development of lateral tap roots, adventitious roots, lateral roots, root length and root biomass, etc. are improved by addition of melatonin or increased endogenous production of phytomelatonin. Melatonin also plays a role in rhizosphere interactions and signaling. Molecular level study of plant responses in relation to melatonin has shown that increased expression of melatonin in plants gives rise to increased root length and biomass as compared to wild-type plants. Furthermore, in a study on apple trees, it was reported that root expression in seedlings was increased due to enhanced expression of MdWOX11 gene. This gene controls the expression of adventitious roots in response to indole acetic acid (IAA). Therefore, it is considered that melatonin plays almost same role as played by IAA in root development. Melatonin causes the lupin hypocotyl root primordia, thus controlling the time and space of lateral root development. Melatonin regulates the mineral absorption of roots along with the other nutrients. Melatonin is also responsible for healthy relation of roots and soil bacteria (Park and Back 2012; Zhang et al. 2017a; Erland and Saxena 2018; Asif et al. 2020; Mao et al. 2020; Yang et al. 2021).

9.4.1.2 Seedling Growth

Exogenous application of melatonin and internal level of phytomelatonin have been also reported to positively affect the seedling growth. This is mainly due to the antioxidant nature of the melatonin that helps the seedling to protect its proteins, photosynthetic pigments, and plastid membranes. Therefore, in presence of phytomelatonin, photosynthetic pigments are enhanced, photosynthetic activity is improved, and assimilation rate of carbon dioxide is increased. Thus melatonin positively affects the growth of seedlings and plants even in the presence of stress factors like drought, salt, or pest stress. Stomatal conductance is also reported to be improved, and a decrease in cellular carbon dioxide content has been reported in the presence of melatonin. Such growth conditions give rise to more photosynthetic products and plant biomass. Other than priming, spraying of melatonin has also shown positive effects on seedling growth. In monocots like maize, melatonin can increase the gene expression of hexokinase and other genes related to photosynthetic enzymes. This leads to the increased sucrose, starch, and glucose metabolism along with enhanced phloem loading (Wang et al. 2013; Arnao and Hernández-Ruiz 2014; Zhao et al. 2015; Sharif et al. 2018a, b).

9.4.1.3 Role of Melatonin in Leaf Senescence, Fruit Maturation, and Postharvest Quality

Leaf senescence is an important and final stage in life cycle of plants and is determined by a number of factors including cellular plant growth regulator level, physiological age of plant, genetics of plant, its nutritional status, and external stress factors. Interaction of all these factors determines the onset of leaf senescence. Research has proven that melatonin is not going to directly affect the leaf senescence but interacting with other growth regulators like ABA and cytokinin. It delays the leaf senescence by enhancing the health of chlorophyll content and increasing the rate of photosynthesis. Melatonin and cytokinin show an additive effect on decreasing the rate of leaf senescence induced by stress, while melatonin and ABA work in an antagonistic way to reduce leaf aging in plants where melatonin inhibits the biosynthesis of ABF-mediated ABA synthesis (Ma et al. 2018; Tan et al. 2019; Ahmad et al. 2020).

Fruit ripening is also affected by melatonin concentration where melatonin decreases the content of proteins related to senescence. Melatonin is also known to play a role in increasing the apoptosis inhibitor proteins and anthocyanin-accumulating proteins during fruit ripening. It is also known to increase fruit quality by increasing the total soluble sugars, ascorbic acid content, and hardness and decrease the fruit rot in plants like peaches and tomatoes. In fruits, melatonin has a role in postharvest ripening of fruits. Melatonin is considered responsible for increasing the antioxidant activity and ATP content in fruits that results in accumulation of anthocyanins and phenolics in them. This leads to reduced postharvest decay and increased quality of harvested fruits. In some fruits like grape, melatonin along with ethylene plays its role in fruit ripening (Gao et al., 2016, b; Sun et al., 2016; Aghdam and Fard 2017; Xu et al., 2018).

Various fruits have variable limits of shelf life. Their shelf life depends upon their susceptibility to various pathogens, their structure, and their physiology and on the biochemical they have in their pulp and in the pericarp. Various phytoconstituents present in the cell sap help to increase the shelf life by decreasing the pathogen and decreasing the oxidative pressure. Melatonin also enhances the postharvest quality of various fruits. Litchi is one of such fruits that have lesser shelf life and market life due to postharvest browning. But application of melatonin caused a decrease in the expression of genes of antioxidant enzymes. It reduced the oxidative stress from litchi fruits and reduced its browning while maintaining its freshness (Marak et al., 2024). Liu et al. (2018b) also reported that strawberry fruits were treated with melatonin, and their shelf life was increased due to decreased accumulation of hydrogen peroxide and lesser lipid peroxidation. In fact, melatonin is reported to be reducing postharvest senescence by regulating ROS species (Gao et al., 2016, b; Bal, 2021).

9.4.1.4 Floral Transition

Recently, the role of melatonin has also been identified in floral development as well. It indirectly controls the floral development by controlling the floral transition phase (conversion from vegetative to reproductive phase). Melatonin interacts with DELLA proteins in a way that it stops the transcription factors of flowering. This is accomplished by the cross talk of other signaling molecules with melatonin. These signaling molecules include strigolactones and nitrogen oxide like molecules (Shi et al., 2016; Zhang et al., 2019; Arnao and Hernández-Ruiz, 2020).

9.4.1.5 ROS Scavenging

When plants grow in natural environment, they have to face various stress situations and have to compete difficult internal and external problems. During such stress conditions, reactive oxygen species (ROS) are produced. These ROS include hydrogen peroxide, singlet oxygen, superoxide radicle, and hydroxyl radical. Therefore, in order to survive in such a stressful and adverse situation, plants have to develop some specific systems because this enhanced content of ROS causes cellular oxidative damage (Ramasamy et al., 2023). This may lead to protein damage, DNA damage, enzyme inhibition, necrosis, and early leaf senescence. Plant may fight with these ROS by using its enzymatic and nonenzymatic antioxidant systems. Expression of these systems is also enhanced by the enhanced melatonin content of plants because melatonin is known to be a pleiotropic signaling molecule. Production of melatonin by plants is directly proportional to the stress faced by plants. Furthermore, melatonin is known to activate the ABA-degrading enzymes. Melatonin also plays its role in detoxification of ROS by interfering with the AsA-GSH cycle or Asada-Halliwell pathway (ascorbate-glutathione pathway). Activity of glutathione peroxidase is also enhanced by melatonin. This results in reduced amount of MDA in plant cells, ultimately preventing the cell from oxidative damage (Gu et al., 2022; Garcia-Caparros et al., 2021; Liu et al., 2021; Zhang et al., 2022a, b; Roy et al., 2022).

9.5 Melatonin Levels and Biosynthesis in Plants Under Stress Conditions

As we have discussed earlier, biosynthesis of melatonin starts from tryptophan and involves four main enzymes as the foll
owing:

1. N-Acetylserotonin methyltransferase (ASMT).
2. N-Acetylserotonin acetyltransferase (SNAT).
3. Tryptamine 5-hydroxylase (T5H).
4. Tryptophan decarboxylase (TDC).

SNAT and ASMT has a very low catalytic efficiency in normal conditions of plants. But during abiotic and biotic stresses like salt stress, drought stress, cadmium stress, and pathogen stresses, expression of SNAT and ASMT gene is enhanced. Study of these gene expression in transgenic plants has shown that melatonin is overexpressed in such situation. HIOMT (o-methyl transferase) activity is also affected by melatonin (Mukherjee et al., 2014; Back 2021; Tan and Reiter 2020). Even the biosynthetic pathways for melatonin are also different in plants in normal and in stressed conditions (Ye et al., 2019). In vitro studies of melatonin and study of its isoforms have shown that there is a great variability in melatonin component in plants with various environments (Zhang et al., 2019). Different intermediates of melatonin metabolism are studied in various cellular compartments including chloroplast, endoplasmic reticulum, mitochondria, and cytoplasm. It enables plants to be more flexible toward various environmental stresses (Back et al., 2016).

9.6 Melatonin-Mediated Stress Tolerance

Many studies have shown that exogenous melatonin can enhance the tolerance of plants to abiotic stress; therefore, it is considered as an effective means to alleviate environmental stress.

9.6.1 Drought Stress

Water-stressed environments greatly affect plants and their metabolism at molecular and cellular level. This results in very much reduced growth of plants. Water stress alters the efficiency of Rubisco enzyme and reduces its activity abnormally. Therefore, it is known to badly affect photosynthetic activity as studied in various plants. In such situation, melatonin is known to help plants by protecting the photosynthetic enzymes and photosynthetic pigments of the affected plants. Melatonin causes reduction in melatonin degradation and increases the rate of transpiration, respiration, and photosynthesis. It enhances the efficiency of photosystem II by increasing the rate of electron transport. This is one way of its action in drought stress (to increase photosynthesis rate). It also maintains the cell turgor pressure and leaf water content, thus creating a balance for water exchange. Its controls the stomatal conductance that ultimately controls the movement of water vapors and carbon dioxide and directly and indirectly controlling the rate of transpiration and photosynthesis. If water-stressed plants are treated with melatonin, they have lesser cell membrane and cell wall damage, lesser electrolyte leakage, and increased content of antioxidant systems (Cui et al., 2017; Hussain et al., 2018; Liang et al., 2019; Anitha et al., 2022).

9.6.2 Salinity Stress

Salt stress is another alarming indicator for a huge part of agricultural lands and causes huge economic losses every year. There are a number of treatments suggested for treating soils and plants to reduce the salt stress effects from them. One of those suggested solutions is the use of melatonin as melatonin has been used in maize, cucumber, wheat, rice, and tomato to treat salt stress successfully (Liang et al., 2015; Zhou et al., 2016; Chen et al., 2018; Ke et al., 2018).

Melatonin is known to cross talk with other signaling molecules as well as to treat salt stress. This is usually done by altering the glutathione reductase activity. Furthermore, when melatonin is applied to the salt stress-affected plants, then K^+/Na^+ homeostasis is mediated, and integrity of cellular membrane is maintained. During this whole process, energy turnover is increased, triglyceride metabolism is increased, and accumulation of sodium and chloride ions is reduced. Therefore, it is believed that melatonin alleviates the effect of salinity stress by increasing the energy production in the cells (Kaur and Bhatla, 2016; Li et al., 2017; Zhang et al., 2017b).

9.6.3 Temperature Stress

Like other stresses, temperature stress (heat and cold stress) causes a change in redox status of the plant. This damage is quite similar to the damage caused by other stresses and hinders the rate of development of the plants. Various studies on plants like wheat, rice, tomato, *Lupinus*, and maize has shown that temperature stress badly affects the yield of above said plants. Increased ROS and peroxidation of lipids during temperature stress can cause severe damage to the plants particularly at seedling stage, but melatonin can reverse it. Melatonin controls plant during temperature stress by controlling the expression of enzymatic and nonenzymatic antioxidants. An enhanced level of 54 secondary metabolite contents has been recorded in cold stress that also helped plant to grow better. Expression of some stress-related genes is controlled by melatonin including C-repeat-binding factors (CBFs)/drought response element binding factors (DREBs); the cold-responsive gene COR15a; CAMTA1, encoding a transcription factor involved in freezing and drought-stress tolerance; and ZAT10 and ZAT12, encoding transcriptional activators of ROS-related antioxidant genes (Zhao et al., 2017; Guo et al., 2018; Liu et al., 2018a; Wang et al., 2018). Therefore, melatonin is known to induce thermotolerance in plants through various pathways. Melatonin is also known to balance temperature stress from plants by altering the biosynthesis of NO and polyamine. Melatonin are also involved in heat shock factors (HSFA2 and HSA32) and heat shock proteins (HSP90 and HSP101) based on thermotolerance. It also helps the cells to remove the heat stress-damaged proteins by the process of autophagy (Shi et al., 2015; Xu et al., 2016; Ahammed et al., 2019; Jahan et al., 2019).

9.6.4 Heavy Metal Stress

Heavy metal stress is another main issue that is causing yield reduction in main cash crops and causing toxicity in consumers as well. They also cause toxicity in plants by creating ROS stress leading to various damages to plant cells at cell, tissue, and organ level (Chen et al., 2017). Application of external melatonin and enhancement of endogenous melatonin can help plants to fight against heavy metal stress. There are a number of proposed pathways for explaining these responses of plants to internal and external melatonin. Melatonin can help plants to accumulate nutrients ions; to chelate heavy metal ions, like lead and copper; to enhance active oxygen scavenging; to promote the carbon metabolism; etc. It is reported that melatonin can also help plant in alleviating the heavy metal stress by NO signaling. Melatonin is also known to trigger the synthesis of other phytochelatins in the plant cells, thus chelating the cadmium. This results in accumulation of heavy metals like cadmium into the cell wall and vacuoles, thus reducing their concentration in cell sap and improving the metabolism (Hasan et al., 2015; Cao et al., 2018; Kaya et al., 2019). Therefore, it can be concluded that melatonin reduces the effect of heavy metal stress by channelizing the ROS and by helping the plants in exclusion or sequestration of heavy metals from main plant metabolism (Sun et al., 2020a).

9.6.5 UV-Radiation Stress

Melatonin has also shown a protective effect against UV rays in plants because UV rays are not good for plants and cause damage to them, even leading to tissue injury or whole plant death. It has been noted that melatonin levels are higher in ozone-resistant species naturally. Furthermore, melatonin content of plants grown in UV-exposed areas like Mediterranean and Alpine areas is also higher as compared to plants of low UV-exposed areas. Some experiments on plants have also shown that on exposure to UV rays, survived plants have shown an enhancement in the melatonin content showing the probable role of melatonin in alleviating the effect of UV stress in them (Sharma et al., 2019). Similarly, in another experiment, a transgenic plant was used as an experimental organism which had a gene for melatonin synthesis. On exposure to UV-B rays, these transgenic plants showed lesser damage to DNA as compared to plants of normal or control group (Nawaz et al., 2020).

9.7 Melatonin Cross Talk with Other Hormones

Plant growth regulators are responsible for plant growth and physiology with their own organ of production, target organs, mode of action, and specific metabolism. These include a range of chemicals like auxins, cytokinins, gibberellins, ethylene, jasmonic acid, and many more. As we have discussed earlier, melatonin is resembling to IAA in its structure and both share same precursor, that is, tryptophan (Arnao and Hernández-Ruiz, 2018; Zhang et al., 2022a, b). Therefore, it also shows activities and physiological control of plant metabolism like IAA. It has shown that it has a vital role in root development and vegetative growth of various plants (Arnao and Hernández-Ruiz, 2018). Even it can develop roots of a plant in an IAA-independent pathway in *Arabidopsis* (Pelagio-Flores et al., 2012).

There are also reports of cross talks of melatonin with other growth regulators in controlling the plant metabolism. For example, melatonin can mediate the synthesis of abscisic acid (ABA) in plants. Therefore, in presence of melatonin, ABA content is reduced in stress-affected plants. There are a number of possible ways of this mediation and one of them is the control at gene level. In apple plant, melatonin is reported to downregulate the MdNCED3 gene that is an essential gene for the biosynthesis of ABA. Same was studied with positive results in rye grass (Li et al., 2015; Zhang et al., 2017a). Melatonin application is known to enhance the endogenous IAA concentration of plants, that is, in *Brassica*. In another report, external application of IAA has also raised the levels of endogenous melatonin with positive impacts on plants (Arnao and Hernández-Ruiz, 2018; Zafar et al., 2020). Melatonin is also known to interfere with ABA signaling, reducing its effects during stress (Fu et al., 2017). Its cross talk with jasmonic acid and salicylic acid has also been reported in alleviating various stresses in plants. During any stressful conditions in plants, this cross talk is very important for normal plant growth that can protect plants from all the possible levels of losses and damages caused by these stresses. Melatonin can interact with a number of plant hormones to defeat drought stress, for example, increasing the gene expression of ABA and jasmonic acid (Sun et al., 2023). Melatonin is also known to interact with exogenous arginine in growth promotion of *Capsicum annuum* L. during salt stress (Usman et al., 2023). In Zucchini, application of melatonin caused an increase in endogenous GABA and proline, increasing the cell wall stability and cell life (Ali et al., 2023).

9.8 Melatonin Regulates Gene Expression

It is mentioned earlier that there are various methods of action of melatonin in plants for improving growth and enhancing the stress tolerance. This includes ROS scavenging and cross talk with other growth regulators, and this may all done by gene expression control as well. There is a lot of data available online explaining the effects of melatonin on gene expression for various characters in plants. Melatonin has been reported to have a role in the expression of chloride channel proteins

(CCLs 1 and 2) during salt stress. It helps the plants by activating the SOS pathway (Li et al., 2017).

Melatonin is also known to regulate the expression of potassium and sodium channel proteins that helps plants to absorb more potassium (Li et al., 2016). It is also reported to upregulate the expression of antioxidant enzyme genes. It also enhances the lignin and gossypol synthetic gene expression and reduces the stress effect of verticillium wilt (Li et al. 2019). During drought stress, melatonin can regulate the expression of genes involved in cuticle thickening and wax production on fruit surface and leaves' surface (Ding et al., 2018). Some detailed studies on RNA sequences have also shown that application of melatonin during cold stress has also enhanced the expression of defense genes (Li et al., 2017). This can be done by upregulating the expression of cold stress response gene (COR15), transcription factor (CBF/DREBs), transcriptional activators (ZAT10, CAMTA1, and ZAT12), etc. (Bajwa et al., 2014).

9.9 Approaches to Inducing Endogenous Melatonin Levels

9.9.1 Transgenic Approaches

Various techniques have been used in plants for increasing the endogenous level of melatonin in plants after finding their positive effects on plants' growth and stress tolerance. One of them is controlling the expression of enzymes involved in the biosynthesis of melatonin including arylalkylamine N-acetyltransferase hydroxyindole-O-methyltransferase (HIOMT), N-acetylserotonin methyltransferase, serotonin N-acetyltransferase, tryptamine5-hydroxylase, and tryptophan decarboxylase (TDC). Overexpression of these enzymes resulted in an increased content of endogenous melatonin (Kim et al., 2013).

Various genes that are overexpressed can increase the content of melatonin include MZASMT1 in *Arabidopsis* (Zou et al., 2014), N-acetylserotonin methyltransferase (ASMT) gene in tomato (Ahammed et al., 2019), HsFA1a gene in tomato (Cai et al., 2017); and MeTDC2, MeASMT2, MeASMT3, and MeWRKY genes in cassava (Wei et al., 2018).

9.9.2 Other Strategies

Transgenic plants or plants with altered genes are one of the options to enhance the endogenous melatonin concentration of plants. One of some other options is applying the melatonin exogenously. There are several methods of applying exogenous melatonin including priming and foliar spray. In stress conditions, application of exogenous melatonin also triggers the overexpression of genes involved in biosynthesis of melatonin (including ASMT, SNAT, T5H, and TDC). Furthermore, some exogenous compounds like benzothiadiazole (BTH) and chitosan (CHT) can also increase the endogenous melatonin content (Umapathi et al., 2018; Sun et al., 2021).

Some researchers report that application of some endophytic bacteria like *Pseudomonas fluorescens*, *Bacillus amyloliquefaciens* SB-9, etc. promote the internal melatonin level of plants like grapes. This treatment helped the plants to survive through salt stress (Jiao et al., 2016; Ma et al., 2017).

There are also some studies reported in mutant plants that proved if plants are deficient in melatonin or they cannot synthesize proper amounts of melatonin due to any reason, they would not survive pathogenic attack. But in such situation, application of melatonin helps plants to survive and improve their defense system (Lee et al., 2015).

A few examples of different experiments are given in Table 9.1.

Table 9.1 **Various methods to increase melatonin** content in plants

Sr. n10.o.	Plant	Method	Effect	References
1	*Trigonella foenum-graecum* L.	Application of rhizospheric bacteria	Endogenous melatonin and growth was increased	Shakeel et al. (2024)
2	*Brassica rapa*	Melatonin foliar spray	Amelioration of drought stress	Hasnain et al. (2023)
3	*Capsicum annuum* L.	Melatonin in combination with arginine	Amelioration of salt stress	Usman et al. (2023)
4	*Prunus persica*	Melatonin in solution form in early growth phases	Amelioration of drought stress by increasing root strength and enhancing photosynthesis	Iqbal et al. (2024)
5	*Zea mays* L.	Exogenous melatonin	Amelioration of salt stress by increasing photosynthetic efficiency and decreasing ROS species	Ahmad et al. (2021)
6	*Medicago sativa* L.	Melatonin foliar spray	Salt tolerance by enhancing antioxidant capacity	Cen et al. (2020)
7	*Zea mays* L.	Melatonin foliar spray	Delaying leaf senescence	Ahmad et al. (2020)
8	*Oryza sativa* L.	Melatonin foliar spray	Drought tolerance	Silalert and Pattanagul (2020)
9	*Capsicum* spp.	Melatonin foliar spray	Drought tolerance	Khosravi et al. (2023)
10	*Brassica juncea* L.	Exogenous methyl jasmonate	Increased melatonin endogenously	Yin et al. (2023)
11	*Zea mays* L.	Nano-ZnO	Increased endogenous melatonin and drought resistance	Sun et al. (2020b)
12	*Prunus persica*	IAA application	Increased endogenous melatonin and cold stress resistance	Zhou et al. (2023)

9.10 Conclusion

Melatonin also called phytomelatonin in plants is a signaling molecule that can be the future of sustainable agriculture. It is a phytochemical having resemblance with indole acetic acid in structure and physiology both with a common precursor, that is, precursor. It can enhance the plant growth by increasing germination, seedling growth, floral transition, etc., whereas it also helps the plants to fight against various stresses like water stress, salt stress, heavy metal stress, etc. Melatonin can also enhance the postharvest quality of agricultural products. It can interact with other phytochemicals and plant growth regulators to enhance the plant growth and mitigate the stress. Melatonin can be applied exogenously like priming and foliar spray. Endogenous melatonin can also be enhanced by adding some other molecules like IAA.

References

Aghdam MS, Fard JR (2017) Melatonin treatment attenuates postharvest decay and maintains nutritional quality of strawberry fruits (*Fragaria ananassa*) by enhancing GABA shunt activity. Food Chem 221:1650–1657

Agozzino P, Avellone G, Bongiorno D, Ceraulo L, Filizzola F, Natoli MC, Livrea MA, Tesoriere L (2003) Melatonin: structural characterization of its non-enzymatic mono-oxygenate metabolite. J Pineal Res 35:269–275

Ahammed GJ, Xu W, Liu A, Chen S (2019) Endogenous melatonin deficiency aggravates high temperature-induced oxidative stress in *Solanum lycopersicum* L. Environ Exp Bot 161:303–311

Ahammed GJ, Mao Q, Yan Y, Wu M, Wang Y, Ren J, Chen S (2020) Role of melatonin in arbuscular mycorrhizal fungi-induced resistance to *Fusarium* wilt in cucumber. Phytopathology 110(5):999–1009

Ahmad S, Su W, Kamran M, Ahmad I, Meng X, Wu X, Han Q (2020) Foliar application of melatonin delay leaf senescence in maize by improving the antioxidant defense system and enhancing photosynthetic capacity under semi-arid regions. Protoplasma 257:1079–1092

Ahmad S, Cui W, Kamran M et al (2021) Exogenous application of melatonin induces tolerance to salt stress by improving the photosynthetic efficiency and antioxidant defense system of maize seedling. J Plant Growth Regul 40:1270–1283

Ahn HR, Kim YJ, Lim YJ, Duan S, Eom SH, Jung KH (2021) Key genes in the melatonin biosynthesis pathway with circadian rhythm are associated with various abiotic stresses. Plan Theory 10(1):129

Ali S, Nawaz A, Naz S, Ali M, Ejaz S, Azam M, Razzaq K (2023) Exogenous melatonin mitigates chilling injury in zucchini fruit by enhancing antioxidant system activity, promoting endogenous proline and GABA accumulation, and preserving cell wall stability. Postharvest Biol Technol 204:11244

Altaf MA, Shahid R, Ren MX, Mora-Poblete F, Arnao MB, Naz S, Anwar M, Altaf MM, Shahid S, Shakoor A, Sohail H (2021) Phytomelatonin: an overview of the importance and mediating functions of melatonin against environmental stresses. Physiol Plant 172(2):820–846

Anitha K, Senthil A, Kalarani MK, Senthil N, Marimuthu S, Djanaguiraman M, Umapathi M (2022) Exogenous melatonin improves seed germination and seedling growth in greengram under drought stress. Journal of Applied and Natural Science 14(4):1190–1197

Arnao MB (2014) Phytomelatonin: discovery, content, and role in plants. Advances in Botany 2014:815769

Arnao MB, Hernández-Ruiz J (2014) Melatonin: plant growth regulator and/or biostimulator during stress. Trends Plant Sci 19(12):789–797
Arnao MB, Hernández-Ruiz J (2018) Melatonin and its relationship to plant hormones. Ann Bot 121(2):195–207
Arnao MB, Hernández-Ruiz J (2019) Melatonin: a new plant hormone and/or a plant master regulator. Trends Plant Sci 24(1):38–48
Arnao MB, Hernández-Ruiz J (2020) Melatonin in flowering, fruit set and fruit ripening. Plant Reproduction 33:77–87
Asif M, Pervez A, Irshad U, Mehmood Q, Ahmad R (2020) Melatonin and plant growth-promoting rhizobacteria alleviate the cadmium and arsenic stresses and increase the growth of *Spinacia oleracea* L. Plant Soil Environ 66(5):234–241
Aversa S, Pellegrino S, Barberi I, Reiter RJ, Gitto E (2012) Potential utility of melatonin as an antioxidant during pregnancy and in the perinatal period. J Matern Fetal Neonatal Med 25(3):207–221
Back K (2021) Melatonin metabolism, signaling and possible roles in plants. Plant J 105(2):376–391
Back K, Tan DX, Reiter RJ (2016) Melatonin biosynthesis in plants: multiple pathways catalyze tryptophan to melatonin in the cytoplasm or chloroplasts. J Pineal Res 61(4):426–437
Bajwa VS, Shukla MR, Sherif SM, Murch SJ, Saxena PK (2014) Role of melatonin in alleviating cold stress in *Arabidopsis thaliana*. J Pineal Res 56(3):238–245
Bal E (2021) Effect of melatonin treatments on biochemical quality and postharvest life of nectarines. J Food Meas Charact 15(1):288–295
Balabusta M, Szafranska K, Posmyk MM (2016) Exogenous melatonin improves antioxidant defense in cucumber seeds (*Cucumis sativus* L.) germinated under chilling stress. *Frontiers. Plant Sci* 7:575
Bandurski RS, Cohen JD, Slovin J, Reinecke DM, Davies PJ (1995) Auxin biosynthesis and metabolism. In: Plant Hormones: Physiology, Biochemistry and Molecular Biology, 2nd edn. Dordrecht Kluwer Academic Publishers, pp 39–65
Blask DE, Dauchy RT, Sauer LA, Krause JA (2004) Melatonin uptake and growth prevention in rat hepatoma 7288CTC in response to dietary melatonin: melatonin receptor-mediated inhibition of tumor linoleic acid metabolism to the growth signaling molecule 13-hydroxyoctadecadienoic acid and the potential role of phytomelatonin. Carcinogenesis 25(6):951–960
Burkhardt S, Tan DX, Manchester LC, Hardeland R, Reiter RJ (2001) Detection and quantification of the antioxidant melatonin in Montmorency and Balaton tart cherries. J Agric Food Chem 49:4898–4902
Cai SY, Zhang Y, Xu YP, Qi ZY, Li MQ, Ahammed GJ, Zhou J (2017) HsfA1a upregulates melatonin biosynthesis to confer cadmium tolerance in tomato plants. J Pineal Res 62(2):e12387
Cao S, Shao J, Shi L, Xu L, Shen Z, Chen W, Yang Z (2018) Melatonin increases chilling tolerance in postharvest peach fruit by alleviating oxidative damage. Sci Rep 8(1):806
Cen H, Wang T, Liu H, Tian D, Zhang Y (2020) Melatonin application improves salt tolerance of alfalfa (*Medicago sativa* L.) by enhancing antioxidant capacity. Plan Theory 9:220
Chen Y, Liu Y, Li Y, Wu Y, Chen Y, Zeng G, Li H (2017) Influence of biochar on heavy metals and microbial community during composting of river sediment with agricultural wastes. Bioresour Technol 243:347–355
Chen YE, Mao JJ, Sun LQ, Huang B, Ding CB, Gu Y, Yuan M (2018) Exogenous melatonin enhances salt stress tolerance in maize seedlings by improving antioxidant and photosynthetic capacity. Physiol Plant 164(3):349–363
Chen L, Liu L, Lu B, Ma T, Jiang D, Li J, Zhang K, Sun H, Zhang Y, Bai Z, Li C (2020) Exogenous melatonin promotes seed germination and osmotic regulation under salt stress in cotton (*Gossypium hirsutum* L.). PLoS One 15(1):e0228241
Cui G, Zhao X, Liu S, Sun F, Zhang C, Xi Y (2017) Beneficial effects of melatonin in overcoming drought stress in wheat seedlings. Plant Physiol Biochem 118:138–149
Ding F, Wang G, Wang M, Zhang S (2018) Exogenous melatonin improves tolerance to water deficit by promoting cuticle formation in tomato plants. Molecules 23(7):1605

Duan Q, Wang Z, Lu T, Chen J, Wang X (2006) Comparison of 6-hydroxylmelatonin or melatonin in protecting neurons against ischemia/reperfusion-mediated injury. J. Pineal Res. 41:351–357

Dubbels R, Reiter RJ, Klenke E, Goebel A, Schnakenberg E, Ehlers C (1995) Melatonin in edible plants identified by radioimmunoassay and by high performance liquid chromatography-mass spectrometry. J Pineal Res 18:28–31

Erland LA, Saxena PK (2018) Melatonin in plant morphogenesis. In Vitro Cell Dev Biol Plant 54:3–24

Fan J, Xie Y, Zhang Z, Chen L (2018) Melatonin: a multifunctional factor in plants. Int J Mol Sci 19(5):1528

Farooq M, Romdhane L, Al Sulti MK, Rehman A, Al-Busaidi WM, Lee DJ (2020) Morphological, physiological and biochemical aspects of osmopriming-induced drought tolerance in lentil. J Agron Crop Sci 206(2):176–186

Fu J, Wu Y, Miao Y, Xu Y, Zhao E, Wang J, Hu T (2017) Improved cold tolerance in *Elymus nutans* by exogenous application of melatonin may involve ABA-dependent and ABA-independent pathways. Sci Rep 7(1):39865

Fuhrberg B, Balzer I, Hardeland R, Werner A, Luning K (1996) The vertebrate pineal hormone melatonin is produced by the brown alga *Pterygophora californica* and mimics dark effects on growth rate in the light. Planta 200:125–131

Galano A, Tan DX, Reiter RJ (2013) On the free radical scavenging activities of melatonin's metabolites, AFMK and AMK. J Pineal Res 54:245–257

Gao H, Zhang ZK, Chai HK, Cheng N, Yang Y, Wang DN, Cao W (2016) Melatonin treatment delays postharvest senescence and regulates reactive oxygen species metabolism in peach fruit. Postharvest Biol Technol 118:103–110

Garcia-Caparros P, De Filippis L, Gul A, Hasanuzzaman M, Ozturk M, Altay V, Lao MT (2021) Oxidative stress and antioxidant metabolism under adverse environmental conditions: a review. Bot Rev 87:421–466

Gitto E, Tan DX, Reiter RJ, Karbownik M, Manchester LC, Cuzzocrea S, Fulia F, Barberi I (2010) Individual and synergistic antioxidative actions of melatonin: studies with vitamin E, vitamin C, glutathione and desferrioxamine (deferoxamine) in rat liver homogenates. J Pharm Pharmacol 53:1393–1401

Gu Q, Xiao Q, Chen Z, Han Y (2022) Crosstalk between melatonin and reactive oxygen species in plant abiotic stress responses: an update. Int J Mol Sci 23(10):5666

Guo X, Liu D, Chong K (2018) Cold signaling in plants: insights into mechanisms and regulation. J Integr Plant Biol 60(9):745–756

Gupta R (2023) Melatonin: a promising candidate for maintaining food security under the threat of phytopathogens. Plant Physiol Biochem 198:107691

Hacışevki A, Baba B (2018) An overview of melatonin as an antioxidant molecule: a biochemical approach. Melatonin Mol Biol Clin Pharm Approach 2018:59–85

Hara M, Iigo M, Ohtani-Kaneko R, Nakamura N, Suzuki T, Reiter RJ, Hirata K (1997) Administration of melatonin and related indoles prevents exercise-induced cellular oxidative changes in rats. Neurosignals 6:90–100

Hardeland R (2015) Melatonin in plants and other phototrophs: advances and gaps concerning the diversity of functions. J Exp Bot 66(3):627–646

Hardeland R (2016) Melatonin in plants—diversity of levels and multiplicity of functions. Front Plant Sci 7:198

Hardeland R, Fuhrberg B (1996) Ubiquitous melatonin: presence and effects in unicells, plants and animals. Trends Comp Biochem Physiol 2:25–45

Hardeland R, Tan DX, Reiter RJ (2009) Kynuramines, metabolites of melatonin and other indoles: the resurrection of an almost forgotten class of biogenic amines. J Pineal Res 47:109–126

Hasan MK, Ahammed GJ, Yin L, Shi K, Xia X, Zhou Y, Zhou J (2015) Melatonin mitigates cadmium phytotoxicity through modulation of phytochelatins biosynthesis, vacuolar sequestration, and antioxidant potential in *Solanum lycopersicum* L. Front Plant Sci 6:601

Hasnain Z, Zafar S, Usman S, Zhang L, Elansary OH (2023) Elucidating role of melatonin foliar spray in ameliorating adverse effects of drought stress on growth and physio-biochemical attributes of Brassica rapa plants. Sci Hortic 323:112336

Horstman JA, Wrona MZ, Dryhurst G (2002) Further insights into the reaction of melatonin with hydroxyl radical. Bioorg Chem 30:371–382

Hussain MB, Mahmood S, Ahmed N, Nawaz H (2018) Rhizobial inoculation for improving growth physiology, nutrition and yield of maize under drought stress conditions. Pak J Bot 50(5):1681–1689

Iqbal S, Hayat F, Hussain M (2024) Melatonin supplementation alleviates drought stress in peach (*Prunus persica*) seedlings by improving photosynthesis, root morphological traits, and antioxidant defense system. Acta Physiol Plant 46:18

Jahan MS, Shu S, Wang Y, Chen Z, He M, Tao M, Guo S (2019) Melatonin alleviates heat-induced damage of tomato seedlings by balancing redox homeostasis and modulating polyamine and nitric oxide biosynthesis. BMC Plant Biol 19(1):1–16

Jiao J, Ma Y, Chen S, Liu C, Song Y, Qin Y, Liu Y (2016) Melatonin-producing endophytic bacteria from grapevine roots promote the abiotic stress-induced production of endogenous melatonin in their hosts. Front Plant Sci 7:1387

Kaur H, Bhatla SC (2016) Melatonin and nitric oxide modulate glutathione content and glutathione reductase activity in sunflower seedling cotyledons accompanying salt stress. Nitric Oxide 59:42–53

Kaya C, Okant M, Ugurlar F, Alyemeni MN, Ashraf M, Ahmad P (2019) Melatonin-mediated nitric oxide improves tolerance to cadmium toxicity by reducing oxidative stress in wheat plants. Chemosphere 225:627–638

Ke Q, Ye J, Wang B, Ren J, Yin L, Deng X, Wang S (2018) Melatonin mitigates salt stress in wheat seedlings by modulating polyamine metabolism. Front Plant Sci 9:914

Khosravi S, Haghighi M, Mottaghipisheh J (2023) Effects of melatonin foliar application on hot pepper growth and stress tolerance. Plant Stress 9:100192

Kim TK, Kleszczyński K, Janjetovic Z, Sweatman T, Lin Z, Li W, Reiter RJ, Fischer TW, Slominski AT (2013) Metabolism of melatonin and biological activity of intermediates of melatoninergic pathway in human skin cells. FASEB J 27(7):2742

Kołodziejczyk I, Posmyk MM (2016) Melatonin—a new plant biostimulator? J Elem 21(4):1187–1198

Lee HY, Byeon Y, Tan DX, Reiter RJ, Back K (2015) Arabidopsis serotonin N-acetyltransferase knockout mutant plants exhibit decreased melatonin and salicylic acid levels resulting in susceptibility to an avirulent pathogen. J Pineal Res 58(3):291–299

Leon J, Escames G, Rodríguez MI, López LC, Tapias V, Entrena A, Camacho E, Carrión MD, Gallo MA, Espinosa A (2006) Inhibition of neuronal nitric oxide synthase activity by N 1-acetyl-5-methoxykynuramine, a brain metabolite of melatonin. J Neurochem 98:2023–2033

Li C, Tan DX, Liang D, Chang C, Jia D, Ma F (2015) Melatonin mediates the regulation of ABA metabolism, free-radical scavenging, and stomatal behaviour in two Malus species under drought stress. J Exp Bot 66(3):669–680

Li J, Liu J, Zhu T, Zhao C, Li L, Chen M (2019) The role of melatonin in salt stress responses. Int J Mol Sci 20(7):1735

Li X, Tan DX, Jiang D, Liu F (2016) Melatonin enhances cold tolerance in drought-primed wild-type and abscisic acid-deficient mutant barley. J Pineal Res 61(3):328–339

Li H, Chang J, Chen H, Wang Z, Gu X, Wei C, Zhang X (2017) Exogenous melatonin confers salt stress tolerance to watermelon by improving photosynthesis and redox homeostasis. Front Plant Sci 8:295

Li GZ, Liu J, Chen SJ, Wang PF, Liu HT, Dong J, Kang GZ (2021) Melatonin promotes potassium deficiency tolerance by regulating HAK1 transporter and its upstream transcription factor NAC71 in wheat. J Pineal Res 70(4):e12727

Liang C, Zheng G, Li W, Wang Y, Hu B, Wang H, Chu C (2015) Melatonin delays leaf senescence and enhances salt stress tolerance in rice. J Pineal Res 59(1):91–101

Liang D, Ni Z, Xia H, Xie Y, Lv X, Wang J, Luo X (2019) Exogenous melatonin promotes biomass accumulation and photosynthesis of kiwifruit seedlings under drought stress. Sci Hortic 246:34–43

Liu J, Zhai R, Liu F, Zhao Y, Wang H, Liu L, Xu L (2018a) Melatonin induces parthenocarpy by regulating genes in gibberellin pathways of 'Starkrimson' pear (*Pyrus communis* L.). *Frontiers*. Plant Sci 9:946

Liu C, Zheng H, Sheng K, Liu W, Zheng L (2018b) Effects of melatonin treatment on the postharvest quality of strawberry fruit. Postharvest Biol Technol 139:47–55

Liu C, Zhou H, Zhou J (2021) The applications of nanotechnology in crop production. Molecules 26(23):7070

Ljung K, Hull AK, Kowalczyk M, Marchant A, Celenza J, Cohen JD, Sandberg G (2002) Biosynthesis, conjugation, catabolism and homeostasis of indole-3-acetic acid in *Arabidopsis thaliana*. Plant Mol Biol 50:309–332

López-Burillo S, Tan DX, Rodriguez-Gallego V, Manchester LC, Mayo JC, Sainz RM, Reiter RJ (2003) Melatonin and its derivatives cyclic 3-hydroxymelatonin, N1-acetyl-N2-formyl-5--methoxykynuramine and 6-methoxymelatonin reduce oxidative DNA damage induced by Fenton reagents. J Pineal Res 34:178–184

Ma Y, Jiao J, Fan X, Sun H, Zhang Y, Jiang J, Liu C (2017) Endophytic bacterium *Pseudomonas fluorescens* RG11 may transform tryptophan to melatonin and promote endogenous melatonin levels in the roots of four grape cultivars. Front Plant Sci 7:2068

Ma X, Zhang J, Burgess P, Rossi S, Huang B (2018) Interactive effects of melatonin and cytokinin on alleviating drought-induced leaf senescence in creeping bentgrass (*Agrostis stolonifera*). Environ Exp Bot 145:1–11

Mannino G, Caradonna F, Cruciata I, Lauria A, Perrone A, Gentile C (2019) Melatonin reduces inflammatory response in human intestinal epithelial cells stimulated by interleukin-1β. J Pineal Res 67:e12598

Mannino G, Pernici C, Serio G, Gentile C, Bertea CM (2021) Melatonin and phytomelatonin: chemistry, biosynthesis, metabolism, distribution and bioactivity in plants and animals—an overview. Int J Mol Sci 22(18):9996

Mao J, Niu C, Li K, Chen S, Tahir MM, Han M, Zhang D (2020) Melatonin promotes adventitious root formation in apple by promoting the function of MdWOX11. BMC Plant Biol 20(1):1–11

Marak K, Mir H, Singh P, Siddiqui MW, Ranjan T, Singh DR, Siddiqui MH, Irfan M (2024) Exogenous melatonin delays oxidative browning and improves postharvest quality of litchi fruits. Sci Hortic 323:112408

McHill AW, Phillips AJ, Czeisler CA, Keating L, Yee K, Barger LK, Garaulet M, Scheer FA, Klerman EB (2017) Later circadian timing of food intake is associated with increased body fat. Am J Clin Nutr 106:1213–1219

Mukherjee S, David A, Yadav S, Baluška F, Bhatla SC (2014) Salt stress-induced seedling growth inhibition coincides with differential distribution of serotonin and melatonin in sunflower seedling roots and cotyledons. Physiol Plant 152(4):714–728

Murch SJ, Simmons CB, Saxena PK (1997) Melatonin in feverfew and other medicinal plants. Lancet 350:1598–1599

Nawaz K, Chaudhary R, Sarwar A, Ahmad B, Gul A, Hano C, Abbasi BH, Anjum S (2020) Melatonin as master regulator in plant growth, development and stress alleviator for sustainable agricultural production: current status and future perspectives. Sustain For 13(1):294

Nazir M, Asad-Ullah M, Mumtaz S, Siddiquah A, Shah M, Drouet S, Abbasi BH (2020) Interactive effect of melatonin and UV-C on phenylpropanoid metabolite production and antioxidant potential in callus cultures of purple basil (*Ocimum basilicum* L. var purpurascens). Molecules 25(5):1072

Normanly J, Slovin J, Cohen JD, Davies PJ (2004) Auxin biosynthesis and metabolism. In: Plant Hormones: Physiology, Biochemistry and Molecular Biology, 3rd edn. Dordrecht Kluwer Academic Publishers, pp 36–62

Omer RA, Koparir P, Ahmed L, Koparir M (2021) Computational and spectroscopy study of melatonin. Indian J Chem B 60:732–741

Park S, Back K (2012) Melatonin promotes seminal root elongation and root growth in transgenic rice after germination. J Pineal Res 53(4):385–389

Pelagio-Flores R, Muñoz-Parra E, Ortiz-Castro R, Lopez-Bucio J (2012) Melatonin regulates Arabidopsis root system architecture likely acting independently of auxin signaling. J Pineal Res 53(3):279–288

Qi W, Reiter RJ, Tan DX, Manchester LC, Siu AW, Garcia JJ (2000) Increased levels of oxidatively damaged DNA induced by chromium (III) and H_2O_2: protection by melatonin and related molecules. J Pineal Res 29:54–61

Rajora N, Vats S, Raturi G, Thakral V, Kaur S, Rachappanavar V, Deshmukh R (2022) Seed priming with melatonin: a promising approach to combat abiotic stress in plants. Plant Stress 4:100071

Ramasamy K, Karuppasami KM, Alagarswamy S, Shanmugam KP, Rathinavelu S, Vellingiri G, Kanagarajan S (2023) Role of melatonin in directing plant physiology. Agronomy 13(9):2405

Raza A, Charagh S, García-Caparrós P, Rahman MA, Ogwugwa VH, Saeed F, Jin W (2022) Melatonin-mediated temperature stress tolerance in plants. GM Crops and Food 13(1):196–217

Reina M, Martínez A (2018) A new free radical scavenging cascade involving melatonin and three of its metabolites (3OHM, AFMK and AMK). Comput Theor Chem 1123:111–118

Ressmeyer AR, Mayo JC, Zelosko V, Sáinz RM, Tan DX, Poeggeler B, Antolín I, Zsizsik BK, Reiter RJ, Hardeland R (2003) Antioxidant properties of the melatonin metabolite N1-acetyl-5-methoxykynuramine (AMK): scavenging of free radicals and prevention of protein destruction. Redox Rep 8:205–213

Roy R, Sultana S, Begum N, Fornara D, Barmon M, Zhang R, Rabbany MG (2022) Exogenous melatonin reduces water deficit-induced oxidative stress and improves growth performance of *Althaea rosea* grown on coal mine spoils. Environ Sci Pollut Res 29(41):61550–61560

Sakano K, Oikawa S, Hiraku Y, Kawanishi S (2004) Oxidative DNA damage induced by a melatonin metabolite, 6-hydroxymelatonin, via a unique non-o-quinone type of redox cycle. Biochem Pharmacol 68:1869–1878

Semak I, Naumova M, Korik E, Terekhovich V, Wortsman J, Slominski A (2005) A novel metabolic pathway of melatonin: oxidation by cytochrome *c*. Biochemistry 44:9300–9307

Shakeel M, Tufail A, Rashid M, Noreen Z, Basharat N, Tanveer F, Al-Qthanin RN, Alomran MM, Alomrai AO (2024) Effect of rhizospheric bacteria on growth, physiological attributes and melatonin contents of *Trigonella foenum-graecum* L. Sci Hortic 323:112552

Sharif R, Xie C, Zhang H, Arnao MB, Ali M, Ali Q, Li Y (2018a) Melatonin and its effects on plant systems. Molecules 23(9):2352

Sharif R, Xie C, Zhang H, Arnao MB, Ali M, Ali Q, Muhammad I, Shalmani A, Nawaz MA, Chen P, Li Y (2018b) Melatonin and its effects on plant systems. Molecules 23(9):2352

Sharma S, Kataria S, Joshi J, Guruprasad KN (2019) Antioxidant defense response of fenugreek to solar UV. Int J Veg Sci 25(1):40–57

Shi H, Tan DX, Reiter RJ, Ye T, Yang F, Chan Z (2015) Melatonin induces class A1 heat-shock factors (HSFA 1s) and their possible involvement of thermotolerance in *Arabidopsis*. J Pineal Res 58(3):335–342

Shi H, Wei Y, Wang Q, Reiter RJ, He C (2016) Melatonin mediates the stabilization of DELLA proteins to repress the floral transition in Arabidopsis. J Pineal Res 60(3):373–379

Silalert P, Pattanagul W (2020) Foliar application of melatonin alleviates the effects of drought stress in rice (*Oryza sativa* L.) seedlings. Notulae Botanicae Horti Agrobotanici Cluj-Napoca 49(3):12417

Simlat M, Ptak A, Skrzypek E, Warchoł M, Morańska E, Piórkowska E (2018) Melatonin significantly influences seed germination and seedling growth of *Stevia rebaudiana* Bertoni. PeerJ 6:e5009

Simopoulos AP, Tan DX, Manchester LC, Reiter RJ (2005) Purslane: a plant source of omega-3 fatty acids and melatonin. J Pineal Res 39:331–332

Sun Q, Zhang N, Wang J, Cao Y, Li X, Zhang H, Guo YD (2016) A label-free differential proteomics analysis reveals the effect of melatonin on promoting fruit ripening and anthocyanin accumulation upon postharvest in tomato. J Pineal Res 61(2):138–153

Sun C, Lv T, Huang L, Liu X, Jin C, Lin X (2020a) Melatonin ameliorates aluminum toxicity through enhancing aluminum exclusion and reestablishing redox homeostasis in roots of wheat. J Pineal Res 68(4):e12642

Sun L, Song F, Guo J, Zhu X, Liu S, Liu F, Li X (2020b) Nano-ZnO-induced drought tolerance is associated with melatonin synthesis and metabolism in maize. Int J Mol Sci 21(3):782

Sun C, Liu L, Wang L, Li B, Jin C, Lin X (2021) Melatonin: a master regulator of plant development and stress responses. J Integr Plant Biol 63(1):126–145

Sun H, Jia M, Wang Y, Lu H, Wang X (2023) The complexity of melatonin and other phytohormones crosstalk with other signaling molecules for drought tolerance in horticultural crops. Sci Hortic 323:112348

Tan DX, Reiter RJ (2020) An evolutionary view of melatonin synthesis and metabolism related to its biological functions in plants. J Exp Bot 71(16):4677–4689

Tan DX, Hardeland R, Manchester LC, Poeggeler B, Lopez-Burillo S, Mayo JC, Sainz RM, Reiter RJ (2003) Mechanistic and comparative studies of melatonin and classic antioxidants in terms of their interactions with the ABTS cation radical. J Pineal Res 34:249–259

Tan D, Reiter R, Manchester L, Yan M, El-Sawi M, Sainz R, Mayo J, Kohen R, Allegra M, Hardelan R (2005) Chemical and physical properties and potential mechanisms: melatonin as a broad spectrum antioxidant and free radical scavenger. Curr Top Med Chem 2:181–197

Tan DX, Manchester LC, Terron MP, Flores LJ, Reiter RJ (2007) One molecule, many derivatives: a never-ending interaction of melatonin with reactive oxygen and nitrogen species? J Pineal Res 42:28–42

Tan XL, Fan ZQ, Kuang JF, Lu WJ, Reiter RJ, Lakshmanan P, Shan W (2019) Melatonin delays leaf senescence of Chinese flowering cabbage by suppressing ABFs-mediated abscisic acid biosynthesis and chlorophyll degradation. J Pineal Res 67(1):e12570

Umapathi M, Kalarani MK, Srinivasan S (2018) Optimization of melatonin to mitigate cadmium stress at seedling level in tomato. Madras Agric J 105:471

Usman S, Yaseen G, Noreen Z, Rizwan M, Noor H, Elansary HO (2023) Melatonin and arginine combined supplementation alleviate salt stress through physiochemical adjustments and improved antioxidant enzymes activity in *Capsicum annuum* L. Sci Hortic 323:112270

Van TDL, Roberts N, O'Neill SD (1995) Melatonin from higher plants: isolation and identification of N-acetyl-5-methoxytryptamine. Plant Physiol 108:101

Wang P, Sun X, Li C, Wei Z, Liang D, Ma F (2013) Long-term exogenous application of melatonin delays drought-induced leaf senescence in apple. J Pineal Res 54(3):292–302

Wang L, Feng C, Zheng X, Guo Y, Zhou F, Shan D, Kong J (2017) Plant mitochondria synthesize melatonin and enhance the tolerance of plants to drought stress. J Pineal Res 63(3):e12429

Wang YP, Reiter RJ, Chan ZL (2018) Phytomelatonin: A universal abiotic stress regulator. J Exp Bot 69(5):963–974

Wei Y, Liu G, Chang Y, Lin D, Reiter RJ, He C, Shi H (2018) Melatonin biosynthesis enzymes recruit WRKY transcription factors to regulate melatonin accumulation and transcriptional activity on W-box in cassava. J Pineal Res 65(1):e12487

Xu W, Cai SY, Zhang Y, Wang Y, Ahammed GJ, Xia XJ, Zhou J (2016) Melatonin enhances thermotolerance by promoting cellular protein protection in tomato plants. J Pineal Res 61(4):457–469

Xu L, Yue Q, Xiang G, Bian FE, Yao Y (2018) Melatonin promotes ripening of grape berry via increasing the levels of ABA, H_2O_2, and particularly ethylene. Hortic Res 5:41

Yang L, You J, Li J, Wang Y, Chan Z (2021) Melatonin promotes *Arabidopsis* primary root growth in an IAA-dependent manner. J Exp Bot 72(15):5599–5611

Ye T, Yin X, Yu L, Zheng SJ, Cai WJ, Wu Y, Feng YQ (2019) Metabolic analysis of the melatonin biosynthesis pathway using chemical labeling coupled with liquid chromatography-mass spectrometry. J Pineal Res 66(1):e12531

Yin Y, Wang C, Cheng C, Yang Z, Fang W (2023) Exogenous methyl jasmonate promotes the biosynthesis of endogenous melatonin in mustard sprouts. Plant Physiol Biochem 203:108055

Zafar S, Akhtar M, Perveen S, Hasnain Z, Khalil A (2020) Attenuating the adverse aspects of water stress on wheat genotypes by foliar spray of melatonin and indole-3-acetic acid. Physiol Mol Biol Plants 26:1751–1762

Zhang HJ, Zhang NA, Yang RC, Wang L, Sun QQ, Li DB, Guo YD (2014) Melatonin promotes seed germination under high salinity by regulating antioxidant systems, ABA and GA 4 interaction in cucumber (*Cucumis sativus* L.). J Pineal Res 57(3):269–279

Zhang J, Shi Y, Zhang X, Du H, Xu B, Huang B (2017a) Melatonin suppression of heat-induced leaf senescence involves changes in abscisic acid and cytokinin biosynthesis and signaling pathways in perennial ryegrass (*Lolium perenne* L.). Environ Exp Bot 138:36–45

Zhang R, Sun Y, Liu Z, Jin W, Sun Y (2017b) Effects of melatonin on seedling growth, mineral nutrition, and nitrogen metabolism in cucumber under nitrate stress. J Pineal Res 62(4):e12403

Zhang Z, Hu Q, Liu Y, Cheng P, Cheng H, Liu W, Chen F (2019) Strigolactone represses the synthesis of melatonin, thereby inducing floral transition in *Arabidopsis thaliana* in an FLC-dependent manner. J Pineal Res 67(2):e12582

Zhang M, Gao C, Xu L, Niu H, Liu Q, Huang Y, Lv G, Yang H, Li M (2022a) Melatonin and indole-3-acetic acid synergistically regulate plant growth and stress resistance. Cells 11(20):3250

Zhang T, Wang J, Sun Y, Zhang L, Zheng S (2022b) Versatile roles of melatonin in growth and stress tolerance in plants. J Plant Growth Regul 41:507–523

Zhao H, Su T, Huo L, Wei H, Jiang Y, Xu L, Ma F (2015) Unveiling the mechanism of melatonin impacts on maize seedling growth: sugar metabolism as a case. J Pineal Res 59(2):255–266

Zhao H, Zhang K, Zhou X, Xi L, Wang Y, Xu H, Zou Z (2017) Melatonin alleviates chilling stress in cucumber seedlings by up-regulation of CsZat12 and modulation of polyamine and abscisic acid metabolism. Sci Rep 7(1):4998

Zheng X, Tan DX, Allan AC, Zuo B, Zhao Y, Reiter RJ, Kong J (2017) Chloroplastic biosynthesis of melatonin and its involvement in protection of plants from salt stress. Sci Rep 7(1):41236

Zhou C, Liu Z, Zhu L, Ma Z, Wang J, Zhu J (2016) Exogenous melatonin improves plant iron deficiency tolerance via increased accumulation of polyamine-mediated nitric oxide. Int J Mol Sci 17(11):1777

Zhou Q, Bao Z, Yu Y, Chen W, Yang Z, Cao S, Shi L (2023) IAA regulated levels of endogenous phytohormones in relation to chilling tolerance in cold-stored peaches after harvest. Postharvest Biol Technol 205:112490

Zuo B, Zheng X, He P, Wang L, Lei Q, Feng C, Kong J (2014) Overexpression of MzASMT improves melatonin production and enhances drought tolerance in transgenic *Arabidopsis thaliana* plants. J Pineal Res 57(4):408–417

Tools and Techniques to Quantify PGR Content in Plants

10

Havza Imtiaz, Yamshi Arif, and Shamsul Hayat

Abstract

Plant growth regulators play a pivotal role in finely tuning plant growth, development, and responses to environmental cues. Advancing our understanding requires insights into hormone structure, distribution at various levels, and the genes governing the hormone process. Qualitative analysis of plant hormones in minute quantities poses challenges, leading to the development of sensitive techniques. Modern mass spectrometry (MS)-based methodologies have replaced early bioassays and immunoassays. While MS presents challenges like matrix interference, triple-quadrupole mass spectrometers (QQQ) excel in targeted quantitative analysis, allowing multiple reaction monitoring (MRM) or selected reaction monitoring (SRM). Gas chromatography (GC) and liquid chromatography (LC) coupled with MS, such as GC-MS and LC-MS/MS, quantify PGRs. HPLC-MS/MS is preferred for its high sensitivity and specificity. Challenges like matrix interference are addressed by QQQ, making it suitable for complex samples. Gas chromatography is limited to volatile and thermostable PGRs. This chapter delves into techniques like HPLC-MS, GC, LC, QQQ, and the tobacco syringe agroinfiltration assay for PGR quantification. Updated protocols for detecting and quantifying PGRs in plant tissues are also highlighted, providing comprehensive insights into PGR analysis.

Keywords

Plant growth regulators · Quantification · Chromatography · Spectrometry

H. Imtiaz · Y. Arif · S. Hayat (✉)
Plant Physiology Section, Department of Botany, Faculty of Life Sciences, Aligarh Muslim University, Aligarh, India

© The Author(s), under exclusive license to Springer Nature Singapore Pte Ltd. 2024

M. Faizan, S. Hayat (eds.), *Plant Growth Regulators: Resilience for Sustainable Agriculture*, https://doi.org/10.1007/978-981-97-2918-0_10

10.1 Introduction

Plant growth regulators (PGRs) play crucial roles in regulating plant growth, development, and responses to several environmental stimuli, both biotic and abiotic. These hormones exhibit structural diversity and are categorized into several major classes. Each of class hormones like auxin, cytokinin (CK), abscisic acid (ABA), gibberellin (GA), brassinosteroids (BR), salicylic acid (SA), jasmonic acid (JA), ethylene (ET), and recently discovered plant hormones like strigolactone (SL) exhibits distinct biological effects, and the coordination of numerous plant hormones frequently modulates development and stress responses through additive, synergistic, or antagonistic actions (EL Sabagh et al. 2022). Advancing our comprehension of the regulation of plant growth and development necessitates insights into plant hormone structure and function, the distribution of hormones at various organizational levels (organ, cellular, and sub-cellular), and detailed study on the genes and their by-products that govern hormone production, utilization, and breakdown. Plant hormones are usually found in minute quantities in plants, rendering qualitative analysis challenging (Wang et al. 2020). The precise identification of plant hormones necessitates advancing susceptible and efficient analytical techniques for their analysis. Bioassays and immunoassays were extensively employed in the early stages of plant hormone research (Wang et al. 2020). However, unlike MS-based methods, these approaches lack sensitivity and specificity and cannot simultaneously detect multi-class plant hormones (Oklestkova et al. 2017). As a result, they have been supplanted by more modern techniques.

MS is an analytical technique that quantifies nearly small organic molecules by measuring the ratio of mass to charge for ions in a gaseous state. However, the MS possesses specific challenges in quantifying PGRs, such as matrix interference, standards of PGRs, and sample preparations. Triple-quadrupole mass spectrometers (QQQ) offer high sensitivity and selectivity, accuracy, and precision by allowing multiple reaction monitoring (MRM) or selected reaction monitoring (SRM). Moreover, quantifying complex samples by reducing the sample matrix interference, thus QQQ excels in targeted quantitative analysis of PGRs (Almeida Trapp et al. 2014). Gas chromatography (GC) and liquid chromatography (LC) separate the compounds with different polarities, molecular weights, and electronic charges. Chromatography and mass spectrometry, such as GC-MS and LC-MS, are used to quantify PGRS. GC-MS is applied to quantify acidic PGRs such as GA, ABA, and IAA; nevertheless, this technique is confined to only volatile and thermostable PGRs (Chiwocha et al. 2003). LC-MS/MS, a combination of HPLC (high-performance liquid chromatography) and tandem mass spectrometry, is the preferred method for analyzing PGRs due to its superior sensitivity and specificity. HPLC is used for separating and identifying individual components within a sample. Subsequently, tandem mass spectrometry (MS/MS) provides comprehensive details about the molecular structure and concentration of PGRs. To date, ABA, CKs, and SLs have been quantified through HPLC-MS (Pan et al. 2010; Xin et al. 2020). Chromatographic methods and MS systems efficiently analyze multi-class plant hormones within the plant matrix (excluding plant hormones). GC or LC with distinct MS detectors creates a swift, sensitive, and efficient method for quantitative measurement of plant hormones (Li et al. 2017; Rawlinson et al. 2015). PGR biosynthesis, signal transduction, and transport across cells and organs are incomprehensible. The transporter activity of PGRs, fundamental characterization of new phytohormone transporter substrates, or the extensive screening of novel transporters targeting a specific phytohormone is studied with the assistance of tobacco syringe agroinfiltration assay (Zhao et al. 2021). This chapter discusses HPLC-MS, GC, LC, QQQ, tobacco syringe agroinfiltration techniques, and MS for quantifying PGRs. This chapter also highlights the updated protocols for detecting and quantifying PGRs in plant tissues.

10.2 High-Performance Liquid Chromatography Coupled with Tandem Mass Spectrometry (HPLC-MS/MS)

HPLC separates mixtures in complex samples at a low concentration level. Nevertheless, the structural details of substances separated through HPLC have little or no structural information. The shortcomings of HPLC can be overcome by the coupling of HPLC with MS (Hollender et al. 2017). Tandem mass spectrometry, also known as MS, is an advanced analytical technique that provides detailed information about the molecular structure of a sample. It is recognized as the third most potent spectral technique for unveiling the structure of unknown organic compounds. Nonetheless, it holds specific advantages over X-ray diffraction and NMR spectroscopy, particularly in bioanalytical assays. It boasts exceptional sensitivity and minimal sample usage, and the necessary information can be efficiently acquired by carefully choosing the ionization technique and mass analyzer (Holčapek et al. 2008). Due to these considerations, HPLC-MS/MS is significant in examining plant PGRs. This sophisticated technique allows for a comprehensive and accurate assessment of ten PGRs, namely, 6-furfurylaminopurine (6-KT), paclobutrazol (PBZ), indole-2-acetic acid (IAA), and indole-3-butyric acid (IBA) that are examined in positive ion mode. Conversely, the negative ion mode is employed for analyzing gibberellin A3 (GA3), 2,4-dichlorophenoxy acetic acid (2,4-D), 4-chlorophenoxyacetic acid (4-CPA), forchlorfenuron (FCF), thidiazuron (TDZ), and 6-benzyl adenine (6-BA) within plant samples, contributing to a more thorough understanding of plant hormonal dynamics and responses (Cai et al. 2023). The detailed outline for the quantification of PGRs is discussed below.

10.2.1 Standard Solution Preparation

- Dissolve 0.0250 g of sample in acetonitrile (MeCN) in a 25 mL volumetric flask for the stock solution of standard plant growth regulator 1.0 mg/mL and stored at 4 °C.
- PGR compound stock solutions at 10.0 μg/mL were prepared by combining 1 mL each of 1.0 mg/mL standard stock solutions and diluting to 100 mL with

MeCN. Further, 100.0 ng/mL mixed stock solutions were obtained by diluting 1 mL of the 10.0 μg/mL stock in 100 mL MeCN. All solutions were stored at 4 °C.

10.2.2 Sample Preparation

10.2.2.1 Sample Extraction

- Combine 5.0 g of plant samples, 20 μL formic acid, and 20 mL MeCN in a 50 mL centrifuge tube.
- Cap the tube and vortex for 1 min, and then treat in an ultrasonic water bath for 5 min at 25 °C.
- Centrifuge the resulting suspension at 9400 g for 5 min, and decant the top layer to another 50 mL centrifuge tube.
- Add 2.0 g NaCl to the solution, vortex for 2 min, and centrifuge at 9400 g for 2 min.
- Transfer 2 mL of the upper MeCN layer to a 15 mL centrifuge tube, and concentrate to dryness with nitrogen flow.
- Reconstitute the residue with 0.2 mL MeOH and vortex mix.
- Add 40 mM aqueous HCl (3 mL), transfer the mixture to another 15 mL centrifuge tube, cap, and centrifuge at 14,000 g for 5 min.
- Load the top layer on an activated MCS SPE column for further purification.

10.2.2.2 Sample Purification

- Activate the MCS column with sequential addition of MeOH (5 mL), water (5 mL), and 40 mM aqueous HCl (5 mL).
- Load the supernatant obtained from sample extraction onto the activated MCS column and wash it with 5 mL water.
- Vacuum-dry the column for 5 min.
- Elute target analytes with MeOH (5 mL) and collect as elution I.
- Add 5% ammoniated MeOH (5 mL) to the column and collect as elution II.
- Concentrate both fractions by flowing nitrogen and reconstitute each in MeOH (0.1 mL) ultrasonically.
- Add 0.9 mL of 10% MeOH in water to each fraction.
- Individually filter both fractions through a 0.22 μm filter membrane (for organic reagent) before LC-MS/MS injection.
- Elution I contains GA3, 2,4-D, 4-CPA, IAA, and IBA.
- Elution II contains FCF, TDZ, 6-BA, PBZ, and 6-KT.

10.2.3 LC-MS/MS Conditions

- Inject 2 μL of sample.
- HPLC-MS was carried out on the C18 column at 40 °C.
- Mass spectrometric detection was performed using ESI ionization.

- ESI ionization in positive ion mode for 6-KT, PBZ, IAA, and IBA and negative ion mode for GA3, 2,4-D, 4-CPA, FCF, TDZ, and 6-BA.
- Tandem mass spectrometer operated at unit resolution in multi-reaction monitoring mode (MRM).
- Mass spectrometric conditions optimized through continuous infusion of standard solution at 10 μL/min.
- Quantification based on peak area ratio.
- Data processing is conducted using the software.
- Quantitative and qualitative ions, cone voltage, and collision energy were studied.

10.2.4 Standard Curve Preparation

- Standard FCF, TDZ, 6-BA, PBZ, and 6-KT solutions at concentrations of 1, 2, 5, 10, 20, and 50 ng/mL were prepared.
- They were prepared by adding 0.01 mL, 0.02 mL, 0.05 mL, 0.1 mL, 0.2 mL, and 0.5 mL of 100.0 ng/mL PGR mixed stock solutions and 10% MeOH/water solution to 1 mL.
- Constructed a standard curve through LC-MS/MS analysis based on peak areas and concentrations of the standards.
- Data fitting was accomplished using a linear regression method.
- For the analysis of 5 PGRs (GA3, 2, 4-D, 4-CPA, IAA, and IBA) with blank matrix quantification.
- The internal standard will be added through the extraction solution.
- Constructed standard curves based on LC-MS/MS analysis results.

10.2.5 Data Analysis

- Analyze the data using specialized software to identify and quantify the plant growth regulators.
- Consider factors such as retention times and mass spectra for accurate identification.

10.2.6 Quality Control

- Implement quality control measures to ensure the accuracy and precision of the analysis.
- Monitor instrument performance using appropriate standards.

10.2.7 Reporting

- Present the quantified results, including concentrations of plant growth regulators in the analyzed samples.
- It is crucial to emphasize that the procedural specifics may vary depending on the nature of the samples and the targeted plant growth regulators. Standard operating procedures and method validation are crucial for obtaining reliable and reproducible results.

10.3 Triple-Quadrupole Mass Spectrometers

The triple-quadrupole mass spectrometer, or QQQ or QTRAP, was developed by Christie G. Enke and Richard A. Yost in the late 1970s, revolutionizing analytical chemistry. This type of tandem mass spectrometer is renowned for its remarkable sensitivity and selectivity (Yost 2022). The instrument comprises three quadrupole mass analyzers arranged sequentially. In its fundamental configuration, two linear quadrupole analyzers are linked by a third quadrupole and scanned similarly to a single-quad instrument. This scanning process generates an MS/MS spectrum, allowing for the confirmation of analyte identities. Alternatively, the third quadrupole can be configured to monitor specific ions exclusively, significantly enhancing the analyzer's sensitivity and specificity and being particularly beneficial for quantitative analysis (Yost 2022).

PGRs' classification is based on chemical structure or their role in different plant activities; although these molecules are found in low concentration and play a significant role in metabolism, their quantification is not an easy task; thus, to quantify PGRs, a triple-quadrupole mass spectrometers' technique is used due to their high sensitivity and selectivity. Nevertheless, certain limitations hinder the technique from reaching its full potential, such as ineffective extraction methods because of low amounts, generally at ng g − 1 or even pg g − 1 level with a diverse range of concentrations in plants. Another factor impacting the analysis of plant hormones is the matrix effect (Du et al. 2012). The removal of interferences is crucial to obtaining precise information about the metabolism and functions of plant hormones. Apart from these characteristics, plant hormones exhibit instability and extreme sensitivity to environmental changes, including temperature, humidity, and light variations, owing to their chemical heterogeneity (Forcat et al. 2008). Advancements in sample pretreatment and detection methods are necessary to tackle these challenges. These methods should exhibit high selectivity, sensitivity, efficient enrichment, ease of handling, and an expanded dynamic range of detections. In the initial phases of plant hormone research, nonchromatographic analytical methods like bioassays and immunoassays were extensively employed. However, compared to MS-based techniques, these methods exhibit lower sensitivity and specificity and lack the capability to detect multiple classes of plant hormones simultaneously. Consequently, they have been replaced by more advanced and contemporary approaches (Almeida Trapp et al. 2014). Modern approaches have superseded these methods, with QQQ detection coupled with chromatographic techniques emerging as the commonly utilized analytical method for simultaneously analyzing multi-class plant hormones. The integration of QQQ (mass analyzer) with mainly GC or LC has found widespread application in the analysis of PGRs, showcasing their capability for precise quantitative analysis and efficient separation (Cao et al. 2016). A systematic protocol involves quantifying plant growth hormones using triple-quadrupole mass spectrometers with chromatography. Here is a detailed outline of the process:

10.3.1 Sample Pretreatment

This is the first step, which includes homogenization, extraction, and purification:

(a) A fine powder was obtained by grinding 200 mg of fresh weight using a mortar and pestle in liquid nitrogen, followed by an overnight extraction at 4 °C in 80% MeOH. The resulting supernatant was gathered, and the residue underwent a second extraction with 80% MeOH for 60 min. Purification is done through the microextraction method. It is the most advanced method over the classical method because of its high extraction efficiency, device miniaturization, and enrichment capacity. There are many extraction methods, but mainly liquid-liquid extraction (LLE) and liquid phase microextraction (LPME) are used, and it is primarily for acidic plant growth regulators (IAA, SA, ABA, Gas, and JA).

(b) A binary solid-phase extraction using commercial polymer anion exchange resin (PAX) and polymer cation exchange resin (PCX) was used to purify acidic and alkaline phytohormones from crude extracts. A 4.0 mL portion of the crude extract was applied to a PAX cartridge using gravity flow. Subsequently, it was washed with 2 mL of water and 2 mL of MeOH. The PAX cartridge underwent two elution steps using a 2 mL MeOH solution containing 1% formic acid. The resulting eluates were collected and concentrated at 40 °C under a gentle nitrogen flow. The concentrated material was redissolved in 0.5 mL of 20% MeOH and filtered through a 0.22 μm polytetrafluoroethylene membrane. The final sample solution obtained was injected into an LC-MS/MS system to analyze PGRs and their metabolites comprehensively. An additional 4.0 mL portion of the crude extract, with the addition of 40 μL formic acid, was loaded onto a PCX cartridge. This loaded cartridge underwent washing with 2 mL of 0.1% formic acid and 2 mL of MeOH. Subsequently, it was eluted twice with 2 mL of MeOH containing 0.5% NH4OH. Following the elution, the subsequent steps (concentration, redissolution, and filtration) remained consistent with the previously described process. The resulting sample solution was then injected into the LC-MS/MS system to analyze alkaline PGRs and their metabolites comprehensively (Cao et al. 2016).

10.3.2 Chromatographic Separation

(a) The HPLC system was used, the samples were kept at 10 °C within the autosampler tray, and each injection involved 5 μL.
(b) Phytohormones and their metabolites underwent separation on a ZORBAX Extend-C18 column (100 × 2.1 mm, 1.8 μm; Agilent, Shanghai, China). The mobile phase consisted of 5 mM ammonium formate and MeOH. The gradient elution program was configured as follows: 0–30 min, 10–45% MeOH; 30–35 min, 45–95% MeOH; and 35.1 min, 10% MeOH. The optimized flow rate was set at 150 μL min–1, and the column temperature was maintained at 30 °C.

10.3.3 Mass Spectrometry Analysis

(a) Quantification and confirmation were performed on QQQ equipped with an electrospray ionization source. The tuning parameters were fine-tuned for each compound by infusing standard solutions (1–10 μg mL–1) and adjusting them based on their respective intensities. The mobile phase consisted of a mixture of MeOH and 5 mM ammonium formate in a 50:50 ratio (v/v), with a flow rate set at 150 μL min–1.
(b) The electrospray ion source was set at its operating condition, and all the analytes were tracked utilizing the selected reaction monitoring (SRM) mode.

10.3.4 Quantification

(a) Instrument control, data acquisition, and processing were executed through Xcalibur and software.

This protocol ensures a systematic and accurate quantification of plant growth hormones using triple-quadrupole mass spectrometers with chromatography, providing valuable insights into the hormonal regulation of plant growth and development.

10.4 Gas Chromatography and Liquid Chromatography

Chromatography is a phase difference-based separation technique. It involves the sample interacting with two immiscible phases: mobile and stationary. These could be a solid and a liquid, a gas and a liquid, or a liquid and another liquid. The mobile phase moves the sample through a stationary or immobile phase. Based on the physical state of the mobile phase, chromatography is divided into categories: mobile phase gas, GC, and mobile phase, liquid, LC (Coskun 2016). Chromatography is the primary tool for separating and quantifying compounds within a sample. GC and LC are the two most commonly used techniques in quantifying plant growth regulators. These techniques enable identifying and measuring the concentration of plant hormones and other regulators present in a given sample. Most plant hormones are analyzed by GC or LC combined with mass spectrophotometry. Still, the choice depends upon the availability of the instrument and the limitation of gas chromatography, for example, the chemical nature of the target compound, heat liability, high polarity, and difficulty in derivatization (Kumari et al. 2022).

In gas chromatography, the stationary phase could be a solid adsorbent or liquid. Gas-liquid chromatography is usually shortened to gas chromatography. It is confined to volatile analytes in the gas phase and should be thermally stable. The only type of chromatography does not utilize the mobile phase to interact with the analyte, that is, only the stationary phase interacts with the analyte (Falaki 2019; Stauffer et al. 2008). The analyte in GC makes the boundary between a gaseous mobile phase and a liquid stationary phase on the overside of a nonreactive solid packing or walls of the capillary tube. Mobile phase gas and carrier gas are of different types, such as helium, nitrogen, hydrogen, argon, and so on, but they must be dry, free of oxygen, and chemically inert (Falaki 2019).

Liquid chromatography is an exceptionally potent and extensively employed method for separating and analyzing multicomponent mixtures encompassing molecular weights ranging from 50 to several million. This includes simple molecules, macromolecules, ions, and even viruses. The separation process in LC occurs under gentle conditions, making it suitable for analyzing unstable compounds (Yashin and Yashin 2012). The chromatographic process involves the following steps:

10.4.1 Sample Preparation

The quantification process begins with extracting plant growth regulators from the plant tissues or samples. This extraction step is pivotal in ensuring the accurate representation of the regulators present. Once removed, the sample is introduced into the chromatograph for separation and analysis. Samples for gas chromatography can be prepared by extraction, cleanup, derivatization, transfer to vapor phase, and concentration (Su et al. 2017). Meanwhile, for liquid chromatography, solid-phase extraction, solid-phase microextraction, liquid-phase microextraction, and related microextraction techniques have been used (Kataoka 2023).

10.4.2 Loading the Sample to the Chromatograph

In GC, the extracted plant growth regulator sample was vaporized to load the sample in the gas chromatograph. Inside the instrument, the sample travels through a long, thin column, where each plant growth regulator interacts with a stationary phase, causing them to separate based on their different properties. In the case of an LC sample containing plant growth regulators injected into the chromatograph, it passes through a column packed with a stationary phase. Within this column, the plant growth regulators separate based on their interaction with the stationary and mobile phases, which are liquid rather than gas. Finally, the separated regulators are detected by a sensitive detector, which produces signals used for quantification.

Quantifying plant growth regulators is necessary to justify their unclear role, metabolism, transport, and molecular mechanism in plants. Various techniques include biological (bioassay and immunoassays), physical (chromatographic), and chemical analytical methods. They all are at their best level but must be more effective for concurrently analyzing plant hormones belonging to multiple classes. Mass spectrophotometry is introduced to satisfy the needs, and the high-throughput analysis is coupled with chromatography, which has already been discussed in the mass spectrometry section.

10.5 Tobacco Syringe Agroinfiltration Assay

Tobacco syringe agroinfiltration assay or *Agrobacterium*-mediated transient expression is the most widely used technique known for its flexibility and lack of necessity for specialized equipment. As the name suggests, in *Nicotiana tabacum* or tobacco leaves, the suspension of *Agrobacterium tumefaciens* containing the gene(s) of interest is introduced using a needle-free syringe (Yang et al. 2000). This assay is detailed in the following steps:

10.5.1 Plasmid Construction

Constructing a plasmid for the tobacco syringe agroinfiltration assay involves selecting a plant expression vector, incorporating a plant-specific promoter and terminator, and inserting the gene of interest (GOI). Optionally, include tags and selectable markers and verify the construct by sequencing. At last, transform the plasmid into *Agrobacterium tumefaciens* and confirm via polymerase chain reaction, restriction digestion, plasmid sequencing, and colony PCR.

10.5.2 *Agrobacterium Tumefaciens* Culture

- Agrobacterium colony with plasmid cultured for 48 h at 28 °C in LC medium (50 µg/mL kanamycin, 50 µg/mL rifampin).
- The culture was centrifuged at OD600 = 1.5, and the supernatant was removed.
- Pellet reconstituted in infiltration medium (10 mM MgC12, 10 mM MES pH 5.6, 150 µM acetosyringone) to OD600 = 0.5.
- Two-hour incubation at room temperature.
- Suspension was transferred to tobacco leaves for further experimentation.

10.5.3 *Nicotiana Benthamiana* Plant

- Select tobacco plants aged 4–5 weeks in good health and disease-free.
- Incise leaves on the lower side to expose the epidermis.
- Employ a needle-less 1-mL plastic syringe for the infiltration process.
- Prepare a bacterial suspension; administer 100 μL per spot.
- Introduce the bacterial suspension into the intercellular spaces of fully expanded leaves for optimal results.

10.5.4 Post Infiltration

After infiltration, allow the treated plants to incubate under conditions of 22–18 °C with a light/dark cycle of 16 h light and 8 h dark for a period ranging from one to four days. Subsequently, conduct microscopic analysis and immunoblotting.

In recent times, several studies have been reported. Zhao et al. (2021) investigated the transporter functions of three established hormone transporters: AtABCG25 for abscisic acid, AtABCG16 for jasmonic acid, and AtPUP14 for cytokinin hormones. The research findings indicated that an unidentified maize transporter, ZmABCG43, preferred cytokinins as substrates. The transporter's efficacy in handling cytokinins was confirmed by complementing the gene in tobacco's cytokinin long-distance transport mutant (atabcg14). As a result, the TSAL method showcases promise for the comprehensive analysis of new phytohormone transporters or the systematic screening of transporters specialized in a specific phytohormone on a larger scale. Rico et al. (2010) observed reduced abscisic acid and enhanced salicylic acid levels after infiltrating the challenged leaf of *Agrobacterium tumefaciens* with *Pseudomonas syringae* in the tobacco plant. Plant promoters and transcription factors have been studied through in vivo assays with the help of this technique; after treatment of salicylic acid -87 35S promoter, which carries as-1 element, it increases 8 times whereas tobacco mosaic virus infection increases 13 times. However, no inducible activity lacking an as-1 element in the −49 35S promoter was noted.

10.6 Mass Spectrometry

Mass spectrometry is an analytical technique to ascertain the mass-to-charge ratio (m/z) of one or multiple molecules within a sample. This method is valuable for determining the precise molecular weight of sample components and identifying isotopes, thereby indirectly confirming the identity of compounds. In this process, ions are generated within the ionization source by inducing a change in charge, such as electron ejection, protonation, or deprotonation. Once ions are formed in the gaseous state, they are electrically directed into a mass analyzer, where they undergo separation based on mass before reaching the detection stage. As a result, a mass spectrometer consists of three essential components, the ionization source, the mass analyzer, and the detector, as outlined by Urban (2016). The operation of the mass spectrometer involves the following stages:

1. Generating the sample in an ionized form in the gas phase: A sample molecule needs to undergo the ionization process, transforming it into charged particles in the gas phase. Various techniques are employed to convert the sample into ions. These methods encompass (1) electron ionization, (2) chemical ionization, (3) desorption ionization, and (4) electrospray ionization.
2. The ions are accelerated in an electric field, each emerging at a velocity proportional to its mass-to-charge ratio (m/z).
3. The ions traverse through a region without an electric field.
4. The detection involves capturing the arrival times of the ions, where the time-of-flight serves as an indicator of the mass-to-charge ratio of the ions.

10.6.1 Mass Analysis

An electric field propels the ion beam forward following ionization, leading it into the mass analyzer. In this region, ions undergo sorting and separation based on their mass-to-charge (m/z) ratio. Various mass analyzers are available, including the time-of-flight (TOF), quadrupole mass analyzer, magnetic/electrostatic double sector mass analyzer, quadrupole ion trap mass analyzers, and ion cyclotron resonance. Each type has specific trade-offs related to operational speed, separation resolution, and other technical criteria. It is noteworthy that the mass analyzer typically collaborates with the ion detection system, as discussed in studies by Luo et al. (2018), Lu et al. (2015), and Okazaki et al. (2013).

The presence of plant hormones in plants is usually in trace amounts, posing challenges for qualitative and quantitative analysis. Quantitative analysis of PGRs by mass spectrometry-based methods has emerged in recent years, revolutionizing the quantitative analysis of plant hormones by enhancing speed, convenience, and sensitivity. Here is a comprehensive outline of the process:

10.6.1.1 Sample Preparation

The quantification process begins with extracting and purifying plant growth regulators from the plant tissues or samples. Once extracted with the ideal solvent like methanol, methanol-water mixture, propanol-water, methylene chloride, methylene chloride-propanol, diethyl ether, dichloromethane, and neutral or acid pH buffer, the sample is purified or separated through chromatography either gas chromatography, liquid chromatography, thin layer chromatography (TLC), solvent-solvent partition, solid-phase extraction, and ion exchange, and column chromatography depends upon the availability of the instrument or depends on nature of the analytes. Recently, GC or LC has mainly been used in purification because the GC provides a separation step and concentrates the analyte into a defined peak. At the same time, the LC avoids the time-consuming chemical derivatization step (Markey 1981).

10.6.1.2 Instrument Calibration

Calibrate the mass spectrometer using known standards of the target PGRs to create a calibration curve.

10.6.1.3 Ionization

Introduce the prepared sample into the mass spectrometer. Ionize PGRs using electrospray ionization (ESI) or atmospheric pressure chemical ionization (APCI). ESI is widely favored among researchers due to its balanced ionization efficiency across different chemical compound classes. As an ionization method, ESI transforms ions present in a solution into the gas phase for MS analysis. The initial phase of ESI involves the electrophoretic migration of both anions and cations within the solution, occurring under the influence of an electric field, generating ions. It is considered a soft ionization method, making it suitable for the analysis of polar compounds. Unlike other methods, ESI does not necessitate the rapid vaporization of solvents, allowing ions to be generated directly in the solution. This feature is particularly beneficial for the analysis of polar plant hormones. The characteristics of ESI make it well-suited for the characterization, identification, and quantification of plant hormones in plant tissues (Kumar 2017; Cui et al. 2015).

10.6.1.4 Mass Analysis

(a) Direct ions into the mass analyzer for sorting based on their mass-to-charge ratio (m/z).
(b) Choose a specific mass analyzer, such as triple-quadrupole (QQQ), time-of-flight (TOF), or quadrupole time-of-flight (QTOF).
(c) Utilize the selected mass analyzer to segregate ions according to their mass before identification through detection.
(d) Employ multiple QQQ mass spectrometers for enhanced qualitative capabilities.
(e) Maintain the original quantitative capabilities of the quadrupole mass spectrometer.
(f) Access various modes in QQQ mass spectrometry, including full scan mode (scan mode), selected ion monitor mode (SIM mode), multiple reaction monitor mode (MRM mode), and selected reaction monitor mode (SRM mode) (Grebe and Singh 2016).

Note that the use of multiple QqQ mass spectrometers preserves the quantitative capabilities of the original quadrupole mass spectrometer.

10.6.1.5 Quantification

(a) Measure ion abundances corresponding to the target PGRs.
(b) Quantify PGR concentrations in the sample utilizing the earlier-established calibration curve.

10.6.1.6 Data Analysis

(a) Analyze mass spectrometry data to determine PGR concentrations in the sample.

(b) Report quantified PGR concentrations in the original plant samples.

10.7 Conclusion

Plant growth regulators are essential for controlling plant growth, development, and responses to environmental stimuli. Various hormones like auxin, cytokinin, abscisic acid, gibberellin, brassinosteroids, salicylic acid, jasmonic acid, ethylene, and strigolactone play distinct roles, often interacting synergistically or antagonistically. Understanding their regulation involves insights into hormone structure, distribution at different levels, and genes governing their processes. Identifying these hormones is challenging due to their minute quantities. Modern techniques like MS offer high sensitivity and selectivity. Triple-quadrupole MS excels in targeted quantitative analysis, overcoming challenges like matrix interference. Chromatography coupled with MS, such as HPLC-MS/MS, is widely used for PGR quantification. While GC-MS is limited to volatile PGRs, LC-MS/MS provides high sensitivity and specificity. This chapter details various techniques, including HPLC-MS, GC, LC, QQQ, and tobacco syringe agroinfiltration, for PGR quantification. It also outlines updated protocols for detecting and quantifying PGRs in plant tissues.

Acknowledgments Havza Imtiaz gratefully acknowledges the financial assistance rendered by the University Grant Commission, New Delhi, India, as a non-net fellowship.

References

Almeida Trapp M, De Souza GD, Rodrigues-Filho E, Boland W, Mithöfer A (2014) Validated method for phytohormone quantification in plants. Front Plant Sci 5:417. https://doi.org/10.3389/fpls.2014.00417

Cai C, Yao F, Li C, Xiang Y, Wu P, Hu Z et al (2023) Determination of ten Plant growth regulators in bean sprouts by mixed solid phase extraction columns separation and LC-MS/MS. PRO 11(9):2586. https://doi.org/10.3390/pr11092586

Cao ZY, Sun LH, Mou RX, Zhang LP, Lin XY, Zhu ZW, Chen MX (2016) Profiling of phytohormones and their major metabolites in rice using binary solid-phase extraction and liquid chromatography-triple quadrupole mass spectrometry. J Chromatogr A 1451:67–74. https://doi.org/10.1016/j.chroma.2016.05.011

Chiwocha SD, Abrams SR, Ambrose SJ, Cutler AJ, Loewen M, Ross AR, Kermode AR (2003) A method for profiling classes of plant hormones and their metabolites using liquid chromatography-electrospray ionization tandem mass spectrometry: an analysis of hormone regulation of thermodormancy of lettuce (Lactuca sativa L.) seeds. Plant J 35(3):405–417

Coskun O (2016) Separation techniques: Chromatography. North Clin Istanb 3(2):156–160. https://doi.org/10.14744/nci.2016.32757

Cui K, Lin Y, Zhou X, Li S, Liu H, Zeng F et al (2015) Comparison of sample pretreatment methods for the determination of multiple phytohormones in plant samples by liquid chromatography–electrospray ionization-tandem mass spectrometry. Microchem J 121:25–31. https://doi.org/10.1016/j.microc.2015.02.004

Du F, Ruan G, Liu H (2012) Analytical methods for tracing plant hormones. Anal Bioanal Chem 403:55–74. https://doi.org/10.1007/s00216-011-5623-x

EL Sabagh A, Islam MS, Hossain A, Iqbal MA, Mubeen M, Waleed M et al (2022) Phytohormones as growth regulators during abiotic stress tolerance in plants. Front Agron 4:765068. https://doi.org/10.3389/fagro.2022.765068

Falaki F (2019) Sample preparation techniques for gas chromatography. IntechOpen. https://doi.org/10.5772/intechopen.84259

Forcat S, Bennett MH, Mansfield JW, Grant MR (2008) A rapid and robust method for simultaneously measuring changes in the phytohormones ABA, JA and SA in plants following biotic and abiotic stress. Plant Methods 4:1–8. https://doi.org/10.1186/1746-4811-4-16

Grebe SK, Singh RJ (2016) Clinical peptide and protein quantification by mass spectrometry (MS). TrAC Trends Anal Chem 84:131–143. https://doi.org/10.1016/j.trac.2016.01.026

Holčapek M, Kolářová L, Nobilis M (2008) High-performance liquid chromatography–tandem mass spectrometry in the identification and determination of phase I and phase II drug metabolites. Anal Bioanal Chem 391:59–78. https://doi.org/10.1007/s00216-008-1962-7

Hollender J, Schymanski EL, Singer HP, Ferguson PL (2017) Nontarget screening with high resolution mass spectrometry in the environment: ready to go? https://doi.org/10.1021/acs.est.7b02184

Kataoka H (2023) Sample preparation for liquid chromatography. In: Liquid chromatography. Elsevier, pp 1–48

Kumar BR (2017) Application of HPLC and ESI-MS techniques in the analysis of phenolic acids and flavonoids from green leafy vegetables (GLVs). J Pharm Anal 7(6):349–364. https://doi.org/10.1016/j.jpha.2017.06.005

Kumari VC, Patil SM, Ramu R, Shirahatti PS, Kumar N, Sowmya BP et al (2022) Chromatographic techniques: types, principles, and applications. In: Analytical techniques in biosciences. Academic Press, pp 73–101

Li D, Guo Z, Liu C, Li J, Xu W, Chen Y (2017) Quantification of near-attomole gibberellins in floral organs dissected from a single Arabidopsis thaliana flower. Plant J 91(3):547–557. https://doi.org/10.1111/tpj.13580

Lu Q, Zhang W, Gao J, Lu M, Zhang L, Li J (2015) Simultaneous determination of plant hormones in peach based on dispersive liquid–liquid microextraction coupled with liquid chromatography–ion trap mass spectrometry. J Chromatogr B 992:8–13. https://doi.org/10.1016/j.jchromb.2015.04.014

Luo XT, Cai BD, Yu L, Ding J, Feng YQ (2018) Sensitive determination of brassinosteroids by solid phase boronate affinity labeling coupled with liquid chromatography-tandem mass spectrometry. J Chromatogr A 1546:10–17. https://doi.org/10.1016/j.chroma.2018.02.058

Markey SP (1981) Quantitative mass spectrometry. Biomed Mass Spectrom 8(9):426–430. https://doi.org/10.1002/bms.1200080913

Okazaki Y, Kamide Y, Hirai MY, Saito K (2013) Plant lipidomics based on hydrophilic interaction chromatography coupled to ion trap time-of-flight mass spectrometry. Metabolomics 9:121–131. https://doi.org/10.1007/s11306-011-0318-z

Oklestkova J, Tarkowská D, Eyer L, Elbert T, Marek A, Smržová Z et al (2017) Immunoaffinity chromatography combined with tandem mass spectrometry: A new tool for the selective capture and analysis of brassinosteroid plant hormones. Talanta 170:432–440. https://doi.org/10.1016/j.talanta.2017.04.044

Pan X, Welti R, Wang X (2010) Quantitative analysis of major plant hormones in crude plant extracts by high-performance liquid chromatography–mass spectrometry. Nat Protoc 5(6):986–992. https://doi.org/10.1038/nprot.2010.37

Rawlinson C, Kamphuis LG, Gummer JP, Singh KB, Trengove RD (2015) A rapid method for profiling of volatile and semi-volatile phytohormones using methyl chloroformate derivatisation and GC–MS. Metabolomics 11:1922–1933. https://doi.org/10.1007/s11306-015-0837-0

Rico A, Bennett MH, Forcat S, Huang WE, Preston GM (2010) Agroinfiltration reduces ABA levels and suppresses pseudomonas syringae-elicited salicylic acid production in Nicotiana tabacum. PLoS One 5(1):e8977. https://doi.org/10.1371/journal.pone.0008977

Stauffer E, Dolan JA, Newman R (2008) Gas chromatography and gas chromatography—mass spectrometry. In: Fire Debris Analysis, pp 235–293

Su Y, Xia S, Wang R, Xiao L (2017) Phytohormonal quantification based on biological principles. In: Hormone metabolism and signaling in plants, vol 13, pp 431–470
Urban PL (2016) Quantitative mass spectrometry: an overview. Philos Trans R Soc A Math Phys Eng Sci 374(2079):20150382. https://doi.org/10.1098/rsta.2015.0382
Wang L, Zou Y, Kaw HY, Wang G, Sun H, Cai L et al (2020) Recent developments and emerging trends of mass spectrometric methods in plant hormone analysis: a review. Plant Methods 16(1):1–17. https://doi.org/10.1186/s13007-020-00595-4
Xin P, Guo Q, Li B, Cheng S, Yan J, Chu J (2020) A tailored high-efficiency sample pretreatment method for simultaneous quantification of 10 classes of known endogenous phytohormones. Plant Commun 1(3):100047. https://doi.org/10.1016/j.xplc.2020.100047
Yang Y, Li R, Qi M (2000) In vivo analysis of plant promoters and transcription factors by agroinfiltration of tobacco leaves. Plant J 22(6):543–551. https://doi.org/10.1046/j.1365-313x.2000.00760.x
Yashin YI, Yashin AY (2012) Liquid chromatography. In: Chemical Analysis of Food: Techniques and Applications, vol 3, pp 285–310
Yost RA (2022) The triple quadrupole: innovation, serendipity and persistence. J Mass Spectrom Adv Clin Lab 24:90. https://doi.org/10.1016/j.jmsacl.2022.05.001
Zhao J, Ju M, Qian J, Zhang M, Liu T, Zhang K (2021) A tobacco syringe agroinfiltration-based method for a phytohormone transporter activity assay using endogenous substrates. Front Plant Sci 12:660966. https://doi.org/10.3389/fpls.2021.660966

Implications in Phytohormone Research: From Lab to Field

11

Samia Muhammad Arif, Ghazala Mustafa, and Kejun Cheng

Abstract

Phytohormones like auxins, gibberellins, and ascorbic acid play very crucial part in the normal growth and development of plants. They are responsible for a variety of physiological functions. The most important aspect is its involvement in stress signaling and responses. With the changing climatic conditions, plants are at high risk of facing adverse environmental influences. In recent years, excessive studies conducted on the activity of phytohormones under stress in laboratories have provided the scientific world with some probable credible solutions. The transition from laboratory-based explications to real-world applications is quite challenging. This review chapter dives into a timeline of the transition from basic research, mostly done in lab settings, to actual agricultural operations. The immense potential in question, such as improved productivity of crops, resiliency, and stress mitigation measures, is contrasted with the difficulties encountered during this transition, such as scalability and ecological unpredictability. Phytohormone research and developing technologies have the potential to lead to novel and sustainable agricultural practices in the future. In the pursuit of a future with safe access to food, this review emphasizes the significance of closing the gap between laboratory research and practical implementation.

S. M. Arif
Department of Plant Sciences, Quaid-i-Azam University, Islamabad, Pakistan

G. Mustafa (✉)
Department of Plant Sciences, Quaid-i-Azam University, Islamabad, Pakistan

Chemical Biology Center, Lishui Institute of Agriculture and Forestry Sciences, Lishui, China
e-mail: mghazala@qau.edu.pk

K. Cheng (✉)
Chemical Biology Center, Lishui Institute of Agriculture and Forestry Sciences, Lishui, China

© The Author(s), under exclusive license to Springer Nature Singapore Pte Ltd. 2024

M. Faizan, S. Hayat (eds.), *Plant Growth Regulators: Resilience for Sustainable Agriculture*, https://doi.org/10.1007/978-981-97-2918-0_11

Keywords

Plant physiology · Strigolactones · Phytohormone quantification · Enzyme-linked immunosorbent assay · CRISPR/Cas9

11.1 Introduction

Phytohormones are organic compounds which affect physiological operations within plants and are essential for supporting development, growth, and metabolic activities as well as acting as buffers against environmental stressors (Arif et al. 2020). These hormones improve plants' ability for adaptation to harsh settings by controlling important physiological processes in both stress and normal situations (Llanes et al. 2018). Plant activities and functions are significantly influenced by the kind and concentration of phytohormones present in the plant (Nadeem et al. 2016). According to Fenn and Giovannoni (2021), they are essential to the control of the development of fruit and its maturation. Furthermore, several phases of plant growth and development, such as the extension of cells, cell division, tissue differentiation, and apical dominance, are influenced by phytohormones.

Since plants are sessile beings, they display plasticity, which is a complicated process incorporating coordinated interactions among several signaling pathways, phytohormones' spatiotemporal inclusion, and environmental cues (Lymperopoulos et al. 2018). Through the regulation of gene expression, these signaling molecules govern the growth and development of plants and regulate cellular processes like elongation, division of cells, differentiation, organ formation, structure formation, and reproduction, adapting the plant to biotic and abiotic stressors (Seif El-Yazal et al. 2015).

Plants employ the regulation of molecular processes, which involves the interaction of these signaling molecules to activate defense responses if faced with stress conditions (Javid et al. 2011). When under stress, phytohormones such as auxins, cytokinins, gibberellins, salicylic acid, jasmonic acid, ethylene (ET), abscisic acid, and brassinosteroids react by acting in antagonistic and synergistic ways, a phenomenon known as signaling cross talk. These hormones work in unison to respond to environmental and developmental stimuli, enabling plants to mount complex and dynamic defensive mechanisms (Checker et al. 2018). Plants adjust to changes in their environment and maintain homeostasis by precisely combining the correct proportion of phytohormones.

Phytohormones come in several forms. Among the classic phytohormones identified by the middle of the twentieth century are auxins, cytokinins, gibberellins, abscisic acid, and ethylene (Seif El-Yazal et al. 2015). Other substances, such as a class of steroidal plant hormones, have also been identified as significant phytohormones in more recent years. Among these include salicylates, jasmonates, strigolactones, and brassinosteroids (Iqbal et al. 2014). Plant physiological, biochemical, and morphological processes are coordinated by the interaction of these phytohormones, which create source/sink transitions and regulate nutrient allocation (Fahad et al. 2015). Their interplay emphasizes how intricately regulated they are and is essential for coordinating tissue expansion in response to environmental changes.

11.2 Role in Regulating Plant Physiology

11.2.1 Auxins: Conductors of Plant Development

According to Enders and Strader (2015), the term "auxin" comes from the Greek word "auxein," which means "to grow." Went (1928) initially extracted it as indol-3-acetic acid (IAA) while researching the tropic activity of *Avena sativa* coleoptiles. According to Sauer et al. (2013), auxins are a class of low molecular weight organic acids that possess an aromatic ring and a carboxylic acid side chain around 0.55 Å of each other. It enables the molecules to have biological activity. Auxin is a crucial regulator of the development and growth of plants. It controls cell division, extension and distinction, the development of embryos, root and stem tropisms, apical dominance, and transition to flowering. The primary auxin found abundantly in plants is indole-3-acetic acid (IAA). With the exception to IAA, indole-3-butyric acid (IBA), 4-chloroindole-3-acetic acid (4-Cl-IAA), and phenylacetic acid (PAA) are also found in both conjugated and free forms in plants (Ma et al. 2018). Despite having an identical structure and role in the growth and development of plants, indole-3-butyric acid (IBA) is classified as an IAA progenitor since it can only trigger auxin sensing and response upon conversion to IAA (Damodaran and Strader 2019). Additionally, a variety of synthetic compounds known as "synthetic auxin analogues" have been developed with the nature and function of endogenous auxins in mind. Examples of these compounds include 1-naphthaleneacetic acid (NAA), 2-(2,4-dichlorophenoxy) propionic acid (2,4-DP), 2,4,5-trichlorophenoxyacetic acid (2,4,5-T), 2,4-dichlorophenoxyacetic acid (2,4-D), dicamba, picloram, quinclorac, and so forth. These synthetic auxin analogues are extensively employed as chemical tools in both scientific and agricultural operations (Cao et al. 2019).

According to Geisler et al. (2014), levels of auxin are greater in populations of undifferentiated cells and decrease during organ formation and tissue differentiation. Polar auxin transport (PAT), which is mediated by auxin transport proteins, is responsible for achieving this variable auxin distribution. The elongation and expansion of root hairs are instances of auxin distribution through the establishment of planar polarity (Ikeda et al. 2009). The gradient in root hairs is established by intercellular auxin transport, which uses AUX1/LAX auxin influx and PIN2 auxin efflux carriers. Auxin is thereby redistributed from the root meristem, where it reached its highest level of localized biosynthesis, into the region of differentiation zone (Velasquez et al. 2016). Plants have four main families of auxin transporters: AUXIN1/LIKE-AUX1s, PIN-LIKES, ATP-binding cassette family B (ABCB), and PIN-FORMED (PIN). These families comprise proteins that contribute to auxin influx, efflux, or both through the apoplast toward the cell or through the cytosol across the ER (endoplasmic reticulum) compartment. These proteins can be found at the plasma membrane or in the ER (Balzan et al. 2014). The PIN family of auxin carriers in plants is the most explored.

Additionally, auxin is essential for leaf patterning because it converts adaxial-abaxial polarity into medio-lateral polarity, which causes the leaf to flatten. MP, an ARF activator, mediates auxin signaling, which affects the expression of WOX1 and PRS. The asymmetric shape of the initial primordium is attributed to enhanced development of the leaf margin and WOX expression in the center domain (Zhao et al. 2019). Blade extension, leaf margin setting, and cell proliferation are all aided by WOX1 and PRS (Coen and Kennaway 2018). Phyllotaxy is also influenced by auxin production, which is carried out by YUCCA (YUC) flavin monooxygenases. Strong phyllotaxy is maintained by YUC4 expression on the abaxial side (Galvan-Ampudia et al. 2018). Auxin regulates the characteristics of the cell wall by causing the wall to loosen, which promotes cell elongation. The changes and enlargement of cell walls were linked to auxin, also known as indole-3-acetic acid (IAA), almost 40 years ago. According to Majda and Robert (2018), IAA treatment stimulates genes linked to cell wall elements, which causes pectin polymerization, enhanced pectin viscosity, and XyG depolymerization. Additionally, it promotes the production of proton pumps, which causes apoplast acidification (Perrot-Rechenmann 2010). According to Arsuffi and Braybrook (2018), auxin stimulates the H+-ATPases in the plasma membrane, causing hyperpolarization, potassium channel stimulation, and the absorption of water for cell wall expansion. Additionally, auxin stimulates vesicle trafficking harboring new cell wall material and works on the cytoskeleton via RHO OF PLANTS (ROP) GUANOSINE-5′-TRIPHOSPHATASES (GTPases) (Majda and Robert 2018).

Moreover, it has been discovered that IAA modulates host physiology and suppresses resistance in a variety of ways during interactions among bacterial plant pathogens and their hosts. According to Kunkel and Harper (2018), elevated auxin levels in infected host tissue encourage pathogenesis processes such as epiphytic colonization and host defense suppression. Auxin-producing pathogens consider it a virulence factor that influences illnesses brought on by plant infections that induce tumors (Duca et al. 2014).

IAA biosynthesis entails numerous routes and is far more intricate. IAA biosynthesis generally happens via both tryptophan-dependent and tryptophan-independent routes, while the primary pathway for auxin production in plants is tryptophan-dependent, an independent pathway employing indole-3-glycerol phosphate has been as well demonstrated to function in maize and *Arabidopsis* (Sharma et al. 2015). In accordance with their primary intermediates, tryptophan-dependent pathways can be divided into four groups: tryptamine (TAM), indole-3-acetaldoxime (IAOx), indole-3-acetamide (IAM), and indol-3-pyruvic acid (IPA). Even though other routes for IAA biosynthesis have been identified (Di et al. 2016), auxin production in higher plants is primarily attributed to IPA.

Using IPA as the intermediate, the IPA route converts tryptophan in two steps to produce IAA. The second stage of the process is controlled by the YUCCA protein family. In *Arabidopsis*, 11 YUCCA variants have been identified, and the precise functions of a few YUCCAs are gradually coming to light. According to Chen et al.

(2014), the YUCCAs can be primarily classified as root- or shoot-active proteins. Because YUCCAs are not highly expressed in the cortex and epidermis under normal circumstances, auxin levels in these cells mostly depend on transport. Since auxin transport is significantly impacted by salt stress as well, auxin distribution during stress is anticipated to be influenced by epidermal biosynthesis, which could have an impact on growth responses (Korver et al. 2018).

IAM is an intermediate used by plant pathogens such as *Agrobacterium* to manufacture IAA. Tryptophan is converted to IAM by tryptophan-2-monoxygenase (iaaM), and then IAM undergoes hydrolysis into IAA by iaaH in a two-step process. Numerous plant species, such as *Arabidopsis*, maize, rice, and tobacco, have IAM (Novák et al. 2012). The IAOx pathway is unique to *Brassica* and is primarily known for its function in secondary metabolism, where it generates camalexin and indole glucosinolates (Julkowska et al. 2017). Nevertheless, in more recent times, a number of studies have suggested that this pathway may be involved in the local synthesis of IAA, particularly under stressful environments. It was demonstrated by Sugawara et al. (2009) that 40% of the 13-C atoms in IAOx that were fed to plants was integrated into IAA, indicating that IAA may be generated from IAOx.

11.2.2 Gibberellins: Controlling Plant Elevation and Blooming

Previously recognized solely as fungus-derived chemicals, gibberellins (GAs) constitute a class of more than 136 structurally similar natural plant components. Gibberellin is the term given to a broad class of diterpenoid carboxylic acids based on structural classification. They were initially discovered in the rice-causing fungus *Gibberella fujikuroi*. The illness known as "bakanae," or "foolish seedling," had been reported by the rice growers of Orient for two centuries. It results in less grain production and supraoptimal lengthening of seedlings (Hedden and Thomas 2016). GAs have an expanded role in higher plants, where they promote organ growth via boosted cell elongation and division and, in many species, activate developmental procedures like the sprouting of seeds, maturation, and stimulation of flowering (Sponsel 2016). In lower plants, like lycophytes and ferns, GAs participate in reproductive growth (Miyazaki et al. 2018).

All things considered, GAs have an impact on sex expression, flowering, extension of stems, germination, dormancy removal, flooding responses, inducement of enzymes, and fruit and leaf withering. In early stems, GA has been found to encourage elongation without affecting the number of nodes. The relaxing of the cell wall leads to increased cell elongation, which in turn promotes internode development. The root-shoot ratio is impacted by GA's promotion of leaf expansion and inhibition of root growth (Bagale et al. 2022). GA has an intricate impact on flowering; based on the species, it can have neutral, restricting, or encouraging effects (Coelho et al. 2018). Application of GA can cause some long-day plants to bolt and flower, indicating a potential role for GA in flowering induction (Leijten et al. 2018).

Prior to the identification of GAs, auxin was believed to be the only factor regulating growth in elongation. GA was later discovered to increase auxin production, which in turn promotes stem elongation. According to recent research, GA is necessary for the transfer of auxin as well (Li et al. 2015). Plant height is influenced by mutations in signaling cascades, metabolism, and GA production. GA3 or gibberellic acid is one of the widely used commercial gibberellin. It is produced on a large scale by microbiology, which makes it easier to utilize in agriculture, especially for fruit production (Singh et al. 2018). Synthetic inhibitors of GA biosynthesis, which function as dwarfing agents, also have significant applications (Sponsel 2016).

Plants can also connect with gibberellin-producing bacterial and fungal species as symbionts or pathogens. According to Lu et al. (2015) and McAdam et al. (2018), GAs in these situations seem to have no developmental function in the generating organism but instead act on the plant host to promote infection by reducing immunity or, in the scenario of nitrogen-fixing rhizobium bacteria, control the development of nodules.

Functioning as growth repressors, DELLA proteins are essential parts of GA signaling. By engaging with transcription factors like PIFs, DELLA proteins control the expression of certain genes (Cardenas 2021). Growth inhibition is relieved when GA binds to GID1 and encourages DELLA breakdown. In shoot tissue, GA transport is nonpolar, moving equally in both acropetal and basipetal planes. GAs have cellular specificity and can move via xylem and phloem. However, the physiological significance of GA transfer from leaves and roots needs more research. GAs also affect stem elongation and floral initiation at the shoot apex during long-distance transport (Hedden and Sponsel 2015).

The process of biosynthesizing the primary gibberellins (GAs) begins in plastics with the tetracyclic diterpene hydrocarbon, ent-kaurene. Membrane-associated cytochrome P450 monooxygenases then oxidize this ent-kaurene to create GA12 and GA53 on the endoplasmic reticulum. The soluble 2-oxoglutarate-dependent dioxygenases GA 20-oxidase and GA 3-oxidase subsequently oxidize these byproducts to produce the active C19-GA end products, GA4 and GA1, respectively (Hedden 2016). Ent-kaurene is generated in plants, primarily through the methylerythritol 4-phosphate (MEP) pathway, with little assistance from mevalonate (MVA) pathway as well. According to Hedden (2020), the formation of ent-kaurene occurs in juvenile plastids lessening the competition from the carotenoid and chlorophyll pathways.

Geranylgeranyl pyrophosphate (GGPP) acts as a precursor of ent-kaurene. There are 10 GGPPS genes active in *Arabidopsis*, with GGPPS11 primarily being confined to plastid. GGPP channeling is indicated by GGPPS11's interactions with enzymes that contribute to the manufacture of carotenoid pigment plastoquinone and chlorophyll. In plastids, the two-step process that transforms GGPP to ent-kaurene is catalyzed by the enzymes CPS and KS (Sankari et al. 2018). Important stages include metal-dependent fragmentation of ent-copalyl diphosphate to ent-kaurene and forward modulation of CPS activity. Particularly in angiosperms connected to plant defense, CPS and KS enzymes exhibit variety among plant species, exhibiting gene expansion and functional variation. Cytochrome P450 monooxygenases then convert ent-kaurene into relevant GAs (Hedden 2020).

11.2.3 Cytokinins: Regulators of Differentiation and Division of Cells

Cytokinins are phytohormones that have a variety of molecular configurations. Carlos Miller and Folke Skoog found the first cytokinins in the 1950s based on their stimulatory properties (Thomas 2016). These are adenine derivatives replaced with N6, which have significant impacts on plant development and mitogenesis. The cytokinins fall into three categories: the most common type are the isoprenoid cytokinins, which include Z and iP; the second category are extremely influential synthetic diphenylureas, which include CPPU (4-chlorfenuron); and the final category are the naturally generated adenine derivatives with fragrant substituents or topolins (Madzikane-Mlungwana et al. 2017).

The three forms of cytokinins are nucleotide, nucleoside, and nucleobase. The biologically active variants that interact to cytokinin receptors are thought to be the nucleobase forms (*trans*-, *cis*-, and dihydro-zeatin (Z)) and isopentenyl adenine (iP). Root-to-shoot coordination is thought to be facilitated by the movement of iP types through the phloem and Z-type cytokinins through the process of transpiration (Schaller et al. 2014).

On the molecular scale, enzymes involved in cytokinin biosynthesis (IPT), stimulation (Lonely Guy (LOG)), deterioration (CKX), reversible deactivation via the association by zeatin O-glucosyl transferases (ZOG), revival by β-glucosidases (GLUs), and perpetual N-glycosylation by UDP glycosyltransferases (UGTs) maintain cytokinin equilibrium (Jameson and Song 2016). These genes are found in multigene families, and the expression of each gene family member varies both temporally and geographically.

The biosynthesis of cytokinins begins when the enzyme isopentenyltransferase (IPT) catalyzes the incorporation of a prenyl moiety from dimethylallyl diphosphate to ATP/ADP to create N 6-isopentenyladenine (iP) ribotides. The cytochrome P450 enzymes CYP735A1/CYP735A2 hydroxylate this isoprenoid side chain of the iP ribotides produced by IPT. It converts them into *trans*-zeatin (tZ)-type cytokinins (Schaller et al. 2015). The LONELY GUY (LOG), class of cytokinin nucleoside 5′ monophosphate phosphoribohydrolases, catalyzes the synthesis of free bases. These free bases are the active forms of cytokinins (Oshchepkov et al. 2020).

Cytokinins control many physiological and developmental processes, including cell division, growth, and proliferation. They play a variety of roles in responses to stress, holding off senescence, fruit and seed formation, shoot advancement, and root suppression. Cytokinins restrict the beginning of lateral roots, influence the development of root vascular tissue, control cell division in growing leaves, and affect photomorphogenic maturation at the tissue and organ level (Mok 2019). They stimulate auxin maxima, drive organ expansion, facilitate undifferentiated cell growth, and improve cell division in the shoot apical meristem (SAM).

The significant role that cytokinins play in seed development is associated with nuclear and cell divisions. They affect the G1/S transition, activate CDKA, and influence the RBR protein, all of which are necessary for the cell cycle to proceed (Dante et al. 2014). Cytokinins have a crucial role in sink size establishment, especially during fast nuclear and cell division in endosperm. These hormones are essential for seed morphogenesis, which is associated with the change from cell division to storage processes. By delaying senescence and increasing sink activity, cytokinins help draw nutrients to the intended site. Their interaction with CWINV, a cell wall invertase, shows that cytokinins and sink activity are related (Velappan et al. 2017).

Cytokinins are important for carpel identity, carpel regeneration, and flower meristem determination in the framework of flower formation. The gene that determines carpel identity, AGAMOUS (AG), is expressed in response to B-type ARRs. The development of gynoecium is mediated by a network of genes, including SPATULA (SPT), ARR1/ARR10/ARR12, and TCS. AHP6's suppression of cytokinin signaling is essential for the proper growth of the tissues that surround the gynecium (Jameson and Song 2016). Changes in cytokinin levels impact the number of stomata via influencing the proliferation of cells in the stomatal lineage. They are also essential for root growth because they control radial extension and cell proliferation, as well as the preservation of bilateral symmetry in vascular tissues. Cytokinins suppress auxin action in the transition zone, which regulates the growth of the root meristem. Their signaling is also connected to the formation of root nodules in plants that interact with microorganisms that fix nitrogen (Wybouw and De Rybel 2019).

Three components make up the cytokinin signal transduction route, which is comparable to bacterial two-component phosphorelays: a histidine-containing phosphotransfer (AHP) protein, a distinct RR, and a "hybrid" receptor kinase that combines the recipient and histidine-kinase motifs in one molecule. Phosphate is moved from amino acid to amino acid in the following sequence: His to Asp to His to Asp, in these multistep phosphorelays (Jameson and Song 2016). With the CHASE domain pointing into the ER, cytokinin receptors are mostly found in the ER membrane of plants, indicating that the lumen of the ER is the primary location of cytokinin affinity in plants (Schaller et al. 2015).

Conjugation to glucose can lower the amount of active cytokinin. Cytokinin is hypothesized to undergo irreversible N-glycosylation, which mainly happens at purine ring locations N-7 or N-9. O-Glycosylation happens at the oxygen on the exterior chains of zeatin or dihydrozeatin and is probably reversible. Cytokinin oxidases, which are copper-dependent amine oxidase enzymes that break the N 6-side chains of tZ- and iP-type cytokinins to permanently inactivate them, are another enzyme that regulates cytokinin levels. Most plant species possess cytokinin oxidase encoded by a multigene family, each of which exhibits unique enzyme characteristics, subcellular sites, and variants in expression (Wybouw and De Rybel 2019).

11.2.4 Abscisic Acid: Stress Responses' Primary Controller

Abscisic acid (ABA) is a significant plant hormone, which is known as an important regulator of the abiotic stress response. According to Hong et al. (2013), it is essential for the synchronization of multiple signal transduction pathways that participate in the plant response to environmental conditions such as high temperatures, low temperatures, drought, and excessive salinity, all of which are known to raise endogenous ABA levels. It is sesquiterpenoid synthesized from C40 carotenoids (Marco et al. 2019). According to Lopez-Climent (2015), zeaxanthin epoxidase, 9-*cis*-epoxycarotenoid dioxygenase (NCED), and abscisic aldehyde oxidase (AAO) catalyze the three steps in the synthesis of ABA.

The metabolic route leading to the buildup of ABA starts in the chloroplasts when β-carotene is hydroxylated to zeaxanthin. This compound is subsequently transformed into violaxanthin via the xanthophyll cycle. Neoxanthin is then isomerized from 9-*trans* to 9-*cis* (neoxanthin) when neoxanthin synthase transforms violaxanthin to neoxanthin (Cardoso et al. 2020). 9-*Cis*-epoxycarotenoid dioxygenase (NCED) further catalyzes the oxidative breakdown of 9-*cis*-neoxanthin and/or 9-*cis*-violaxanthin to xanthoxin, which is the final step in the metabolic pathway that takes place in the chloroplasts. According to Sussmilch et al. (2017), NCED is the primary rate-limiting enzyme in the biosynthesis of ABA and oversees initiating the pathway's initial nonreversible step. After forming xanthoxin, it leaves the chloroplasts and enters the cytoplasm, where it is changed into abscisic aldehyde and then into ABA. In addition to ABA production being activated, plants also accumulate ABA through the hydrolysis of its glycosyl ester, or ABA-GE, which can be used to keep plant concentrations of ABA high amid drought (Han et al. 2020).

There is a huge signaling network for ABA in plants. According to Lumba et al. (2014), the core ABA signaling network consists of 5 basic leucine zipper activators of transcription referred to as ABA-responsive element binding factors (ABFs), 9 A-type protein phosphatase 2Cs (PP2Cs), 3 sucrose nonfermenting-1-related protein kinase 2 s (SNRK2s), and 13 pyrabactin resistance 1-like (PYL) receptors. Still, a network of nearly 500 interactions between 138 distinct proteins in the ABA signaling system has been discovered by modern systems biology techniques. When the ABFs phosphorylate, they attach to cis regions in target gene promoters called ABREs (ABA-responsive elements), which activate gene expression and control the transcriptional stimulation of downstream genes responsive to ABA (Cruz et al. 2014).

Being linked to stress, abscisic acid (ABA) quickly builds up in response to drought stress and is essential for plants' resistance (Hong et al. 2013). Through the inhibition of PP2C phosphatases by ABA receptors (PYR/PYL/RCAR proteins), the ABA signaling pathway activates SnRK2 protein kinases. These kinases, especially SnRK2.6, phosphorylate transcription factors such as AREB/ABFs to regulate stomatal aperture and activate genes that respond to ABA (Huang et al. 2018). Enhanced foliar ABA levels during drought trigger stomatal closure, which helps angiosperm species maximize their water use efficiency (Mega et al. 2019). Under drought stress, ABA and auxin interact to regulate root growth in a way that has a complex effect on root elongation. The "acid growth theory" suggests that basipetal auxin transport is stimulated by ABA, which may lead to the elongation of primary root cells (Xie et al. 2020).

Additionally, ABA is involved in the primary and secondary metabolic pathways. Its signaling affects genes involved in the metabolism of carbohydrates, controlling the metabolism of sucrose and encouraging the breakdown of starch to release compatible solutes to maintain membrane integrity (Cardoso et al. 2020). Under drought stress, ABA-induced stem soluble carbohydrates may support the maintenance of xylem hydraulic integrity and function as a means of energy. Additionally, according to Bandurska et al. (2017), it plays a role in the control of proline accumulation, an essential osmolyte that adjusts osmotic pressure, stabilizes structures, and functions as a signaling molecule. According to Brunetti et al. (2018), ABA has an impact on secondary metabolites, especially flavonoids, which may have functions in regulating the reactive oxygen species-signaling cascade and mitigating the metabolic changes brought on by stress.

Apart from its impact on growth, ABA also modifies the hydraulic conductivity of roots, with different results based on concentration. In the roots of maize and soybeans, micromolar ABA concentrations increase hydraulic conductivity while larger quantities decrease it. These hydraulic effects are mediated by tissue-specific aquaporin modulation, highlighting the complex function of ABA in hydraulic responses (Rosales et al. 2019). Furthermore, there is proof that ABA promotes senescence, and its participation in this process is well documented. ABA's positive regulatory effect in leaf senescence is highlighted by molecular data, such as the involvement of receptor-like kinase 1 (RPK1) (Song et al. 2016).

11.2.5 Ethylene: The Gaseous Hormone in the Plants

Ethylene is a multipurpose gaseous plant hormone that governs both growth and senescence. According to Bakshi et al. (2015), C2H4 is crucial to plant's physiology during the duration of a plant's life cycle, from seedling to mature plant. As a crucial part of the ripening process, C2H4 can cause senescence and abscission. Additionally, it has been linked to responses to both biotic and abiotic stressors. Ethylene can work in three different ways: it can connect to a receptor and perform a vital function; it can oxidize to a component that is needed and then be integrated into tissue; and it can act as a cofactor in a response (Sisler 2018).

All components of higher plants, which involve the foliage, stems, roots, flowers, fruits, tubers, and seeds, are used to create ethylene. During specific growth phases, including germination, fruit maturation, leaf abscission, and floral senescence, ethylene production is stimulated in plants (Kenea 2021). Numerous external factors, including mechanical wounds, environmental stressors, and specific molecules like auxin and other regulators, can also increase the production of ethylene. The Yang cycle, which elucidates the ethylene production route, is named for the scientist Shang Fa Yang (Wang et al. 2002). While bacteria and fungi can make and break down C2H4, plants are the primary producer of it. Plants will be significantly impacted if C2H4 concentrations in the soil gas phase surpass specific thresholds (Lang et al. 2023).

Ethylene biosynthesis begins with conversion of methionine by the enzyme Met adenosyltransferase into S-adenosyl-L-methionine (SAM), also known as AdoMet. The enzyme ACC synthase (ACS) then transforms SAM into

1-aminocyclopropane-1- carboxylic acid (ACC) (Kenea 2021). The modulation of ACS is crucial for the biosynthesis of ethylene, as its activity dictates the pace at which ethylene is produced. The last stage, which calls for oxygen, is carried out by the ACC-oxidase (ACO) enzyme, formerly called the ethylene-forming enzyme (EFE). High concentrations of auxins, particularly cytokinins and indole acetic acid (IAA), stimulate the synthesis of ACC (Hussen 2014).

A crucial part of the complex regulation of leaf senescence is played by ethylene. Elevated levels of ethylene occur during early leaf development, decline till maturity, and then rise during the onset of senescence (Thakur et al. 2016). The increase in ACC content, a marker for ethylene, particularly in senescing leaves, follows this same trend. Ethylene aids in the breakdown of chlorophyll, the apparent indications of senescence, and the concerted procedure for leaf abscission in deciduous plants (Peerzada and Iqbal 2021).

The growth and development of leaves are also significantly influenced by ethylene. The effects differ between plant species and depend on its concentration. It seems that different plant species have different correlations between the variability of the ethylene evolution rate and the rate of leaf elongation (Iqbal et al. 2017). Pea plants with increased ACC deaminase activity in the soil show ethylene-induced reductions in leaf growth, stressed ethylene levels in enclosed spaces cause lettuce leaves to become smaller, and ETHYLENE RESPONSE FACTORS (ERF) in *Arabidopsis* enhance leaf growth (Dubois et al. 2015). These results highlight how complex and context-dependent ethylene's effects on leaf formation and growth are in a variety of plant species.

By altering genetic circuits, ethylene plays a critical role in controlling the time of blooming. The intricacy of ethylene's functions is demonstrated by the fact that mutants that overproduce ethylene display early flowering while ethylene-insensitive mutants display delayed flowering (Dar et al. 2021). Certain floral tissues express genes involved in ethylene production, suggesting that these genes are involved in reproductive functions. Different plant species exhibit different expression levels of ethylene receptors, such as ETR2, during the flowering stage. It also interacts with homeotic genes to establish a genetic network that controls floral morphogenesis (Iqbal et al. 2017).

Ethylene also becomes apparent during fruit ripening as a crucial regulator of several biochemical and developmental pathways. It controls how fruits mature, especially climacteric fruits, in terms of color, texture, nutritional value, and aroma. Processes like the decrease of chlorophyll, the synthesis of volatile organic compounds (VOCs), the rise in carotenoids or anthocyanins, and sugar levels are all closely regulated by ethylene (Gao et al. 2020). This hormone increases the attraction of fruit to frugivores by having a close relationship with VOC production. Fruit fragrance is lessened, and less volatile organic compounds are produced when ethylene biosynthesis is inhibited. Genes related to cell wall breakdown, ethylene receptors, and fruit-specific enzymes are all impacted by ethylene. It has been demonstrated that modifying ethylene levels or action can effectively extend the shelf life of climacteric fruits, highlighting its usefulness in fruit quality management and postharvest storage (Iqbal et al. 2017).

11.2.6 Salicylic Acid, Jasmonates, Strigolactones, and Brassinosteroids: The New Era Phytohormones

The plant hormone salicylic acid (SA) is involved in defense and is important in both abiotic and biotic stresses. When a plant experiences biotic stress, it develops resistance against several microbial pathogens, including bacteria, fungi, viruses, and oomycete. It is widely known that endogenous SA levels in plants are positively correlated with resistance mechanisms against both biotrophic and hemibiotrophic diseases (Ding and Ding 2020). It has only within the last 30 years that the role of SA in plant-defensive response has become clear. Along with many other phenolic secondary metabolites, SA was thought to be a rather inconsequential product before it was identified as an essential plant hormone.

According to Khan et al. (2015), SA in plants not only controls the ability to adapt to different abiotic stressors but also provides resistance to biotic fluctuations. The processes that underlie SA-induced abiotic stress tolerance include buildup of osmolytes, which can support the maintenance of osmotic homeostasis; management of mineral nutrition absorption; increased capacity to scavenge reactive oxygen species; improved manufacturing of secondary metabolites, including terpenes, phenolics, and compounds containing nitrogen like alkaloids, cyanogenic glucosides, and nonprotein amino acids but also sulfur-containing compounds; and monitoring of other hormonal processes (Koo et al. 2020).

According to Pérez-Llorca et al. (2019), SA is produced via the shikimate pathway. The first step involves chorismate synthase converting shikimate into chorismate. Two distinct routes, namely, isochorismate synthase-dependent and phenylalanine ammonia-lyase-dependent, are used to biosynthesize SA using chorismate (Dempsey and Klessig 2017). Even so, research on SA as a plant hormone controlling disease resistance has been boosted by many findings. SA has found to increase resistance in tobacco plants; when treated with it against tobacco mosaic virus (TMV) infection, it showed a 20-fold rise in endogenous SA concentrations in TMV-infected-resistant tobacco plants (Klessig et al. 2018). A growing body of research indicates that SA is crucial for controlling plant resistance to disease as well as for thermogenesis, DNA damage/repair, fruit output, seed germination, and other processes (Lefevere et al. 2020).

Plant hormones known as jasmonates (JAs) play a variety of roles and include jasmonic acid and its derivatives. According to Chini et al. (2016), they also provide tolerance to abiotic challenges, govern stages of development like root growth, stamen formation, flowering, and leaf senescence and regulate plant defenses versus herbivores and pathogens. JA is produced by the octadecanoid pathway from α-linolenic acid. Through the SCFCOI1 E3 ubiquitin ligase complex, JAZ proteins are degraded, releasing downstream transcription factors including MYC2 to control gene expression and initiating JA signaling (Wasternack and Song 2017).

JAs are essential for protecting plants from abiotic stressors, herbivores, and necrotrophs. They also have an impact on several processes related to development (Yuan and Zhang 2015). Although JA by itself has little biological activity, it can be combined with L-isoleucine to create the biologically active JA-Ile, which is essential for JA-mediated reactions (Han 2017). Furthermore, senescence, development, stress tolerance, reproductive procedures, and connections to other hormones are all impacted by JA and its methyl ester, methyl jasmonates (MeJAs). Numerous physiological processes, such as germination stimuli, buildup of storage proteins, chlorosis, increase of antioxidant enzymes, production of floral nectar, response to herbivory, development of seeds and flowers, systemic resistance, and allelopathy, are associated with JAs (Wasternack and Strnad 2016).

Strigolactones (SLs) are an emerging category of phytohormones that control many facets of plant's growth and development. Although the very first strigolactone (SL), strigol, was extracted from cotton root secretions in 1966, it took almost 40 years to determine that SLs belong to a novel group of phytohormones (Xu et al. 2021). Strigol is made up of a methylbutenolide D-ring and an ABC-ring joined by an enol-ether bridge. More than 25 SLs have been found throughout the plant kingdom. These SLs are classified as canonical or noncanonical depending on whether the entire ABC-ring system is present or absent (Bürger and Chory 2020).

SLs originate from β-carotene. Three of the following genes have been partially elucidated concerning their biosynthesis in several plant species: cytochrome P450s (MAX1) homologues, DWARF27 (β-carotene isomerase), and CAROTENOID CLEAVAGE DIOXYGENASE 7 and 8 (CCD7 and CCD8). Strigolactones (SLs) react to a variety of environmental cues and govern several developmental processes. A category of phytohormones called strigolactones shapes the general makeup of plants. They regulate root morphology, lateral development, and shoot branching (Wu et al. 2022). To foster symbiosis with arbuscular mycorrhizal fungi (AMF), SLs also function as signaling molecules secreted from roots into the rhizosphere (Machin et al. 2020). SLs also cause hyphal branching, a morphological alteration only seen in the proximity of host roots (Yoneyama and Brewer 2021).

Plant hormones known as brassinosteroids (BRs) are an essential class of hormones that regulate numerous facets of plant growth and development, allowing plants to adjust to their surroundings (Manghwar et al. 2022). The ability of BRs to promote growth in *Brassica napus* pollen led to their discovery, and in 1979, brassinolide (BL), the most active BR, was isolated.

Steroid hormones known as brassinosteroids (BRs) are found in many different parts of plants, such as the leaves, shoot, and roots, flowers, pollen, anthers, and seeds. Although it is found in all developing tissues of higher plants, fruit, pollen, and seeds have considerably larger amounts of the substance (Li and He 2020). Together, these hormones, known as "brassins," regulate various biological and developmental processes that affect the growth and maturation of plants. BRs affect the elongation of different cell types, vascular development, stem cell restoration, cell division, and the floral transformation (Lv et al. 2018). Furthermore, they contribute to the following processes: xylem development, hypocotyl lengthening, root and shoot expansion, stomata modeling, pollen and seed sprouting, pollen tube development, treachery element differentiation, and senescence (Nolan et al. 2020).

C27-, C28-, or C29-type BRs are produced by three distinct biosynthetic routes (Li et al. 2021). These processes are similar in the early stages of their synthesis, which can happen through the non-MVA or mevalonate (MVA) pathway. In the following steps, where cycloartenol- and cycloartanol-dependent processes are involved, differentiation between the routes takes place (Hafeez et al. 2021). With the identification and characterization of numerous processes, enzymes, and genes involved in BR biosynthesis, the C28-BR biosynthesis pathway has undergone substantial study. The various functions that BRs play in the growth and adaptability of plants highlight their importance in plant physiology (Bajguz et al. 2020).

11.3 Lab Investigations: Genetic and Molecular Approaches

11.3.1 Phytohormone Quantification: Analytical Approaches and Procedures

Understanding plant growth, development, and response to environmental stimuli depends on the quantification of phytohormones. Genetic and molecular approaches are frequently combined in analytical procedures for phytohormone measurement. Here are a few prominent used techniques:

11.3.1.1 The Enzyme-Linked Immunosorbent Assay (ELISA)

ELISA is an immunological technique that measures and identifies phytohormones using antibodies. Target phytohormone-specific antibodies are immobilized on a solid surface in ELISA. When the sample is added, the antibodies bind to the phytohormone, if it is present. Subsequently, an enzyme-linked secondary antibody is introduced, and a substrate is added to provide a detectable signal (Narang and Mohan 2023). Hauvermale et al. (2023) examined the expression of wheat α-amylase transcripts during grain development and germination to identify which proteins were the most suitable targets for the effect of phytohormones for antibody advancement and to help us to design an induced system for antibody evaluation.

11.3.1.2 Gas Chromatography–Mass Spectrometry (GC–MS)

GC–MS separates and quantifies phytohormones on the base of chemical properties and biomass. After being removed from plant tissues and converted into a volatile form, phytohormones are injected into a gas chromatograph. Mass spectrometry is used to identify and measure the separated components (Sheikh et al. 2023). According to an investigation of Song et al. (2022), pyrazine and 2,5-dimethylpyrazine suppress plant growth by measuring phytohormone content abscisic acid (ABA), salicylic acid, and gibberellic acid in plant tissues by using GC–MS techniques.

11.3.1.3 Liquid Chromatography–Mass Spectrometry (LC–MS)

LC–MS separates and quantifies phytohormones in liquid phase before analysis by mass spectrometry phytohormones that are extracted, and the liquid sample is passed through a chromatography column like the GC–MS. Mass spectrometry is then used for detection and quantification (Yao et al. 2023). The current work presents a fresh method for quantifying endogenous CKs in plants, which was not possible with the previous study. For the major CK metabolites, the overall run duration is less, and the matrix effect is less than with the traditional UHPLC–MS/MS. This strategy makes it possible to target plant hormonomics, including other plant hormone families, using UHPSFC–MS/MS apparatus. Ultra-high performance liquid chromatography–tandem mass spectrometry (UHPLC–MS/MS) is the basis of modern CK measurement methods since it allows the separation of CK isomeric forms tht occur naturally in plants (Petřík et al. 2023).

11.3.1.4 Assays Based on Fluorescence and Luminescence Assays

Assays based on fluorescence and luminescence assays use fluorescent or luminous markers to find phytohormones. Tags that are fluorescent or luminous are affixed to molecules that have a specific binding affinity for phytohormones. Next, the amount of light emitted is measured to determine the phytohormone content (Novotná 2023). According to Chen et al. (2023), fluorescent probes can be used to image and detect plant hormones and their receptors. Understanding how plants adapt to their surroundings requires a minimally invasive investigation of plant behavior at the cellular level.

11.3.1.5 Quantitative Polymerase Chain Reaction (qPCR)

qPCR is an expression of genes involved in signaling pathways, or phytohormone production is measured using qPCR. Plant tissues are used to extract RNA, which is subsequently transformed into cDNA and amplified using qPCR and target gene-specific primers. The quantity of phytohormones is directly correlated with the level of gene expression (Tan and colleagues 2023). Prior research revealed that 54 genes were consistently expressed in three of the screening methods. Based on their functional annotations, 11 genes related to plant hormone signal transduction were chosen for further examination. Three genes (Prupe.7G234800, Prupe.8G079200, and Prupe.8G082100) were identified as potential genes for single fruit weight features when combined with the evolving phenotypic trend. Auxin plays a critical function in plant growth and development, and all three genes are involved in auxin signal transduction (Bie et al. 2023).

11.3.1.6 Reporter Gene Assays

According to Freitas et al. (2019), this genetic method makes use of reporter genes whose expression is regulated by phytohormone-responsive promoters. A promoter region that responds to a specific phytohormone is linked to a reporter gene, such as luciferase or green fluorescence protein (GFP). The amount of phytohormone is then estimated by measuring the expression of the reporter gene (Jia et al. 2020).

11.3.2 New Perspectives on Phytohormone Communication: Current Developments and Up-and-Coming Technologies

Developing understanding of the complex communication network of phytohormones is essential in the dynamic field of plant physiology. This section explores the most recent developments and cutting-edge technology that are changing how we think about phytohormone communication. By investigating state-of-the-art research and novel approaches, we hope to reveal new aspects of plant signaling pathways.

11.3.2.1 Single-Cell Transcriptomics in Phytohormone Response

Our ability to evaluate gene expression at the individual cell level is being revolutionized by cutting-edge techniques in single-cell transcriptomics (Zenda et al. 2021). Utilizing single-cell RNA sequencing, scientists are deciphering the complexities of phytohormone-induced gene expression patterns in many cell types, offering never-before-seen insights into the cellular heterogeneity found in plant tissues (Ali 2023).

11.3.2.2 Quantitative Proteomics for Hormone Signaling Networks

Researchers can trace the dynamic changes in protein abundance and posttranslational modifications in response to phytohormones through advancements in mass spectrometry and quantitative proteomics (Chen et al. 2022). By identifying important regulatory proteins and their functions in signal transduction cascades, this method enables a systems-level knowledge of hormone signaling networks (Salem 2018).

11.3.2.3 CRISPR/Cas9 Genome Editing for Hormone

Plant genomes can now be precisely modified due to the deployment of CRISPR/Cas9 technology in receptor studies, which is revolutionizing the area of molecular biology (Zhu and Gao 2020). Using CRISPR/Cas9, plants can be engineered to have altered hormone receptor proteins. This allows for the analysis of receptor-specific responses and yields useful data for focused agricultural improvement (Santo Domingo Martínez 2023).

11.3.2.4 In Vivo Imaging of Phytohormone Dynamics

The real-time dynamics of phytohormones in living plant tissues are visualized using modern imaging technologies including bioluminescence imaging and fluorescence resonance energy transfer (FRET) (Struck et al. 2019). This method reveals the subtleties of hormone flow within plants by providing spatiotemporal information on hormone distribution and transport.

11.3.2.5 Integration of Multi-Omics Data for Systems Biology Insights

To completely understand phytohormone communication, an approach based on systems biology is being made easier by the integration of data from genomes, transcriptomics, proteomics, and metabolomics (Yang et al. 2021). To discover important regulatory nodes and decipher the intricacy of the interactions between various signaling pathways, computational models are being created to evaluate multi-omics datasets.

11.3.2.6 Nanotechnology for Targeted Phytohormone Delivery

Nanoparticle-based delivery systems are being explored to enhance the targeted delivery of phytohormones, allowing for precise modulation of hormonal responses in specific plant tissues (Wu et al. 2022). This technology opens possibilities for optimizing plant growth, stress responses, and crop yields by finely tuning phytohormone levels at the subcellular scale (Li et al. 2023).

11.4 Moving into Fields: Impacts on Ecosystem

11.4.1 Effects of Phytohormones on Soil Microorganisms and Soil Topology

According to Flemming and Wuertz (2019), the soil is home to a remarkable variety of microorganisms, with bacteria making up the about 95% of these species. The rhizosphere is a significant "hot spot" for these microbes (Moulick et al. 2021). The small area of soil that surrounds the root system and is stimulated chemically and physically by root growth and activity is known as the rhizosphere. Zhu et al. (2022) describe the intricate microsystem composed of soil, microorganisms, and plant roots. Numerous types of bacteria, fungus, invertebrates, nematodes, and other creatures can be found in this nutrient-rich rhizosphere area. This is caused by several variables, including soil pH, moisture, temperature, root exudates, and inorganic nutrients, in addition to the general interactions between plants and microbes (van Dam et al. 2016).

The root exudates offer a steady supply of both energy and nutrients, which drives diversity of microbes and their population. Based on how they affect plants, bacteria that relate to plants are classified as useful, neutral, or detrimental. Therefore, extended root exudate excretion functions as a crucial marker of microbial populations in addition to being an increased nitrogen and carbon-demanding activity (Herms et al. 2022). Out of all the microbes present in soil bacteria are the primary agents in the vital nutrient cycle. They are crucial to the uptake and accumulation of nutrients in plants (Narayanan and Ma 2023).

Plants are under constant stress due to changing environmental conditions. Abiotic stress is impacted differently by diverse soil environments. In the microbial ecosystem, physical, chemical, and biological elements all play fundamental roles. The soil ecology is also influenced by other characteristics such as soil texture, pH, salinity, and electrical conductivity (Fig. 11.1). However, abiotic stress has been found to increase the number of microorganisms to reduce stress by moderating several metabolites that induce alterations to root exudates (Jansson and Hofmockel 2020). By secreting messenger molecules, the plant employs the "asking seeking help" technique to draw in helpful microbes that may reside in the rhizosphere or infiltrate the endosphere. The exudate that draws rhizosphere bacteria is thought to include strigolactone. Leguminous plants, for example, release more flavonoids to draw in N-fixing bacteria when they are exposed to a nitrogen (N) shortage (Soliman et al. 2022). Similar role is played by strigolactones. The rhizosphere microbiome is

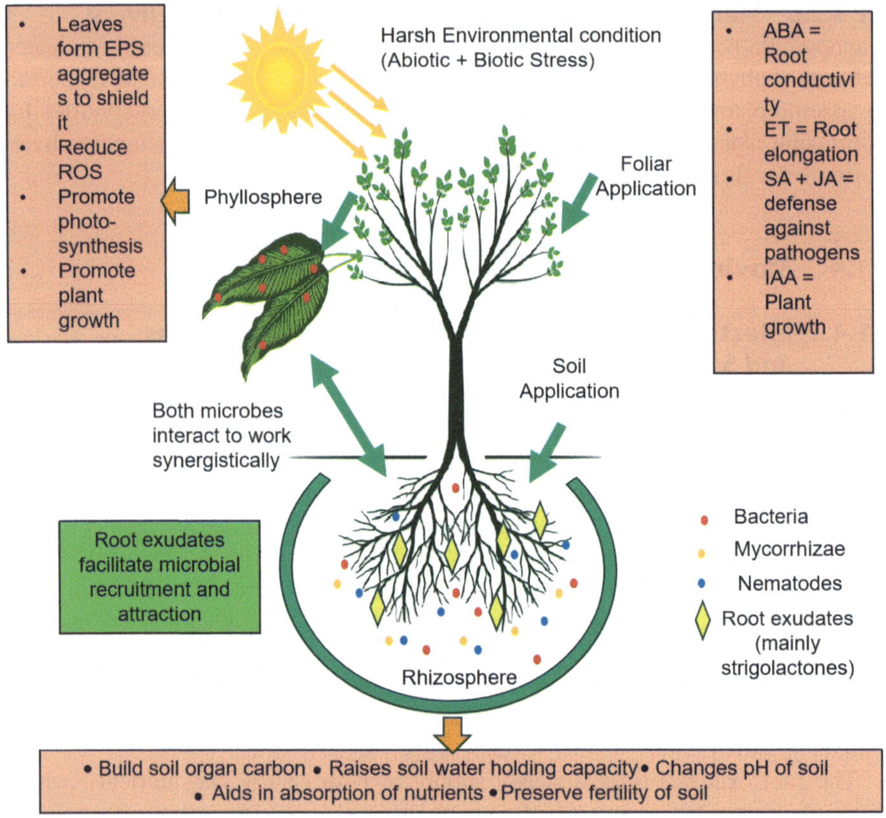

Fig. 11.1 Effects of phytohormones on microbes and soil topology. This schematic diagram represents the effect of exogenously applied and endogenously produced phytohormones on the rhizosphere and phyllosphere. Plants secrete the phytohormones as root exudates mainly strigolactones. This strategy of "asking for help" attracts growth-promoting bacteria, beneficial mycorrhizae, and free-living nematodes in soil. The microbes of the rhizosphere interact with microbes of the phyllosphere bringing changes that improve plant growth under stress conditions. The microbes in rhizosphere build up soil organic carbon which eventually leads to increased soil water-holding capacity. They change the pH of soil which aids in nutrient uptake, thus leading to preserving fertility of soil even under stress conditions like drought, salinity, etc. The microbes in the phyllosphere form aggregates on leaves using EPS (extracellular polymeric substances), which prevents water loss. ROS concentration is reduced, and photosynthesis is promoted, which eventually leads to plant growth. Exogenously applied ABA promotes the conduction of water in roots, ET is involved in root elongation, IAA promotes plant growth, and SA + JA aids in plant defense against parasitic organisms

related to nutrients, particularly the uptake and absorption of phosphate and nitrogen. Numerous more phenolic compounds are also crucial for improving plant tolerance and drawing in a sizable number of microbial species (Jansson and Hofmockel 2020).

Microbial recruitment and attraction are facilitated by the root exudate SLs infiltrating the soil. The diversity of soil microbes is produced by SLs. Many soil bacteria as a response to exudates have evolved defense mechanisms against shifting environmental conditions. They build the soil's organic carbon (SOC), which raises the soil's capacity to hold water (Carvalhais et al. 2019). Improving the ability of the soil to hold water is one physical tactic against the water shortage. Soil microbes and SLs work together to change the chemical makeup of the soil. When these microbes and SLs work together, tolerance processes are facilitated, which helps the plant end the current stress (Libohova et al. 2018).

The bacteria that live on plant components above ground are known as the phyllosphere microbiome, and they are essential for improving plant immunity and protection against abiotic stressors. Phylloplane and phyllotelma, two divisions of the phyllosphere, have an impact on the communities of microorganisms and their interaction with plants (Liu et al. 2020). The phyllosphere microbiome is altered by soil microbes, which impacts how plants react to stress. Strigolactones (SLs) affect fungi and bacteria on leaves, which may influence microbial activities on the phyllosphere (Soliman et al. 2022). SLs may interact with hormones like cytokinin to have an indirect impact on the phyllosphere microbiota.

Microbial communities in the phyllosphere interact with the microbiomes in the roots, and during pathogen infection, beneficial microbes are drawn to the leaves. SLs and flavonoids are two substances that are involved in this communication (Stassen et al. 2021). Microbial diversity and abundance are impacted by plant genotype, which also affects the microbial makeup of the phyllosphere. Plant techniques that make use of the phyllosphere microbiome include secondary metabolites, gene expression, and protein overexpression during stress (Soliman et al. 2022). The microbes on the leaves use extracellular polymeric substances (EPS) and formation of aggregates to shield plants from dehydration during heat stress.

By reducing reactive oxygen species (ROS) degradation during stress and promoting physiological procedures like photosynthesis, phyllosphere microorganisms promote plant growth. Water availability is essential for leaf survival, and plants release chemicals that nourish microorganisms in the phyllosphere. In the communication that occurs among the soil, rhizosphere, endosphere microbes, and root cells, strigolactones play a messenger role (O'Banion et al. 2020). They function as regulators of signals, interface messengers, and hormonal messengers. Rhizosphere bacteria change the pH of the soil, aid in the absorption of nutrients, and exude chemicals into the soil (Soliman et al. 2022).

Plants and arbuscular mycorrhizal fungi (AMF) have a symbiotic connection that improves soil structure, insect tolerance, and nutrient uptake. SLs are involved in the symbiotic relationship that roots have with AMF in times of stress and phosphate deprivation. Cell-to-cell communication between bacteria and plants may be facilitated by N-acyl-homoserine lactones (AHLs) (Soliman et al. 2022). Additionally, SLs are involved in controlling plant development, seed germination, and root-hair extension and acting as messengers for parasitic weeds (Wang et al. 2019).

Among all the abiotic stresses, droughts have been found to have significant impact on microbial community. Plants affected by drought stress show changed hormone buildup in their roots, which increases the outflow of phytohormones toward the rhizosphere (Trinder et al. 2020). Plant-microbial reactions regarding drought are coordinated because of this efflux's impact on soil microbial populations. Abscisic acid (ABA), ethylene, and jasmonates are examples of phytohormones that are involved in plant responses to drought and can influence microbial activity in the soil (Sayer et al. 2021). Root development and hydraulic conductivity are maintained in drying soil by ABA, which is generated in response to a decreasing tissue water status.

Root elongation is influenced by ethylene and its precursor, 1-aminocyclopropane-1-carboxylic acid (ACC), and root development can be stimulated under water stress conditions by bacteria that break down ACC (Naylor and Coleman-Derr 2018). Jasmonates, which are linked to biotic stress, influence rhizosphere microbial populations and plant responses to drought (Fridley et al. 2016). Research on microbial activity mediated by plant hormones in drought-affected areas has mostly concentrated on endophytic or rhizosphere species (Chanclud and Morel 2016). Rhizobacteria that promote plant growth and plant hormone signaling are interdependent, but little is known about the larger effects of phytohormones on the microbial populations and procedures in bulk soils (Sayer et al. 2017).

The most prevalent metazoan on Earth is the nematode (Van Den Hoogen et al. 2019). Most soil-dwelling worms can be categorized into two groups according to their modes of consumption: either they are plant parasitic nematodes (PPN) which survive on tissue from plants, or they are advantageous free-living nematodes which play a significant part in the soil environment by feeding primarily on bacteria, fungi, algae, protozoa, and other parasites (Wilschut and Geisen 2021). Considering defensive mechanisms toward plant parasitic nematodes (PPN), phytohormones are especially important in mediating interactions between plants and nematodes. The two main defense hormones are salicylic acid (SA) and jasmonate acid (JA). SA increases tomato resistance to worms, whereas JA signaling aids in defense toward a variety of nematode species (Kammerhofer et al. 2015). ET works in concert with JA and auxin to trigger systemic defense mechanisms against nematodes. Strigolactones (SLs) have an impact on defense and growth in plants; their function in nematode interactions differs in dicots and monocots.

Abscisic acid (ABA) mediates either beneficial or detrimental defense reactions through complicated interactions with JA, SA, and ET (Sikder et al. 2021). Stronger impacts were observed when JA and SA were disrupted, which is in line with their functions as the primary plant defense hormones (Kyndt et al. 2017). The complex relationships between ABA and various other hormonal signaling routes in plants are highlighted by the fact that while root lesion nematode reproduction was decreased in plants sprayed with ABA, the application of exogenous ABA increased vulnerability to other nematodes (Sikder and Vestergård 2020).

In chickpea, it was found that the plant growth-promoting rhizobacteria (PGPR) and plant hormones interacted synergistically to improve drought tolerance under extreme stressful circumstances in addition to the development of roots and shoots

(Khan et al. 2017). In sandy soils, PGPR are especially essential for preserving fertility and preserving water. To boost development, *Bacillus subtilis* generates plant hormones including indole-3-acetic acid (IAA) and induces stress tolerance in plants (Khan et al. 2018). Rice growth is aided by PGPR from salt-affected soil that produces IAA. When PGPR is injected into maize, auxin and gibberellin accumulate more strongly, and the leaves of the plant accumulate a substantial number of heavy metals. When it comes to how plants react to abiotic stressors, phytohormones and PGPR are essential (Khan et al. 2020).

The PGPR signaling system that confers stress tolerance in plants is revealed by molecular investigations. To enhance stress tolerance, PGPR controls gene expression, osmotic equilibrium, and ion homeostasis (Ilangumaran and Smith 2017). To facilitate cell-to-cell communication and control gene expression and behavior, quorum sensing bacteria generate signal molecules. Certain substances generated from plants function as antiquorum sensing agents (Cosa et al. 2019). The production of indole-3-acetic acid is modulated by PGPR-producing N-acyl-homoserine lactone (AHL). When root exudates are present, *Bacillus* strains suppress plant immune responses, which makes root infection easier (Donoso and Ruz 2017). Plant-associated bacteria may perceive volatile chemicals or secondary metabolites such as surfactin lipopeptides, which could lead to enhanced resilience in reaction to PGPR-induced plant responses (Khan et al. 2020).

11.4.2 Their Mechanisms of Absorption and Transportation in Soil

Phytohormones are applied to plants exogenously via foliar sprays or by directly adding them in soil. Applying phytohormones foliarly allows for immediate delivery to leaves. Some phytohormones can pierce the protective layer known as the cuticle on plants while others may not be able to do so. It has been discovered, for example, that stomata and the cuticle are the pathways by which abscisic acid (ABA) is ingested (Cutler et al. 2010). Roots are capable of absorbing phytohormones that are added to the soil. Soil conditions, phytohormone levels, and root features are some of the elements that may affect the absorption process (Wu et al. 2023). Phytohormones are distributed throughout the plant after absorption via vascular bundles. Phloem permits downward flow, whereas xylem supports upward transport. Phytohormones mostly show polar transmissions (Zhao et al. 2021). Certain phytohormones, such as auxins, are actively transported by cell membrane-bound carrier proteins during polar transport.

11.5 Upgrading Agriculture

11.5.1 Methods of Genetic Modification for Phytohormone Modulation

The process of genetic modification for phytohormone modulation entails modifying the genetic material of the plant to promote growth, development, and stress response features. Essential signaling molecules known as phytohormones control a range of physiological functions in plants. Improvements in crop production, stress tolerance, and overall agricultural productivity can be achieved by adjusting their levels or reactions. The following are some techniques for genetic alteration in the manipulation of phytohormones:

11.5.1.1 Traditional Breeding

To select and cross plants' used desirable phytohormone-related features, traditional breeding techniques can be used; however, they are not genetic alteration approaches in the traditional sense. On the other hand, this procedure takes a long time and might not provide exact control over the desired genes. Traditional breeding techniques were applied in a study by Sheoran et al. (2022) to improve a crop's phytohormone-related qualities, like yield and drought resistance. Their study aimed to develop plants with favorable traits, like high concentrations of the phytohormone abscisic acid, which is linked to the stress response. Their goal was to create a new cultivar that would be more resistant to environmental stresses through a series of carefully crafted crosses and selections made over time.

11.5.1.2 Genetic Engineering (Transgenic Plants)

Plant growth and development can be impacted by the overproduction of certain hormones caused by the introduction of genes encoding enzymes involved in phytohormone biosynthesis (Fenn and Giovannoni 2021), employing gene-silencing methods like RNA interference (RNAi) to suppress the expression of genes involved in the production of phytohormones or their response to them. This may aid in regulating specific hormone levels (Gautam et al. 2022).

11.5.1.3 Synthetic Biology

Creating artificial genetic circuits regulates the synthesis of phytohormones or their response (Andres and Zurbriggen 2019). With this method, specific plant growth and development may be precisely modulated through the engineering of unique routes.

11.5.1.4 Genome-Wide Association Studies (GWAS)

GWAS is applied to identify naturally occurring genetic variations linked to phytohormone-related phenotypes, which are then utilized for breeding or genetic manipulation (Riaz et al. 2021).

11.5.2 Transgenic Crop Achievement Case Studies: Phytohormones in Sustainable Agriculture

Global agricultural productivity must increase significantly to keep up with the continually expanding human population. Nonetheless, several biotic and abiotic stressors are significant variables that restrict the yield of agricultural crops (Fróna et al. 2019). To feed the world's expanding population by 2050, productivity must increase by 70% to accommodate an extra 2.3 billion people (Hemathilake and Gunathilake 2022). The most common and substantial abiotic stressors that reduce agricultural productivity are drought, salt, and extremely high temperatures. The length of the stress response, its nature, the plant's developmental stage, and biotic and abiotic variables can all have an impact (Yadav et al. 2020). Many economically significant crop species now have a fresh means of acquiring abiotic stress resistance thanks to genetic engineering. Recent developments in genetic engineering have elevated this field to the top of the list of effective ways to increase agricultural output in harsh climates.

To improve resistance to abiotic stress, one possible target for alteration is the metabolism of hormones and the signaling pathways that accompany them. But for healthy growth, hormonal balance must be maintained (Meena et al. 2022). Because it is involved in a wide range of processes, ABA is possibly the most sought-after phytohormone for building abiotic stress tolerance in agricultural plants. As a result, efforts have been made to increase the abiotic stress tolerance of economically significant crop species through transgenic manipulation of several of the main enzymes implicated in the ABA biosynthesis pathway (Simmons et al. 2021).

Overexpressing an ABA-responsive stress-related gene enhanced *Arabidopsis*' ability to withstand osmotic stress. Tomatoes with increased drought tolerance were produced through manipulation of C-repeat binding factor (CBF) and dehydration-responsive element binding (DREB) genes, such as DREB1B (Yoon et al. 2020). Increased antioxidative enzyme activity, fewer stomatal openings, lower water loss, and higher ABA levels were all mentioned as contributing factors to improvement of crops. But problems emerged when genes related to catabolic processes or ABA biosynthesis were overexpressed, leading to pleiotropic effects that hindered growth (Meena et al. 2022).

Overexpressed CRK45 is a stress-inducible kinase that controls ABA levels and signaling, to overcome these difficulties. This resulted in transgenics that avoided unintended negative effects while exhibiting enhanced resistance to drought. Like this, stress-inducible promoter-driven overexpression of the IPT gene in *Agrostis stolonifera* led to increased cytokinin (CK) content, antioxidant scavenging, and improved root development, all of which boosted grain yield (Nayak et al. 2024). The stress-inducible SWPA2 promoter-driven amplification of the *Arabidopsis* YUCCA6 gene in poplar trees initiated the tryptophan-dependent IAA biosynthesis pathway. This led to increased root hair creation, slower main root development, faster shoot growth, and a greater ability to withstand drought stress by lowering reactive oxygen species levels.

Gibberellic acid (GA) levels in rice were altered by overexpressing OsGA2ox1. The failure to establish grain and severe dwarfism caused by constitutive expression of this gene suggests that GA is involved in the development of the reproductive system and shoot elongation (Meena et al. 2022). But when OsGA2ox1 was expressed ectopically and controlled by the OsGA3ox2 promoter, the plant produced a semi-dwarf phenotype that had normal grain and flower development. Together, these findings demonstrate how genetic alterations may improve plants' ability to withstand stress, provided that gene targets and regulatory systems are carefully taken into account to prevent unforeseen consequences.

Enhancing plant resistance to environmental stressors and controlling the production of secondary metabolites (SM) can be achieved through genetic engineering of plant hormones and the biosynthesis/regulatory genes linked with them (Sreenivasulu and Fernie 2022). By controlling the lignin and abscisic acid (ABA) signaling pathways, overexpression of the SiMYB56 gene improved drought tolerance in transgenic rice plants. During drought stress, transgenic soybeans overexpressing ABA-responsive element binding factor 3 (ABF3) demonstrated enhanced accumulation of a variety of metabolites, including phenolic compounds, flavonoids, and glucosinolates (GSLs) (Meena et al. 2022). In maize, overexpression of the phosphatidylinositol synthase gene (ZmPIS) boosted drought tolerance and ABA production.

Furthermore, by modifying the pathways for brassinosteroid (BR), jasmonic acid (JA), and ABA, genetic alterations in mustard, apple, and red sage showed enhanced tolerance to a variety of environmental challenges, such as heat, cold, and ultraviolet (UV) radiation. Together, these research highlight how genetic engineering might improve crop stress tolerance and maximize secondary metabolite synthesis, opening the door for more breeding initiatives.

11.6 Conclusion and Future Perspective

In conclusion, it should be noted that phytohormones are essential regulators in the complex dance of plant life, skillfully balancing environmental stressors and responses that are adaptive. With ordinary and stressful conditions, the diverse array of signaling molecules, which includes the well-known auxins, cytokinins, and gibberellins as well as more recent discoveries like salicylic acid, jasmonic acid, ethylene, abscisic acid, strigolactones, and brassinosteroids, collectively orchestrates fundamental physiological processes. The kinds and concentration of phytohormones have a major impact on a range of plant functions, including normal growth and development as well as the difficult tasks presented by environmental hardships.

Plants are a notable example of plasticity because of their complex signaling networks, their spatiotemporal phytohormone distribution, and their sensitivity to external stimuli. Phytohormones exert control over basic features of plant growth through the precise regulation of genes, directing activities such as cell elongation, splitting, tissue differentiation, organogenesis, structural creation, and reproduction. Plants are able to respond nimbly to a wide range of stresses that they face in the biotic and abiotic domains thanks to this regulatory skill. The complex interrelationships between phytohormones provide plants with a dynamic toolkit for assembling complex defensive mechanisms. These hormones orchestrate a complex reaction to environmental and developmental signals because they can function in both antagonistic and complementary ways. By carefully adjusting the ratios of various phytohormones, plants are able to maintain a delicate balance and adapt to the ever-changing conditions in their environment. Physiological, biological, and morphological functions are all regulated by phytohormones, which can take many different forms. Some of the more well-known examples of these include auxins, cytokinins, gibberellins, abscisic acid, and ethylene. Other more recent discoveries include salicylates, jasmonates, strigolactones, and brassinosteroids. The survival and growth of a plant depend on these processes, which include nutrient distribution and source/sink changes. The vast symphony of plant physiology is composed of distinct roles for each type of plant hormones. Honored as master regulators of plant growth, auxins control cell division, extension, differentiation, embryonic development, and a range of tropisms. Gibberellins regulate cell elongation and division to affect organ growth. Gibberellins are the master builders of plant height and flowering. Masterminds of cell differentiation and division, cytokinins control mitogenesis and play a major role in root-to-shoot communication. To sum up, phytohormones play a dual role in plant growth and development, acting as both molecular regulators and architects of stress resilience. Gaining knowledge into the complex roles and interplay of these signaling molecules might help increase crop resilience, boost yields, and ease response to constantly shifting environmental conditions. The intricate regulatory networks controlling phytohormones may be further explored, which could lead to revolutionary developments in plant biotechnology and agriculture. Future study on phytohormones has the potential to be revolutionary. Advanced technologies like single-cell transcriptomics, CRISPR/Cas9 genome editing, and multi-omics integration are coming together, and this could help us understand the complex molecular networks that control the dynamics of phytohormones. Precision agriculture is a promising field that can optimize resource use and reduce environmental impact by targeted treatments and real-time monitoring based on phytohormone data. Using synthetic biology, it is possible to develop crops with unique phytohormone profiles that increase yield and stress tolerance. A thorough knowledge of phytohormone interactions will be possible through the application of holistic systems biology techniques, which include genomics, transcriptomics, proteomics, and metabolomics. Furthermore, phytohormones' ecological uses in regulating nutrient cycling and soil microbial populations highlight their potential for sustainable soil management. The ongoing study of phytohormones puts us in a leading position to pioneer ideas that will transform agriculture, environmental preservation, and ecosystem health as we negotiate the complexity of climate change and global food security.

References

Ali K, Georgiev MI (2023) Omics and its integration: a systems biology approach to understanding plant physiology. Frontiers in Plant Science 14:1324901

Andres J, Blomeier T, Zurbriggen MD (2019) Synthetic switches and regulatory circuits in plants. Plant Physiology 179(3):862–884

Arif Y, Sami F, Siddiqui H, Bajguz A, Hayat S (2020) Salicylic acid in relation to other phytohormones in plant: a study towards physiology and signal transduction under challenging environment. Environ Exp Bot 175:104040

Arsuffi G, Braybrook SA (2018) Acid growth: an ongoing trip. J Exp Bot 69(2):137–146

Bagale P, Pandey S, Regmi P, Bhusal S (2022) Role of plant growth regulator "gibberellins" in vegetable production: an overview. Int J Horticult Sci Technol 9(3):291–299

Bajguz A, Chmur M, Gruszka D (2020) Comprehensive overview of the brassinosteroid biosynthesis pathways: substrates, products, inhibitors, and connections. Front Plant Sci 11:1034

Bakshi A, Shemansky JM, Chang C, Binder BM (2015) History of research on the plant hormone ethylene. J Plant Growth Regul 34:809–827

Balzan S, Johal GS, Carraro N (2014) The role of auxin transporters in monocots development. Front Plant Sci 5:393

Bandurska H, Niedziela J, Pietrowska-Borek M, Nuc K, Chadzinikolau T, Radzikowska D (2017) Regulation of proline biosynthesis and resistance to drought stress in two barley (Hordeum vulgare L.) genotypes of different origin. Plant Physiol Biochem 118:427–437

Bie H, Wang H, Wang L, Li Y, Fang W, Chen C et al (2023) Mining Genes Related to Single Fruit Weight of Peach (Prunus persica) Based on WGCNA and GSEA. Horticulturae 9(12):1335

Brunetti C, Fini A, Sebastiani F, Gori A, Tattini M (2018) Modulation of phytohormone signaling: a primary function of flavonoids in plant–environment interactions. Front Plant Sci 9:1042

Bürger M, Chory J (2020) The many models of strigolactone signaling. Trends Plant Sci 25(4):395–405

Cao X, Yang H, Shang C, Ma S, Liu L, Cheng J (2019) The roles of auxin biosynthesis YUCCA gene family in plants. Int J Mol Sci 20(24):6343

Cardenas J (2021) Elucidate the role of PUX1 in GA signaling through its interaction with the GA receptor GID1 to control cell growth and development. The University of Wisconsin-Madison

Cardoso AA, Gori A, Da-Silva CJ, Brunetti C (2020) Abscisic acid biosynthesis and signaling in plants: key targets to improve water use efficiency and drought tolerance. Appl Sci 10(18):6322

Carvalhais LC, Rincon-Florez VA, Brewer PB, Beveridge CA, Dennis PG, Schenk PM (2019) The ability of plants to produce strigolactones affects rhizosphere community composition of fungi but not bacteria. Rhizosphere 9:18–26

Chanclud E, Morel JB (2016) Plant hormones: a fungal point of view. Mol Plant Pathol 17(8):1289–1297

Checker VG, Kushwaha HR, Kumari P, Yadav S (2018) Role of phytohormones in plant defense: signaling and cross talk. In: Singh A, Singh IK (eds) Molecular aspects of plant-pathogen interaction. Springer, pp 159–184

Chen Q, Dai X, De-Paoli H, Cheng Y, Takebayashi Y, Kasahara H et al (2014) Auxin overproduction in shoots cannot rescue auxin deficiencies in Arabidopsis roots. Plant Cell Physiol 55(6):1072–1079

Chen Y, He XJ, Huang XD et al (2022) Recent Advances in Proteomics and Metabolomics in Plants. Molecular. Horticulture

Chen Y, He B, Hu M, Bao J, Yan W, Han X, Ye Y (2023) Fluorescent probes for imaging and detection of plant hormones and their receptors. Advanced Agrochem

Chini A, Gimenez-Ibanez S, Goossens A, Solano R (2016) Redundancy and specificity in jasmonate signalling. Curr Opin Plant Biol 33:147–156

Coelho LL, Fkiara A, Mackenzie KK, Müller R, Lütken H (2018) Exogenous application of gibberellic acid improves flowering in Kalanchoë. HortScience 53(3):342–346

Coen E, Kennaway R (2018) Early shaping of a leaf. Nat Plants 4(9):618–619

Cosa S, Chaudhary SK, Chen W, Combrinck S, Viljoen A (2019) Exploring common culinary herbs and spices as potential anti-quorum sensing agents. Nutrients 11(4):739
Cruz TM, Carvalho RF, Richardson DN, Duque P (2014) Abscisic acid (ABA) regulation of Arabidopsis SR protein gene expression. Int J Mol Sci 15(10):17,541–17,564
Cutler SR, Rodriguez PL, Finkelstein RR, Abrams SR (2010) Abscisic acid: emergence of a core signaling network. Annu Rev Plant Biol 61:651–679
van Dam NM, Weinhold A, Garbeva P (2016) Calling in the dark: the role of volatiles for communication in the rhizosphere. Deciphering chemical language of plant communication, pp. 175–210
Damodaran S, Strader LC (2019) Indole 3-butyric acid metabolism and transport in Arabidopsis thaliana. Front Plant Sci 10:851
Dante RA, Larkins BA, Sabelli PA (2014) Cell cycle control and seed development. Front Plant Sci 5:493
Dar RA, Nisar S, Tahir I (2021) Ethylene: a key player in ethylene sensitive flower senescence: a review. Sci Hortic 290:110491
Dempsey DMA, Klessig DF (2017) How does the multifaceted plant hormone salicylic acid combat disease in plants and are similar mechanisms utilized in humans? BMC Biol 15:1–11
Di DW, Zhang C, Luo P, An CW, Guo GQ (2016) The biosynthesis of auxin: how many paths truly lead to IAA? Plant Growth Regul 78:275–285
Ding P, Ding Y (2020) Stories of salicylic acid: a plant defense hormone. Trends Plant Sci 25(6):549–565
Donoso R, Ruz G (2017) Quorum-sensing systems in the plant growth-promoting bacterium paraburkholderia phytofirmans PsJN exhibit cross-regulation and are involved in biofilm formation. Mol Plant Microbe Interact 30(7):557–565
Dubois M, Van den Broeck L, Claeys H, Van Vlierberghe K, Matsui M, Inzé D (2015) The ETHYLENE RESPONSE FACTORs ERF6 and ERF11 antagonistically regulate mannitol-induced growth inhibition in Arabidopsis. Plant Physiol 169(1):166–179
Duca D, Lorv J, Patten CL, Rose D, Glick BR (2014) Indole-3-acetic acid in plant–microbe interactions. Antonie Van Leeuwenhoek 106:85–125
Enders TA, Strader LC (2015) Auxin activity: past, present, and future. Am J Bot 102(2):180–196
Fahad S, Hussain S, Matloob A, Khan FA, Khaliq A, Saud S et al (2015) Phytohormones and plant responses to salinity stress: a review. Plant Growth Regul 75:391–404
Fenn MA, Giovannoni JJ (2021) Phytohormones in fruit development and maturation. Plant J 105(2):446–458
Flemming HC, Wuertz S (2019) Bacteria and archaea on earth and their abundance in biofilms. Nat Rev Microbiol 17(4):247–260
Freitas EO, Melo BP, Lourenço-Tessutti IT, Arraes FB, Amorim RM, Lisei-de-Sá ME et al (2019) Identification and characterization of the GmRD26 soybean promoter in response to abiotic stresses: potential tool for biotechnological application. BMC biotechnology 19:1–14
Fridley JD, Lynn JS, Grime JP, Askew AP (2016) Longer growing seasons shift grassland vegetation towards more-productive species. Nat Clim Chang 6(9):865–868
Fróna D, Szenderák J, Harangi-Rákos M (2019) The challenge of feeding the world. Sustain For 11(20):5816
Galvan-Ampudia CS, Cerutti G, Legrand J, Azais R, Brunoud G, Moussu S, et al. (2018) From spatio-temporal morphogenetic gradients to rhythmic patterning at the shoot apex. BioRxiv, 469718
Gao J, Zhang Y, Li Z, Liu M (2020) Role of ethylene response factors (ERFs) in fruit ripening. Food Qual Saf 4(1):15–20
Gautam P, Kumar R, Feroz Z, Vijayaraghavalu S, Kumar M (2022) RNA interference technology in plants: mechanisms and applications in crop improvement. In: Plant genomics for sustainable agriculture. Singapore, Springer Nature Singapore, pp 265–290
Geisler M, Wang B, Zhu J (2014) Auxin transport during root gravitropism: transporters and techniques. Plant Biol 16:50–57

Hafeez MB, Zahra N, Zahra K, Raza A, Batool A, Shaukat K, Khan S (2021) Brassinosteroids: molecular and physiological responses in plant growth and abiotic stresses. Plant Stress 2:100029
Han GZ (2017) Evolution of jasmonate biosynthesis and signaling mechanisms. J Exp Bot 68(6):1323–1331
Han Y, Watanabe S, Shimada H, Sakamoto A (2020) Dynamics of the leaf endoplasmic reticulum modulate β-glucosidase-mediated stress-activated ABA production from its glucosyl ester. J Exp Bot 71(6):2058–2071
Hauvermale AL, Matzke C, Bohaliga G, Pumphrey MO, Steber CM, McCubbin AG (2023) Development of Novel Monoclonal Antibodies to Wheat Alpha-Amylases Associated with Grain Quality Problems That Are Increasing with Climate Change. Plants 12(22):3798
Hedden P (2016) Gibberellin biosynthesis in higher plants. In: Hedden P, Thomas SG (eds) Annual plant reviews, volume 49: gibberellins, the. Wiley, pp 37–72
Hedden P (2020) The current status of research on gibberellin biosynthesis. Plant Cell Physiol 61(11):1832–1849
Hedden P, Sponsel V (2015) A century of gibberellin research. J Plant Growth Regul 34:740–760
Hedden P, Thomas SG (eds) (2016) Annual plant reviews, the gibberellins (Vol. 49). Wiley
Hemathilake DMKS, Gunathilake DMCC (2022) Agricultural productivity and food supply to meet increased demands. In: Bhat R (ed) Future foods. Academic Press, pp 539–553
Herms CH, Hennessy RC, Bak F, Dresbøll DB, Nicolaisen MH (2022) Back to our roots: exploring the role of root morphology as a mediator of beneficial plant–microbe interactions. Environ Microbiol 24(8):3264–3272
Hong JH, Seah SW, Xu J (2013) The root of ABA action in environmental stress response. Plant Cell Rep 32:971–983
Huang X, Hou L, Meng J, You H, Li Z, Gong Z et al (2018) The antagonistic action of abscisic acid and cytokinin signaling mediates drought stress response in Arabidopsis. Mol Plant 11(7):970–982
Hussen S (2014) Ethylene as a postharvest "evil" and its remedies in some horticultural crops. Greener J Plant Breed Crop Sci:34–40
Ikeda Y, Men S, Fischer U, Stepanova AN, Alonso JM, Ljung K, Grebe M (2009) Local auxin biosynthesis modulates gradient-directed planar polarity in Arabidopsis. Nat Cell Biol 11(6):731–738
Ilangumaran G, Smith DL (2017) Plant growth promoting rhizobacteria in amelioration of salinity stress: a systems biology perspective. Front Plant Sci 8:1768
Iqbal N, Umar S, Khan NA, Khan MIR (2014) A new perspective of phytohormones in salinity tolerance: regulation of proline metabolism. Environ Exp Bot 100:34–42
Iqbal N, Khan NA, Ferrante A, Trivellini A, Francini A, Khan MIR (2017) Ethylene role in plant growth, development and senescence: interaction with other phytohormones. Front Plant Sci 8:475
Jameson PE, Song J (2016) Cytokinin: a key driver of seed yield. J Exp Bot 67(3):593–606
Jansson JK, Hofmockel KS (2020) Soil microbiomes and climate change. Nat Rev Microbiol 18(1):35–46
Javid MG, Sorooshzadeh A, Moradi F, Modarres Sanavy SAM, Allahdadi I (2011) The role of phytohormones in alleviating salt stress in crop plants. Aust J Crop Sci 5(6):726–734
Jia L, Mao Y, Ji Q, Dersh D, Yewdell JW, Qian SB (2020) Decoding mRNA translatability and stability from the 5′ UTR. Nature structural & molecular biology 27(9):814–821
Julkowska MM, Koevoets IT, Mol S, Hoefsloot H, Feron R, Tester MA et al (2017) Genetic components of root architecture remodeling in response to salt stress. Plant Cell 29(12):3198–3213
Kammerhofer N, Radakovic Z, Regis JM, Dobrev P, Vankova R, Grundler FM et al (2015) Role of stress-related hormones in plant defence during early infection of the cyst nematode Heterodera schachtii in Arabidopsis. New Phytol 207(3):778–789
Kenea FT (2021) Role of ethylene in horticulture—a review. GP Globalize Res J Chem 81
Khan MIR, Fatma M, Per TS, Anjum NA, Khan NA (2015) Salicylic acid-induced abiotic stress tolerance and underlying mechanisms in plants. Front Plant Sci 6:462

Khan N, Bano A, Babar MA (2017) The root growth of wheat plants, the water conservation and fertility status of sandy soils influenced by plant growth promoting rhizobacteria. Symbiosis 72:195–205
Khan N, Bano A, Zandi P (2018) Effects of exogenously applied plant growth regulators in combination with PGPR on the physiology and root growth of chickpea (Cicer arietinum) and their role in drought tolerance. J Plant Interact 13(1):239–247
Khan N, Bano A, Ali S, Babar MA (2020) Crosstalk amongst phytohormones from planta and PGPR under biotic and abiotic stresses. Plant Growth Regul 90:189–203
Klessig DF, Choi HW, Dempsey DMA (2018) Systemic acquired resistance and salicylic acid: past, present, and future. Mol Plant-Microbe Interact 31(9):871–888
Koo YM, Heo AY, Choi HW (2020) Salicylic acid as a safe plant protector and growth regulator. Plant Pathol J 36(1):1
Korver RA, Koevoets IT, Testerink C (2018) Out of shape during stress: a key role for auxin. Trends in plant science 23(9):783–793
Kunkel BN, Harper CP (2018) The roles of auxin during interactions between bacterial plant pathogens and their hosts. J Exp Bot 69(2):245–254
Kyndt T, Nahar K, Haeck A, Verbeek R, Demeestere K, Gheysen G (2017) Interplay between carotenoids, abscisic acid and jasmonate guides the compatible rice-Meloidogyne graminicola interaction. Front Plant Sci 8:951
Lang V, Schneider V, Puhlmann H, Schengel A, Seitz S, Schack-Kirchner H et al (2023) Spotting ethylene in forest soils—what influences the occurrence of the phytohormone? Biol Fertil Soils 59(8):953–972
Lefevere H, Bauters L, Gheysen G (2020) Salicylic acid biosynthesis in plants. Front Plant Sci 11:338
Leijten W, Koes R, Roobeek I, Frugis G (2018) Translating flowering time from Arabidopsis thaliana to Brassicaceae and Asteraceae crop species. Plan Theory 7(4):111
Li Z, He Y (2020) Roles of brassinosteroids in plant reproduction. Int J Mol Sci 21(3):872
Li G, Zhu C, Gan L, Ng D, Xia K (2015) GA 3 enhances root responsiveness to exogenous IAA by modulating auxin transport and signalling in Arabidopsis. Plant Cell Rep 34:483–494
Li S, Zheng H, Lin L, Wang F, Sui N (2021) Roles of brassinosteroids in plant growth and abiotic stress response. Plant Growth Regul 93:29–38
Li J, Tian Z, Han A, Li J, Luo A, Liu R, Zhang Z (2023) Integrative physiological, critical plant endogenous hormones, and transcriptomic analyses reveal the difenoconazole stress response mechanism in wheat (Triticum aestivum L.). Pesticide Biochemistry and Physiology 197:105688
Libohova Z, Seybold C, Wysocki D, Wills S, Schoeneberger P, Williams C et al (2018) Reevaluating the effects of soil organic matter and other properties on available water-holding capacity using the National Cooperative Soil Survey Characterization Database. J Soil Water Conserv 73(4):411–421
Liu H, Brettell LE, Singh B (2020) Linking the phyllosphere microbiome to plant health. Trends Plant Sci 25(9):841–844
Llanes A, Reginato M, Devinar G, Luna V (2018) What is known about phytohormones in halophytes? A review. Biologia 73:727–742
Lopez-Climent MF (2015) Water stress responses of tomato mutants impaired in hormone biosynthesis reveal abscisic acid, jasmonic acid and salicylic acid interactions. Front Plant Sci 6:997
Lu X, Hershey DM, Wang L, Bogdanove AJ, Peters RJ (2015) An ent-kaurene-derived diterpenoid virulence factor from Xanthomonas oryzae pv. oryzicola. New Phytol 206:295–302
Lumba S, Toh S, Handfield LF, Swan M, Liu R, Youn JY et al (2014) A mesoscale abscisic acid hormone interactome reveals a dynamic signaling landscape in Arabidopsis. Dev Cell 29(3):360–372
Lv B, Tian H, Zhang F, Liu J, Lu S, Bai M et al (2018) Brassinosteroids regulate root growth by controlling reactive oxygen species homeostasis and dual effect on ethylene synthesis in Arabidopsis. PLoS Genet 14(1):e1007144

Lymperopoulos P, Msanne J, Rabara R (2018) Phytochrome and phytohormones: working in tandem for plant growth and development. Front Plant Sci 9:1037
Ma Q, Grones P, Robert S (2018) Auxin signaling: a big question to be addressed by small molecules. J Exp Bot 69(2):313–328
Machin DC, Hamon-Josse M, Bennett T (2020) Fellowship of the rings: a saga of strigolactones and other small signals. New Phytol 225(2):621–636
Madzikane-Mlungwana O, Moyo M, Aremu AO, Plíhalová L, Doležal K, Van Staden J, Finnie JF (2017) Differential responses to isoprenoid, N 6-substituted aromatic cytokinins and indole-3-butyric acid in direct plant regeneration of Eriocephalus africanus. Plant Growth Regul 82:103–110
Majda M, Robert S (2018) The role of auxin in cell wall expansion. Int J Mol Sci 19(4):951
Manghwar H, Hussain A, Ali Q, Liu F (2022) Brassinosteroids (BRs) role in plant development and coping with different stresses. Int J Mol Sci 23(3):1012
Marco F, Busó E, Lafuente T, Carrasco P (2019) Spermine confers stress resilience by modulating abscisic acid biosynthesis and stress responses in Arabidopsis plants. Front Plant Sci 10:972
McAdam EL, Reid JB, Foo E (2018) Gibberellins promote nodule organogenesis but inhibit the infection stages of nodulation. J Exp Bot 69:2117–2130
Meena S, Taria S, Nagar S, Yadav S (2022) Phytohormone engineering: a potential approach for inducing abiotic stress tolerance in crop plants. Multidisciplinary 2022:35
Mega R, Abe F, Kim JS, Tsuboi Y, Tanaka K, Kobayashi H et al (2019) Tuning water-use efficiency and drought tolerance in wheat using abscisic acid receptors. Nat Plants 5(2):153–159
Miyazaki S, Hara M, Ito S, Tanaka K, Asami T, Hayashi KI et al (2018) An ancestral gibberellin in a moss Physcomitrella patens. Mol Plant 11(8):1097–1100
Mok MC (2019) Cytokinins and plant development—an overview. In Cytokinins, pp 155–166
Moulick D, Samanta S, Sarkar S, Mukherjee A, Pattnaik BK, Saha S et al (2021) Arsenic contamination, impact and mitigation strategies in rice agro-environment: An inclusive insight. Sci Total Environ 800:149477
Nadeem SM, Ahmad M, Zahir ZA, Kharal MA (2016) Role of phytohormones in stress tolerance of plants. In: Hakeem KR, Akhtar MS (eds) Plant, soil and microbes: volume 2: mechanisms and molecular interactions. Springer, pp 385–421
Narang J, Mohan H (2023) Biosensors for phytohormone Abscisic acid and its role in humans: A review. Sensors International 4:100234
Narayanan M, Ma Y (2023) Mitigation of heavy metal stress in the soil through optimized interaction between plants and microbes. J Environ Manag 345:118732
Nayak L, Lal MK, Tiwari RK, Kumar R, Kumar D, Altaf MA, Kumar A (2024) Modulation in phytohormone metabolism in plants under stress conditions. In: Ahanger MA (ed) Improving stress resilience in plants. Academic Press, pp 297–319
Naylor D, Coleman-Derr D (2018) Drought stress and root-associated bacterial communities. Front Plant Sci 8:2223
Nolan TM, Vukašinović N, Liu D, Russinova E, Yin Y (2020) Brassinosteroids: multidimensional regulators of plant growth, development, and stress responses. Plant Cell 32(2):295–318
Novák O, Hényková E, Sairanen I, Kowalczyk M, Pospíšil T, Ljung K (2012) Tissue-specific profiling of the Arabidopsis thaliana auxin metabolome. Plant J 72(3):523–536
Novotná, L. (2023). Creating and testing bioluminescence sensors for cytokinins in Arabidopsis thaliana.
O'Banion BS, O'Neal L, Alexandre G, Lebeis SL (2020) Bridging the gap between single-strain and community-level plant-microbe chemical interactions. Mol Plant-Microbe Interact 33(2):124–134
Oshchepkov MS, Kalistratova AV, Savelieva EM, Romanov GA, Bystrova NA, Kochetkov KA (2020) Natural and synthetic cytokinins and their applications in biotechnology, agrochemistry and medicine. Russ Chem Rev 89(8):787
Peerzada YY, Iqbal M (2021) Leaf senescence and ethylene signaling. In: Aftab T, Hakeem KR (eds) Plant growth regulators: signalling under stress conditions. Springer, pp 153–171

Pérez-Llorca M, Muñoz P, Müller M, Munné-Bosch S (2019) Biosynthesis, metabolism and function of auxin, salicylic acid and melatonin in climacteric and non-climacteric fruits. Front Plant Sci 10:136
Perrot-Rechenmann C (2010) Cellular responses to auxin: division versus expansion. Cold Spring Harb Perspect Biol 2(5):a001446
Petřík M, Johnson A, Smith B, Lee C (2023) Ultra-high performance liquid chromatography-tandem mass spectrometry (UHPLC-MS/MS) is the basis of modern CK measurement methods since it allows the separation of CK isomeric forms that occur naturally in plants. Journal of Chromatography A 1567:123456
Riaz A, Kanwal F, Börner A, Pillen K, Dai F, Alqudah AM (2021) Advances in genomics-based breeding of barley: molecular tools and genomic databases. Agronomy 11(5):894
Rosales MA, Maurel C, Nacry P (2019) Abscisic acid coordinates dose-dependent developmental and hydraulic responses of roots to water deficit. Plant Physiol 180(4):2198–2211
Salem MSZ (2018) Biological networks: An introductory review. Journal of Proteomics and Genomics Research 2(1):41–111
Sankari M, Rao PR, Hemachandran H, Pullela PK, Tayubi IA, Subramanian B et al (2018) Prospects and progress in the production of valuable carotenoids: insights from metabolic engineering, synthetic biology, and computational approaches. J Biotechnol 266:89–101
Santo Domingo Martínez, M (2023) Genetic dissection of ETHQV8. 1, a QTL related with climacteric fruit ripening in melon.
Sauer M, Robert S, Kleine-Vehn J (2013) Auxin: simply complicated. J Exp Bot 64(9):2565–2577
Sayer EJ, Oliver AE, Fridley JD, Askew AP, Mills RT, Grime JP (2017) Links between soil microbial communities and plant traits in a species-rich grassland under long-term climate change. Ecology and Evolution 7(3):855–862
Sayer EJ, Crawford JA, Edgerley J, Askew AP, Hahn CZ, Whitlock R, Dodd IC (2021) Adaptation to chronic drought modifies soil microbial community responses to phytohormones. Commun Biol 4(1):516
Schaller GE, Street IH, Kieber JJ (2014) Cytokinin and the cell cycle. Curr Opin Plant Biol 21:7–15
Schaller GE, Bishopp A, Kieber JJ (2015) The yin-yang of hormones: cytokinin and auxin interactions in plant development. Plant Cell 27(1):44–63
Seif El-Yazal A, Samir, Seif El-Yazal MA, Dwidar EF, Rady MM (2015) Phytohormone crosstalk research: cytokinin and its crosstalk with other phytohormones. Curr Protein Pept Sci 16(5):395–405
Sharma E, Sharma R, Borah P, Jain M, Khurana JP (2015) Elucidation of abiotic stress signaling in plants. Springer
Sheikh TMM, Zhou D, Ali H, Hussain S, Wang N, Chen S et al (2023) Volatile Organic Compounds Emitted by the Biocontrol Agent Pythium oligandrum Contribute to Ginger Plant Growth and Disease Resistance. Microbiology Spectrum 11(4):e01510–e01523
Sheoran S, Kaur Y, Kumar S, Shukla S, Rakshit S, Kumar R (2022) Recent advances for drought stress tolerance in maize (Zea mays l.): Present status and future prospects. Frontiers. Plant Science 13:872566
Sikder MM, Vestergård M (2020) Impacts of root metabolites on soil nematodes. Front Plant Sci 10:1792
Sikder MM, Vestergård M, Kyndt T, Kudjordjie EN, Nicolaisen M (2021) Phytohormones selectively affect plant parasitic nematodes associated with Arabidopsis roots. New Phytol 232(3):1272–1285
Simmons CR, Lafitte HR, Reimann KS, Brugière N, Roesler K, Albertsen MC et al (2021) Successes and insights of an industry biotech program to enhance maize agronomic traits. Plant Sci 307:110899
Singh J, Nigam R, Singh R, Kumar A, Kumar A (2018) Effect of gibberellic acid and cycocel on growth, flowering and yield of chrysanthemum (Dendranthema grandiflora ramat) cv. birbal sahni. J Pharmacogn Phytochem 7(1S):2753–2758
Sisler EC (2018) Ethylene-binding components in plants. In: The plant hormone ethylene, pp. 81–99. CRC Press

Soliman S, Wang Y, Han Z, Pervaiz T, El-Kereamy A (2022) Strigolactones in plants and their interaction with the ecological microbiome in response to abiotic stress. Plan Theory 11(24):3499
Song Y, Xiang F, Zhang G, Miao Y, Miao C, Song CP (2016) Abscisic acid as an internal integrator of multiple physiological processes modulates leaf senescence onset in Arabidopsis thaliana. Front Plant Sci 7:181
Song GC, Jeon JS, Sim HJ, Lee S, Jung J, Kim SG et al (2022) Dual functionality of natural mixtures of bacterial volatile compounds on plant growth. Journal of experimental botany 73(2):571–583
Sponsel VM (2016) Signal achievements in gibberellin research: the second half-century. In Annual plant reviews, volume 49: gibberellins, the, pp 1–36
Sreenivasulu N, Fernie AR (2022) Diversity: current and prospective secondary metabolites for nutrition and medicine. Curr Opin Biotechnol 74:164–170
Stassen MJ, Hsu SH, Pieterse CM, Stringlis IA (2021) Coumarin communication along the microbiome–root–shoot axis. Trends Plant Sci 26(2):169–183
Struck M, Álvarez-Venegas R, Fang H, Yeh SC, Panarese V, Zander M et al (2019) Spatio-temporal dynamics of phytohormones in the tomato graft healing process. Journal of Experimental Botany 71(10):2898–2914
Sugawara S, Hishiyama S, Jikumaru Y, Hanada A, Nishimura T, Koshiba T et al (2009) Biochemical analyses of indole-3-acetaldoxime-dependent auxin biosynthesis in Arabidopsis. Proc Natl Acad Sci 106(13):5430–5435
Sussmilch FC, Brodribb TJ, McAdam SA (2017) Up-regulation of NCED3 and ABA biosynthesis occur within minutes of a decrease in leaf turgor but AHK1 is not required. J Exp Bot 68(11):2913–2918
Tan K-E, Lan W, Dou D, Huan J (2023) Detection of Cytoplasmic and Nuclear Circular RNA via RT-qPCR. International Journal of Molecular Sciences 24(23):16924
Thakur N, Sharma V, Kishore K (2016) Leaf senescence: an overview. Indian J Plant Physiol 21:225–238
Thomas B (2016) Encyclopedia of applied plant sciences. Academic Press
Trinder S, Askew AP, Whitlock R (2020) Climate-driven evolutionary change in reproductive and early-acting life-history traits in the perennial grass Festuca ovina. J Ecol 108(4):1398–1410
Van Den Hoogen J, Geisen S, Routh D, Ferris H, Traunspurger W, Wardle DA et al (2019) Soil nematode abundance and functional group composition at a global scale. Nature 572(7768):194–198
Velappan Y, Signorelli S, Considine MJ (2017) Cell cycle arrest in plants: what distinguishes quiescence, dormancy and differentiated G1? Ann Bot 120(4):495–509
Velasquez SM, Barbez E, Kleine-Vehn J, Estevez JM (2016) Auxin and cellular elongation. Plant Physiol 170(3):1206–1215
Wang KLC, Li H, Ecker JR (2002) Ethylene biosynthesis and signaling networks. Plant Cell 14(suppl_1):S131–S151
Wang JY, Haider I, Jamil M, Fiorilli V, Saito Y, Mi J et al (2019) The apocarotenoid metabolite zaxinone regulates growth and strigolactone biosynthesis in rice. Nat Commun 10(1):810
Wasternack C, Song S (2017) Jasmonates: biosynthesis, metabolism, and signaling by proteins activating and repressing transcription. J Exp Bot 68(6):1303–1321
Wasternack C, Strnad M (2016) Jasmonate signaling in plant stress responses and development–active and inactive compounds. New Biotechnol 33(5):604–613
Went FW (1928) Growth material and growth. Recueil des travaux botaniques néerlandais 25(1/2):1–116
Wilschut RA, Geisen S (2021) Nematodes as drivers of plant performance in natural systems. Trends Plant Sci 26(3):237–247
Wu F, Gao Y, Yang W, Sui N, Zhu J (2022) Biological functions of strigolactones and their crosstalk with other phytohormones. Front Plant Sci 13:821563
Wu K, Wang L, Wu Z, Liu Z, Li Z, Shen J et al (2023) Selenite reduced cadmium uptake, interfered signal transduction of endogenous phytohormones, and stimulated secretion of tartaric acid based on a combined analysis of non-invasive micro-test technique, transcriptome and metabolome. Plant Physiol Biochem 206:108107

Wybouw B, De Rybel B (2019) Cytokinin–a developing story. Trends Plant Sci 24(2):177–185
Xie Q, Essemine J, Pang X, Chen H, Cai W (2020) Exogenous application of abscisic acid to shoots promotes primary root cell division and elongation. Plant Sci 292:110385
Xu X, Jibran R, Wang Y, Dong L, Flokova K, Esfandiari A et al (2021) Strigolactones regulate sepal senescence in Arabidopsis. J Exp Bot 72(15):5462–5477
Yadav S, Modi P, Dave A, Vijapura A, Patel D, Patel M (2020) Effect of abiotic stress on crops. In Sustainable crop production, Vol 3
Yang Y, Saand MA, Huang L, Abdelaal WB, Zhang J, Wu Y et al (2021) Applications of multiomics technologies for crop improvement. Frontiers in Plant Science 12:563953
Yao L, Zhang J, Lu J, Chen D, Song S, Wang H et al (2023) Revealing the influence of microbiota on the flavor of kombucha during natural fermentation process by metagenomic and GC-MS analysis. Food Research International 169:112909
Yoneyama K, Brewer PB (2021) Strigolactones, how are they synthesized to regulate plant growth and development? Curr Opin Plant Biol 63:102072
Yoon Y, Seo DH, Shin H, Kim HJ, Kim CM, Jang G (2020) The role of stress-responsive transcription factors in modulating abiotic stress tolerance in plants. Agronomy 10(6):788
Yuan Z, Zhang D (2015) Roles of jasmonate signalling in plant inflorescence and flower development. Curr Opin Plant Biol 27:44–51
Zenda T, Liu S, Dong A, Li J, Wang Y, Liu X et al (2021) Omics-facilitated crop improvement for climate resilience and superior nutritive value. Frontiers in Plant Science 12:774994
Zhao F, Du F, Oliveri H, Zhou L, Ali O, Chen W et al. (2019) A microtubule-mediated mechanical feedback controls leaf blade development in three dimensions. bioRxiv, 604710
Zhao B, Liu Q, Wang B, Yuan F (2021) Roles of phytohormones and their signaling pathways in leaf development and stress responses. J Agric Food Chem 69(12):3566–3584
Zhu H, Li C, Gao C (2020) Applications of CRISPR–Cas in agriculture and plant biotechnology. Nature Reviews Molecular Cell Biology 21(11):661–677
Zhu X, Wang K, Yan H, Liu C, Zhu X, Chen B (2022) Microfluidics as an emerging platform for exploring soil environmental processes: a critical review. Environ Sci Technol 56(2):711–731

The Potential for Plant Growth Regulators to Impact Crop Productivity in Future Agricultural System

12

Urvasha Patyal, Khushboo, Ritu Bala, Mukesh Narwat, and Manpreet Kaur

Abstract

In contemporary agriculture, plant growth regulators are commonly utilised for plant growth promotion. By altering a crop's rate, pattern, responses to the internal and external factors that control development from germination to post harvest through vegetative growth, reproductive development, maturity, and senescence or aging, plant growth regulators can modify a crop. Plant growth-promoting microorganisms are well known to increase plant growth and crop harvest by the production of various secondary metabolites and atmospheric nitrogen conversion into ammonia, through induced systemic resistance and systemic acquired resistance. It enables more sustainable agriculture without impairing ecosystem function. This chapter will focus on the role of microbes as plant growth regulators, stress tolerance, soil health management and sustainable crop production, current progress, and the challenges of the sustainable crop production.

Keywords

PGPs · IAA · Siderophore · Phosphate solubilisation

U. Patyal
Department of Biosciences and Technology, Maharishi Markandeshwar University, Mullana, Ambala, Haryana, India

Khushboo · R. Bala · M. Narwat
Department of Microbiology, School of Bioengineering and Biosciences, Lovely Professional University, Phagwara, Punjab, India

M. Kaur (✉)
Department of Biotechnology, School of Bioengineering and Biosciences, Lovely Professional University, Phagwara, Punjab, India

© The Author(s), under exclusive license to Springer Nature Singapore Pte Ltd. 2024

M. Faizan, S. Hayat (eds.), *Plant Growth Regulators: Resilience for Sustainable Agriculture*, https://doi.org/10.1007/978-981-97-2918-0_12

12.1 Introduction

Plant growth regulators (PGRs) are organic substances that, in minute quantities, encourage, hinder, or alter any biological processes in plants. PGRs are distinct from nutrients, which provide either energy or minerals. Both synthesised molecules, also known as chemical analogues, and organically produced growth-stimulating substances, also known as phytohormones, are included in the term PGR (Bisht et al. 2018). As for "classical" phytohormones, there are five recognised groups: auxins, gibberellins, cytokinins, abscisic acid, and ethylene (Müller and Munné-Bosch 2021). More recently, oligosaccharins, brassinosteroids, jasmonates, salicylates, and polyamines have been identified as a few more substances that can control different aspects of plant growth and development (Creelman and Mullet 1997).

Our knowledge of PGRs' controlled role in plant growth and development has been substantially expanded by the development of advanced techniques for determining their existence and quantitative evaluation, hormone mutants, and potent biological techniques. In order to improve agricultural crops by regulating their developmental phases, from sprouting to growth of vegetative tissue, reproductive growth, maturity, ageing, and post-harvest maintenance, both natural and synthetic growth regulators have been used more and more in horticulture and agriculture (Sabagh et al. 2021). The main emphasis in the field of synthetic bioregulators has been on herbicides and fungicides. The expense and difficulty of finding novel PGRs are estimated to be several times higher than those of herbicides.

Therefore, the various applications of PGRs in agriculture have been surpassed by the weed-killing aspects of plant development control. Increasing concerns about the environment have made it harder to make investments in research and development and register new PGRs, which has limited the enormous scope of using PGRs effectively to improve productivity effectiveness of the world's key agricultural crops (Gupta et al. 2023). In addition to facilitating the production and development of novel PGR compounds, a thorough grasp of their biological functions and mechanisms of action can help anticipate and avoid any negative side effects (Yan et al. 2023). Therefore, it is very appropriate to critically evaluate this topic, covering both fundamental and applied features.

An optimal agricultural system preserves and enhances the quality of the soil, benefits producers and consumers, and safeguards the environment and public health. Additionally, they provide food security and produce enough food to feed the globe's human population. Farmers may be able to achieve these goals at a minimal cost with a sustainable agricultural technique. Using a biological method reduces the need for fertilisers and pesticides. This approach increases agricultural output while being economical and sustainable (Gamage et al. 2023).

The use of microbial consortium decreased pollution and the loss of organic matter in the soil. Thus, rhizobacteria that promote plant development provide an economical and ecologically friendly method of raising agricultural yields. Numerous research works have proven the efficacy of PGM rhizobacteria in both typical and stressful environments (Dos Santos et al. 2020). PGP mycorrhizae and rhizobacteria respond differently to various environmental circumstances (Raklami et al. 2019).

As the most developed industry, agriculture utilises a variety of chemical fertilisers to boost output and feed the world's expanding population. Both the environment and human health are seriously threatened by fertilisers that include chemicals (Bisht and Singh Chauhan 2021). Eco-friendly methods are being used to increase crop yield in order to solve this issue. One strategy that might lead to the next agricultural green revolution is the use of beneficial bacteria as bio-pesticides and fertilisers in the agricultural sector (Mishra et al. 2020). Beneficial fungi that are related to plants are crucial to their development and growth because they solubilise a variety of insoluble and inaccessible nutrients and produce plant growth regulators, among other processes (Begum et al. 2019). Additionally, fungi aid in the reduction of biotic stressors such as plant pests and pathogens as well as abiotic stressors such as temperature, salt, drought, and heavy metals (Wahab et al. 2023). One new area in the agricultural industry for development and improving crop productivity for environmentally friendly farming is the use of fungi as biofertilisers (Odoh et al. 2020).

By lowering the demand for artificial agricultural chemicals and water, plant growth regulators (PGRs) enable farmers to adopt sustainable farming techniques like precision farming (Dhanaraju et al. 2022). It also makes it possible for crops to develop well with minimal irrigation. Growth-regulating agents can be beneficial to certain agricultural practices (such as crop preservation, canopy management, and seed planting) (Dhanaraju et al. 2022), though there are countless potential applications for growth regulators in the future.

It is well known that certain crop plant types or varieties are more vulnerable to pests and diseases than others. These variations may be linked to minor variations in the range of endogenous growth regulators, which might alter the attractiveness of plants to infections. Developing protoplasts would be isolated from other plant parts where they would typically obtain the endogenous regulators required for healthy functioning; growth regulators would undoubtedly be important in any system involving protoplast growth (Reed and Bargmann 2021).

12.2 Microbes as Plant Growth Regulators

The most significant contenders for improving the fertility and health of soil are soil microorganisms. Arbuscular mycorrhizae (AM) and plant growth-promoting microorganisms are utilised to increase agricultural crop yields and plant growth in both normal and stressful environments (Begum et al. 2019). Through a variety of processes, it enhances plant growth on a range of physiological parameters in response to external stimuli. Plant growth regulators, the synthesis of various metabolites, the direct and indirect conversion of atmospheric nitrogen to ammonia, and other processes are among the mechanisms behind growth promotion (Backer et al. 2018). Furthermore, it confers resistance to biotic agents, or infections, via means of systemic acquired resistance and induced systemic resistance (Yu et al. 2022). The interaction of plant microbes promotes plant growth, controls disease in a changing environment, and allows for more sustainable agriculture without sacrificing ecosystem function. Plant growth-promoting rhizobacteria (PGPR) and arbuscular fungus are two examples of the positive plant-microbe interactions that are maintained by plant growth regulators (Sagar et al. 2021). Rhizospheric soil's microbial variety keeps the soil productive and healthy. Therefore, the widespread usage of rhizobacteria that promote plant development aids in achieving sustainable agricultural output both under normal and stressful conditions (Chandran et al. 2021).

Applying advantageous microorganisms might be a viable substitute for toxic chemical pesticides and fertilisers. Microbes have a significant impact on soil management and crop yield. Soil microorganisms linked with plants are essential for plant growth and development, including nutrient cycling and crop yield. Although the interactions between plants and microorganisms play a significant role in regulating the functioning of ecosystems, these interactions vary widely and are contingent upon the availability of nutrients (Wang et al. 2022). Plant growth-promoting (PGP) microbes are primarily employed to promote plant development through a variety of mechanisms, including nitrogen fixation and plant growth control (Backer et al. 2018). These days, the agricultural sector has to adopt a different approach in order to make agriculture more effective and sustainable. Sustainable agricultural production has to be increased by innovative and promising approaches in order to feed the growing global population. The use of plant growth-promoting bacteria, which is sustainable and advantageous for farmers and soil health, is how agriculture is improved (Pathania et al. 2020).

12.3 Microbial Processes that Promote Plant Growth

Rhizobacteria (PGPR), which is necessary for plant growth promotion, may be used in a variety of ways by microbes. The two main strategies for accelerating plant development are indirect and direct mechanisms (Saeed et al. 2021) (Fig. 12.1). One of the indirect mechanisms is mitigating the adverse effects of infections through other means. For instance, siderophore inhibits the development of infections by decreasing their access to iron. Furthermore, it may offer protection against infections and boost resistance through a process known as induced systemic resistance.

The availability of several beneficial compounds produced by bacteria provides a direct pathway. Similar to this, they promote the development of the host plant by releasing siderophores, solubilising minerals, and increasing nutrient availability (Zhang et al. 2023) (Table 12.1).

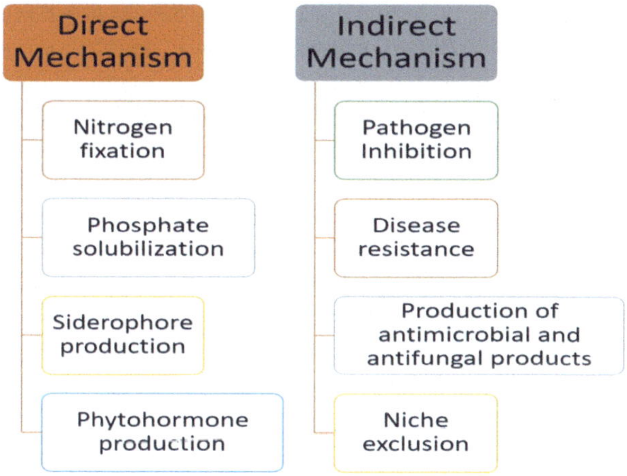

Fig. 12.1 Plant growth promotion by different mechanisms

Table 12.1 Microorganism with their sources and their role in plant growth

Plant growth-promoting microbes	Source	Plant growth regulation	References
P. chlororaphis	*Coffea arabica* L.	Effective utilisation of insoluble phosphate from soil	Muleta et al. (2013)
Bacillus amyloliquefaciens 5113	Wheat	Encourage plant development in drought-prone areas and boost wheat plant enzyme activity	Abd El-Daim et al. (2019)
Bacillus mucilaginosus	*Panax*	Transforms insoluble forms of potassium (K) into soluble forms	Li et al. (2022)
B. thuringiensis	*Lavandula dentate*	Increased potassium and proline levels and decreased glutathione reductase and ascorbate levels because of IAA	Abdelaal et al. (2021)
Serratia nematodiphila PEJ1011	*Capsicum annuum* L.	Enhancing plant development and managing biocontrol to prevent plant illnesses	Kang et al. (2015)

12.4 Direct Mechanism of Plant Growth Promotion

12.4.1 Nitrogen Fixation

As nitrogen (N) makes up 78% of the atmosphere, it is the most important element for plant growth and development (Bloom 2015). Plants may use it even if it is only available to them in molecular form through biological nitrogen fixation to ammonia. The nitrogen fixation process is powered by the oxygen-sensitive nitrogenase enzyme complex. Nitrogenase enzymes use ATP as an energy source to convert atmospheric nitrogen into ammonia (Hanrahan and Chan 2005). One crucial growth factor that influences plant development and output is biological nitrogen fixation.

Numerous microbial species, including those from the genera *Bacillus*, *Azospirillum*, *Pseudomonas*, *Enterobacter*, *Flavobacterium*, *Erwinia*, and *Rhizobium*, are capable of fixing nitrogen (Ahemad and Kibret 2014). There are two types of nitrogen-fixing microbes: symbiotic and non-symbiotic (Bekele and Yilma 2021). Symbiotic bacteria fix nitrogen. Among the symbiotic nitrogen-fixing bacteria are rhizobia, which associate symbiotically with leguminous plants. *Azospirillum*, *Azotobacter*, *Gluconacetobacter diazotrophicus*, and *cyanobacteria* are among the non-symbiotic bacteria. Symbiotic relationships fix the majority of atmospheric nitrogen, with non-symbiotic sources contributing very little. A few of these microorganisms for root nodule with their corresponding host plant root product are important in plant-microbe communication (Bekele and Yilma 2021). Numerous stress factors, such salt and dryness, prevent nodulation, which leads to nitrogen fixation.

12.4.2 Phosphate Solubilisation

The most important micronutrient for the growth and development of plants is phosphorus. Phosphate ions are the form in which plants absorb phosphorus from the soil (Johan et al. 2021). Plant growth-promoting rhizobacteria (PGPR) are microorganisms that may mineralise phosphate and solubilise it, making it accessible to plants. In a natural agricultural system, the biogeochemical cycle is regulated by phosphorus-solubilising bacteria (Figueiredo et al. 2016). But the majority of the phosphorus in the soil is insoluble and inaccessible to plants. Phosphate's bioavailability to plants is dependent on the environment, microorganisms, and plants. The mobilisation of phosphorus in soil and its availability to plants may be improved by the association of plant microorganisms. In nature, rhizobacteria that solubilise phosphate are widely distributed, with varying populations seen in different soil types. It is mobilised by phosphate-solubilising microorganisms (PSM), which include fungus and bacteria, through the production of phosphatase and organic acid (Kalayu 2019). Microbes that solubilise phosphate are described as belonging to several species of bacteria and fungi. As a biofertiliser, phosphate-solubilising microorganisms (PSM) promote plant development and give plants phosphorus in a sustainable manner. The most powerful acid that phosphate-solubilising microorganisms can make is 2-ketogluconic acid. Thus, using phosphate-solubilising microorganisms in an agricultural system might be a sustainable and economical strategy (Timofeeva et al. 2022).

12.4.3 Siderophore Production

Iron is necessary for enzymatic and physiological processes in almost all living things. Microbes can access iron through the siderophore, an iron-chelating molecule secreted by bacteria. Microbial siderophores are the primary mechanism via which plants absorb iron. Siderophore's affinity for iron is greater. The siderophore transports into the bacterial cell and attaches to iron with ease. Plants can directly absorb iron from the bacterial siderophore, or they can use ligand exchange reactions to get iron from the siderophore (Leventhal et al. 2019). The amount of chlorophyll a and chlorophyll b in plant leaves is one of siderophore-iron's main effects. *Aspergillus fumigatus* is susceptible to the siderophore activity of pseudochelin A, a newly discovered siderophore of *Pseudoalteromonas piscicida* S2040 (Sonnenschein et al. 2017).

12.4.4 Phytohormone Production

Indole acetic acid (IAA), a key auxin, regulates a number of physiological processes, including cell signalling, plant development, and the creation of disease resistance. Rhizobacteria that colonise roots efficiently and provide more IAA for plant growth and development are linked to phytohormones (Tang et al. 2023). Another common phytohormone that is essential for both biotic and abiotic stress tolerances is ethylene. Under stressful conditions, ethylene inhibits the growth of root hair and root elongation (Fatma et al. 2022). Additionally, another hormone found in plants, abscisic acid, regulates physiological processes and stress tolerance, particularly in drought situations (Sah et al. 2016). In times of drought stress, PGPR increases the concentration of abscisic acid and increases plant stress tolerance. It is evident from the explanation above that phytohormones influence plants' endogenous mechanisms and promote plant development. These phytohormones primarily stimulate the production of lateral roots and root hairs, which aid in the uptake of nutrients and the absorption of water.

12.5 Indirect Mechanism of Plant Growth Promotion

By a variety of processes, bacteria that promote plant development can either reduce or prevent the harmful effects of pathogens. This is known as the indirect impact. Biocontrol refers to the use of microorganisms to manage plant disease. It is an economical and ecologically friendly way to lessen the harmful effects of infections. Rhizobacteria release a variety of antifungal and antibacterial agents that aid in disease resistance (Saeed et al. 2021). The main indirect effects include niche exclusion, increased systemic resistance, competition for nutrients, and antifungal metabolites. Numerous metabolites, including 2,4-diacetylphloroglucinol, HCN, phenazines, and pyrrolnitrin, which are significant biocontrolling agents, have been discovered to be secreted by rhizobacteria (Wang et al. 2015).

A broad variety of microorganisms are found in PGPR, including bacteria that fix nitrogen and solubilise phosphate, which are beneficial soil microbiota for plant growth and development. These bacteria associate with other microorganisms to produce nodules. Through a variety of activities, including as improving soil nutrient availability, producing plant growth regulators, controlling plant growth characteristics, and maintaining soil health, these bacteria promote plant growth and development (De Andrade et al. 2023).

In addition, biological nitrogen fixation, potassium, and phosphate solubilisation are involved in this connection, all of which support plant growth and development. For the growth of shoots and roots, plants need phytohormones such as cytokines, gibberellin, and indole acetic acid (IAA) (Mukherjee et al. 2022). At various stages of a plant's growth and development, several hormones are responsible. For instance, IAA influences the differentiation of plant tissues and root growth.

Because they increase surface area, IAA-induced root development and improved root hair production aid in the absorption of water and nutrients (Tang et al. 2023). This led to significant growth and a high production of crops. Gibberellins are a key player in the regulation of plant growth, having an impact on flowering, fruit set, stem elongation and development, and seed germination (Castro-Camba et al. 2022). Additionally, it reduces the production of ethylene and controls the quantity of ethylene during stress conditions by synthesising 1-aminocyclopropane-1-carboxylate (ACC) deaminase through PGPR (Etesami et al. 2014).

In agricultural fields, arbuscular mycorrhizae (AM) are the most prevalent type of fungus. Approximately 80% of higher plant forms have a fungus relationship. In the realm of agriculture, it is essential to the cycling of nutrients. Mycorrhizae are also essential for enhancing the physical characteristics of soil, preserving plant nutrient availability, and aiding in soil aggregation (Begum et al. 2019). Under stressful conditions, mycorrhizae create a variety of metabolites, including vitamins, amino acids, phytohormones, and mineralisation. The relationship between PGPR and mycorrhizae fungus is thought to have an impact on temperature, moisture content, nutrient levels, and rhizospheric microorganisms. Helping plants grow and develop both normally and under stress is another of this association's cumulative impacts.

12.6 The Potential for Plant Growth Regulators to Impact Crop Productivity in Future Agriculture System

In the ever-evolving landscape of agriculture, the quest for enhancing crop productivity to meet the growing global demand for food remains a paramount challenge. As traditional farming practices encounter the complexities of climate change, resource scarcity, and population expansion, the agricultural sector seeks innovative solutions to optimise yield and resource utilisation (Hossain et al. 2020).

Plant growth regulators encompass a diverse array of natural and synthetic compounds that regulate plant growth processes, including cell division, elongation, and differentiation (Raza et al. 2019). These regulators play pivotal roles in orchestrating plant responses to environmental stimuli, such as light, temperature, and nutrient availability. In recent years, the potential of harnessing PGRs to modulate crop productivity has garnered considerable attention within the scientific and agricultural communities. Plant growth regulators (PGRs) play a pivotal role in enhancing the tolerance of plants to abiotic stressors, a critical consideration in the face of changing climatic conditions (Sharma et al. 2020). Abiotic stresses, such as drought, salinity, extreme temperatures, and nutrient deficiencies, pose significant challenges to crop productivity. PGRs, through having the ability to modulate various physiological and biochemical processes, offer a valuable tool for bolstering plants' resilience in adverse environments. For instance, certain PGRs can regulate stomatal closure, reducing water loss during drought conditions. Additionally, PGRs may influence the expression of stress-responsive genes, enabling plants to activate adaptive mechanisms against environmental challenges (Xuan et al. 2020). The intricate interplay between PGRs and stress signalling pathways holds promise for the development of crops with heightened tolerance to adverse conditions, contributing to the establishment of more robust and climate-resilient agricultural systems (Fig. 12.2).

Harnessing the potential of PGRs in this context represents a forward-looking strategy to ensure food security and sustainable agriculture in the face of a changing climate (Chakraborty et al. 2019). In changing weather conditions, different kinds of stresses can make it hard for seeds to grow, seedlings to develop, and plants to thrive. This leads to a serious drop in how much crops grow and how much money they can make (Shabbir et al. 2021). Plants make substances called plant growth regulators (PGRs), like gibberellins, auxin, cytokinin, abscisic acid, ethylene, jasmonic acid, brassinosteroid, nitric oxide, salicylic acid, and strigolactone, to deal with stress (Ahmad et al. 2021). These substances work together, sometimes helping each other and sometimes working against each other, in a complicated way (Aerts et al. 2021). This review looks at recent progress in understanding how PGRs help plants defend against stresses (Basu et al. 2021). It also finds areas where we need to learn more and suggests ideas for future research. The article also talks about certain substances, like polyamines, that act like PGRs against different stresses (Chen et al. 2019).

Fig. 12.2 Overview of abiotic stress and effects on plant growth and development

Soil health management stands as the cornerstone of sustainable crop production, embodying the interconnected relationship between the land and agricultural productivity. Sustainable agriculture seeks to balance the utilisation of natural resources with the preservation of ecosystem integrity, and soil health emerges as a linchpin in this delicate equilibrium. Effective soil health management involves practices that enhance soil structure, fertility, and biodiversity while minimising the use of synthetic inputs. Techniques such as cover cropping, crop rotation, and organic matter incorporation foster soil structure and microbial diversity, promoting nutrient cycling and water retention. By prioritising soil health, farmers can mitigate the impact of soil degradation, erosion, and nutrient depletion, safeguarding the long-term viability of agricultural systems. Sustainable crop production, therefore, is intricately woven into the fabric of soil health management, creating a harmonious cycle where resilient soils foster productive crops, while the crops, in turn, contribute to the overall health and sustainability of the soil (Cardinale et al. 2018). This synergy is essential for ensuring food security, safeguarding environmental health, and meeting the demands of a growing global population in a manner that respects the finite resources of our planet.

The future of agriculture demands sustainable and resilient practices that can navigate the challenges posed by climate variability and resource limitations. PGRs offer a compelling avenue for innovation, holding the promise to optimise resource use efficiency, mitigate stress responses, and bolster overall crop performance. This potential impact on crop productivity is multifaceted, encompassing improved photosynthetic efficiency, enhanced nutrient uptake, and increased resistance to environmental stressors. This article explores the evolving landscape of PGRs in agriculture, emphasising their potential to revolutionise the way we cultivate crops. By delving into the intricate mechanisms through which PGRs influence plant growth and development, we can gain insights into how these regulators might be strategically employed to propel future agriculture systems toward greater sustainability and productivity. As we navigate the intricate web of plant signalling pathways and molecular responses, the utilisation of PGRs emerges as a promising frontier in the pursuit of a more resilient and productive global food supply (Fig. 12.3).

12.7 Soil Health Management and Sustainable Crop Production

Growing conditions that are less than perfect highlight the need of healthy soil in maintaining crop productivity. A crop's ability to endure climatic fluctuations, such as brief but intense precipitation episodes and droughts within a particular season, is enhanced by robust soils. Fields using conventional tillage methods (such as deep ripping and chisel plough) were most severely affected by the variable yield drop in maize and soybean production that the severe drought of 2012 brought to Iowa (Cole et al. 2007) weather patterns that are becoming more and more unpredictable put crops at more risk; thus conservation planning has to be done with more care to

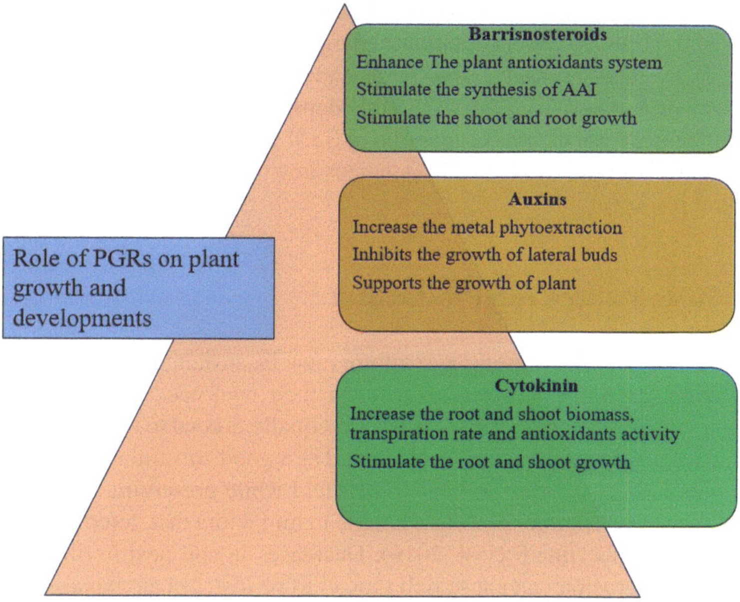

Fig. 12.3 Role of PGRs on plant growth and development

minimise negative effects on crop yield and soil health. Since soil is their main source of mineral nutrients, the great majority of biological entities are nourished by it. A sound soil management system confirms that sufficient nutrients must enter the food chain and that these components must never become dangerous or insufficient for plants. Crop productivity, environmental sustainability, and human well-being are all impacted by soil management, both directly and indirectly.

The primary objective in agriculture is to minimise unfavourable environmental conditions while maintaining a suitable increase in output. The predicted increase in world's population and the consequent need to expand food production make soil management more important than ever in the years to come (Kelley 1983). Achieving a sustainable food supply for future generations while ensuring human well-being will require managing soils in a sustainable manner through adequate nutrient management and soil conservation practices. This will be challenging. Adopting agricultural solutions and sustainable practices will be the only way to make it happen.

Agriculture should be conducted in a sustainable and environmentally sound manner if the natural world is to be sustainable. For the purpose of feeding the world's growing population, enough nutrient-dense food must be produced at affordable rates while also protecting environmental and natural agricultural resources through sustainable agriculture systems and practices (Giller et al. 2021). Achieving food security, addressing issues of poverty, promoting rural and agricultural development, and enhancing nutrition all depend on sustainable soil management, which is essential for effective land management. Since soil is the most widespread terrestrial carbon storage system on Earth, it produces over 95% of the food consumed worldwide. Sustainable development methods (SSM) are a useful tactic for mitigating the effects of climate change and for protecting biodiversity and valuable ecosystem services. Large-scale and small-scale agricultural producers around the world, whose livelihoods directly depend on their soil resources, stand to gain significantly from the broad adoption of SSM practices. The implementation of sustainable agricultural systems and practices is vital to enhance the protection of agricultural resources.

12.8 Sustainable Crop Production

To meet the needs of the present, agriculture must be sustainable without making it more difficult for future generations to meet their own needs. Within given this, attaining financial success in the short term is equally crucial to sustainable agriculture practices that have always been promoted as a good substitute for conventional farming practices in an effort to boost crop yield while preserving soil health. The local soil types and climate, however, are the main factors that determine the benefits of these systems (Imadi et al. 2016). Decreases in soil health, tillage erosion, wind, and water erosion can cause rich topsoil to be lost, as can management practices like over-tilling sensitive soils and removing too much residue. These factors can also lead to reduced soil organic carbon (SOC) content, decreased soil resilience, and decreased crop productivity, phytohormones, and soil bacteria for sustainable crop production. Since it can regulate all aspects of plant growth, the most prevalent, extensively researched, and naturally occurring auxin is indole-3-acetic acid (IAA). A few of these are fruit setting and ripening, apical dominance, vascular tissue differentiation, growth elongation, and lateral root initiation.

IAA synthesis can take place through different pathways such as the following:

1. IAA formation via indole-3-pyruvic acid and indole3-acetic acid aldehyde is present in most of rhizobacteria like *Agrobacterium*, *Azospirillum*, *Bradyrhizobium*, *Rhizobium*, *Enterobacter*, *Erwinia herbicola*, *Pseudomonas*, *Klebsiella*, etc.
2. Conversion of tryptophan into indole-3-acetic aldehyde produces tryptamine, for example, *Azospirillum* and *Pseudomonads*.
3. Biosynthesis of IAA via indole-3-acetamide formation is reported by *Azospirillum*, *A. tumefaciens*, *E. herbicola*, *Rhizobium* spp., *Bradyrhizobium* sp., *Saprophytic Pseudomonads*, etc.
4. In plants, *Cyanobacteria* and *Alcaligenes faecalis* are involved in the production of IAA via indole-3-acetonitrile.
5. *Cyanobacteria* and *Azospirilla* are primarily involved in the tryptophan-independent pathway (Fig. 12.4).

Conservation practices have been promoted in order to increase the SOC content. Conservation tillage is one management strategy for increasing SOC and improving soil health on agricultural sites. Other conservation practices include cover crops,

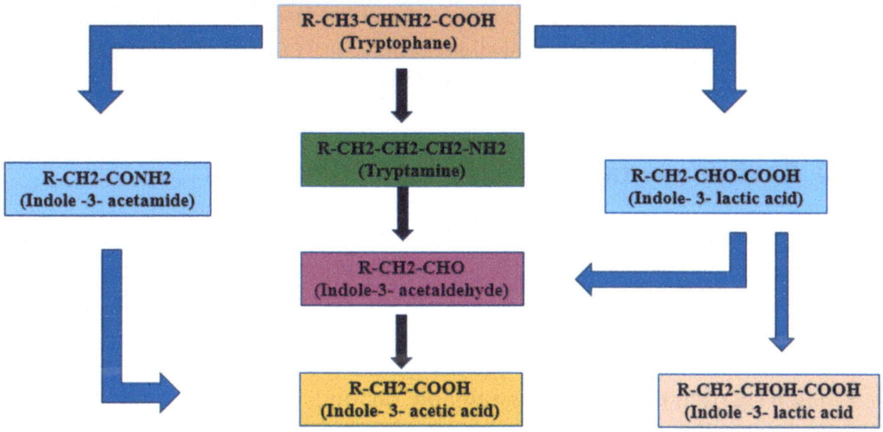

Fig. 12.4 Bacterial IAA production biosynthetic route

organic farming, conservation tillage, grazing management, and waste management. A key indicator of crop productivity and soil health, as well as a major driver of many ecosystem services, the no-till (NT) method, raises soil organic carbon (SOC) concentration (Martínez-Mena et al. 2021). Conversely, rapid erosion and breakdown caused by intensive tillage frequently encourage soil C losses. By creating a favourable environment for the soil biome, sustainable agriculture aims to preserve the health and quality of the soil.

12.9 Sustainable Agricultural Production Principles

1. Maintaining and improving soil is essential since it is the foundation of a thriving ecosystem. For the purpose of nourishing the soil, add compost, organic and green manure, and other natural fertilisers. In comparison to chemical fertilisers, natural fertilisers have more benefits for the land, plants, water, air, and people.
2. Water and water resources: Clean water is the only resource needed for agriculture, just like in real life. Cultivating plants that are adapted to the local environment or that only need water during periods of precipitation is the best way to preserve water in dry locations. Both mulch and green manure work well to keep moisture in the soil. The contour barriers protect the water by preventing it from flowing. There are two other ways to conserve water: using drip irrigation instead of traditional methods and planning irrigation schedules.
3. Control of diseases and pests: Using natural or integrated protection management is advised above chemical control in order to preserve ecological equilibrium. To achieve sustainable agriculture, care should be made in agricultural activities such as selecting hardy types, arranging plants appropriately, and using other methods.

4. Diversification: It is preferable to grow a variety of crops in order to preserve the nutritional balance of the soil and attain food security (Lal 2005). This maintains soil nutrients, ensures agricultural diversification, and yields nutrient-dense food.

12.10 Environmentally Friendly Farming Methods

1. **Crop rotation**: Compared to monoculture, crop rotation improves farming techniques and encourages sustainable agriculture by growing diverse crops in succession. Crop rotation reduces soil depletion, minimises compaction, supplies nitrogen, helps with pest management, lowers farming hazards, stays away from chemical inputs, and increases soil activity.
2. **Using irrigation to promote sustainable farming:** Irrigation is crucial to crop production and uses a significant amount of energy and water (Hartmann and Six 2023). Sustainability maximises plant hydration while consuming the least amount of energy and water. In place of furrow (flood) irrigation, water- and energy-efficient irrigation techniques like drip irrigation should be used.
3. **Cover crops**: Cover crops sown in the off-season help farmers prevent erosion on their farms. By increasing organic matter with cover crop green manure, this method lowers fertiliser costs. Cover cropping maintains soil moisture and suppresses weeds. In addition, because they naturally bloom, cover crops increase the number of honey bees.
4. **Coordinated weed control:** In order to control weeds and protect natural resources, sustainable weed control tries to forgo chemicals and replace them with eco-friendly methods (Sheng 1989). Using insects, birds, cover crops, resilient crops, mechanical and manual weeding, allelopathic plants, crop rotation, and other organic techniques are some examples of integrated weed management.

12.11 Plant Growth Regulators: Current Progress and Future

It is a major milestone that the field of plant growth regulators, or PGRs, has advanced significantly in recent years. These compounds, which also include ethylene, gibberellins, cytokinins, auxins, and abscisic acid, are essential for coordinating many aspects of plant growth and development. The body of knowledge regarding the complex signal transduction pathways and molecular mechanisms that PGRs use to regulate activities including stress response, cell elongation, and flowering is growing as a result of current research. Specifically, the clarification of new PGR interactions has expanded our understanding of the complex networks of regulation controlling plant physiology (Pullaiah et al. 2017).

The investigation of the molecular mechanisms by which PGRs alter gene expression is one area of significant progress. Key genes implicated in PGR-mediated activities have been identified by researchers thanks to the integration of transcriptome and genomic techniques. For example, research has revealed how auxin-responsive genes affect tropic responses and cell elongation. Furthermore, the discovery of certain transcription factors that PGRs regulate has shed important light on the subsequent impacts of these substances on plant development. It looks more and more possible to use targeted genetic manipulation to improve desirable features in crops as we learn more about the genetics of PGR responses.

Despite these noteworthy advancements, problems in the field of plant growth regulators continue to exist and demand the attention of both practitioners and researchers. Understanding the complex interactions between distinct PGRs and their dynamic interplay across different developmental stages is one of the main challenges. Although the roles of individual PGRs are well established, our understanding becomes more complex when multiple signals are integrated in a spatio-temporal manner. To fully utilise PGRs in maximising crop performance and guaranteeing sustainable farming practices, this complexity must be unravelled (Kanwar et al. 2018; Mittal et al. 2023).

Furthermore, there are real-world obstacles to PGR application in agriculture, especially with regard to sustainability and environmental impact. An advanced strategy is needed to strike a balance between the need for ecological responsibility and the goal of higher crop yield. Careful thought must be given to the possible ecological effects of widespread PGR use, such as contaminated soil and water. Collectively, researchers and agricultural practitioners need to devise strategies that optimise PGR benefits while reducing their environmental impact (Dias 2019).

As we celebrate the advancements in PGR research, we must face the problems brought on by climate change. Temperature, precipitation, and the frequency of extreme weather events are all altered by climate change, and these changes have a substantial effect on plant growth. PGRs may be extremely important for improving stress resilience and reducing the effects of climate-related challenges. To maintain global food security in the face of climate uncertainty, future research in this field must concentrate on creating PGR-based strategies that allow crops to flourish under changing climatic conditions (Campos et al. 2023).

In the future, incorporating cutting-edge technologies will continue to be essential to addressing the challenges facing PGR research. High-throughput approaches like mass spectrometry and next-generation sequencing enable thorough systems-level analyses of PGR responses, while computational modelling helps predict complex interactions and optimise PGR applications for specific crops and environments.

In conclusion, examining plant growth regulators represents an approach marked by significant advancement and an awareness of obstacles that need to be overcome. The field of PGR research is changing quickly, from the discovery of synthetic biology to an explanation of molecular pathways (Bhatla and Lal 2023). However, there are still issues with navigating the effects of climate change, addressing environmental concerns, and comprehending PGR interactions. PGR research has a bright future ahead of it. Multidisciplinary research collaborations and cutting-edge technologies will drive us towards resilient and sustainable farming methods, protecting the well-being of our plant-based society.

References

Abd El-Daim IA, Bejai S, Meijer J (2019) Bacillus velezensis 5113 induced metabolic and molecular reprogramming during abiotic stress tolerance in wheat. Sci Rep 9:16282. https://doi.org/10.1038/s41598-019-52567-x

Abdelaal K, AlKahtani M, Attia K, Hafez Y, Király L, Künstler A (2021) The role of plant growth-promoting bacteria in alleviating the adverse effects of drought on plants. Biology 10:520. https://doi.org/10.3390/biology10060520

Aerts N, Pereira Mendes M, Van Wees SC (2021) Multiple levels of crosstalk in hormone networks regulating plant defense. Plant J 105(2):489–504

Ahemad M, Kibret M (2014) Mechanisms and applications of plant growth promoting rhizobacteria: current perspective. J King Saud Univ Sci 26:1–20. https://doi.org/10.1016/j.jksus.2013.05.001

Ahmad Z, Waraich EA, Tariq RMS, Iqbal MA, Ali S, Soufan W et al (2021) Foliar applied salicylic acid ameliorates water and salt stress by improving gas exchange and photosynthetic pigments in wheat. Pak J Bot 53(5):1553–1560

Backer R, Rokem JS, Ilangumaran G, Lamont J, Praslickova D, Ricci E, Subramanian S, Smith DL (2018) Plant growth-promoting Rhizobacteria: context, mechanisms of action, and roadmap to commercialization of biostimulants for sustainable agriculture. Front Plant Sci 9:1473. https://doi.org/10.3389/fpls.2018.01473

Basu S, Clark RE, Bera S, Casteel CL, Crowder DW (2021) Responses of pea plants to multiple antagonists are mediated by order of attack and phytohormone crosstalk. Mol Ecol 30(19):4939–4948

Begum N, Qin C, Ahanger MA, Raza S, Khan MI, Ashraf M, Ahmed N, Zhang L (2019) Role of arbuscular mycorrhizal fungi in plant growth regulation: implications in abiotic stress tolerance. Front Plant Sci 10:1068. https://doi.org/10.3389/fpls.2019.01068

Bekele M, Yilma G (2021) Nitrogen fixation using symbiotic and non-symbiotic microbes: A review article. BMB 6:92. https://doi.org/10.11648/j.bmb.20210604.12

Bhatla SC, Lal MA (2023) Plant growth regulators: an overview. Plant physiology, development and metabolism, pp 391–8

Bisht N, Singh Chauhan P (2021) Excessive and disproportionate use of chemicals cause soil contamination and nutritional stress. In: Larramendy ML, Soloneski S (eds) Soil contamination—threats and sustainable solutions. IntechOpen. https://doi.org/10.5772/intechopen.94593

Bisht TS, Rawat L, Chakraborty B, Yadav V (2018) A recent advances in use of plant growth regulators (PGRs) in fruit crops—a review. Int J Curr Microbiol App Sci 7:1307–1336. https://doi.org/10.20546/ijcmas.2018.705.159

Bloom AJ (2015) The increasing importance of distinguishing among plant nitrogens sources. Curr Opin Plant Biol 25:10–16. https://doi.org/10.1016/j.pbi.2015.03.002

Campos EV, Pereira A d ES, Aleksieienko I, Do Carmo GC, Gohari G, Santaella C, Fraceto LF, Oliveira HC (2023) Encapsulated plant growth regulators and associative microorganisms: nature-based solutions to mitigate the effects of climate change on plants. Plant Sci 331:111,688

Cardinale F, KorwinKrukowski P, Schubert A, Visentin I (2018) Strigolactones: mediators of osmotic stress responses with a potential for agrochemical manipulation of crop resilience. J Exp Bot 69(9):2291–2303

Castro-Camba R, Sánchez C, Vidal N, Vielba JM (2022) Plant development and crop yield: the role of gibberellins. Plan Theory 11:2650. https://doi.org/10.3390/plants11192650

Chakraborty S, Nguyen B, Wasti SD, Xu G (2019) Plant leucine-rich repeat receptor kinase (LRR-RK): structure, ligand perception, and activation mechanism. Molecules 24(17):3081

Chandran H, Meena M, Swapnil P (2021) Plant growth-promoting rhizobacteria as a green alternative for sustainable agriculture. Sustain For 13:10986. https://doi.org/10.3390/su131910986

Chen D, Shao Q, Yin L, Younis A, Zheng B (2019) Polyamine function in plants: metabolism, regulation on development, and roles in abiotic stress responses. Front Plant Sci 9:1945

Cole JJ, Prairie YT, Caraco NF, McDowell WH, Tranvik LJ, Striegl RG, Duarte CM, Kortelainen P, Downing JA, Middelburg JJ, Melack J (2007 Feb) Plumbing the global carbon cycle: integrating inland waters into the terrestrial carbon budget. Ecosystems 10:172–185

Creelman RA, Mullet JE (1997) Oligosaccharins, brassinolides, and jasmonates n:ontraditional regulators of plant growth, development, and gene expression. Plant Cell 9:1211–1223. https://doi.org/10.1105/tpc.9.7.1211

De Andrade LA, Santos CHB, Frezarin ET, Sales LR, Rigobelo EC (2023) Plant growth-promoting Rhizobacteria for sustainable agricultural production. Microorganisms 11:1088. https://doi.org/10.3390/microorganisms11041088

Dhanaraju M, Chenniappan P, Ramalingam K, Pazhanivelan S, Kaliaperumal R (2022) Smart farming: internet of things (IoT)-based sustainable agriculture. Agriculture 12:1745. https://doi.org/10.3390/agriculture12101745

Dias JP (2019) Plant growth regulators in horticulture: practices and perspectives. Biotecnología Vegetal 19(1):3–14

Dos Santos RM, Diaz PAE, Lobo LLB, Rigobelo EC (2020) Use of plant growth- promoting rhizobacteria in maize and sugarcane: characteristics and applications. Front Sustain Food Syst 4:136. https://doi.org/10.3389/fsufs.2020.00136

Etesami H, Mirseyed Hosseini H, Alikhani HA (2014) Bacterial biosynthesis of 1-aminocyclopro pane-1-caboxylate (ACC) deaminase, a useful trait to elongation and endophytic colonization of the roots of rice under constant flooded conditions. Physiol Mol Biol Plants 20:425–434. https://doi.org/10.1007/s12298-014-0251-5

Fatma M, Asgher M, Iqbal N, Rasheed F, Sehar Z, Sofo A, Khan NA (2022) Ethylene signaling under stressful environments: analyzing collaborative knowledge. Plants 11:2211. https://doi.org/10.3390/plants11172211

Figueiredo MDVB, Bonifacio A, Rodrigues AC, De Araujo FF (2016) Plant growth-promoting rhizobacteria: key mechanisms of action. In: Choudhary DK, Varma A (eds) Microbial-mediated induced systemic resistance in plants. Springer Singapore, pp 23–37. https://doi.org/10.1007/978-981-10-0388-2_3

Gamage A, Gangahagedara R, Gamage J, Jayasinghe N, Kodikara N, Suraweera P, Merah O (2023) Role of organic farming for achieving sustainability in agriculture. Farm Syst 1:100005. https://doi.org/10.1016/j.farsys.2023.100005

Giller KE, Hijbeek R, Andersson JA, Sumberg J (2021) Regenerative agriculture: an agronomic perspective. Outlook Agric 50(1):13–25

Gupta S, Bhattacharyya P, Kulkarni MG, Doležal K (2023) Editorial: growth regulators and biostimulants: upcoming opportunities. Front Plant Sci 14:1209499. https://doi.org/10.3389/fpls.2023.1209499

Hanrahan G, Chan G (2005) Nitrogen. In Encyclopedia of analytical science. Elsevier, pp. 191–196. doi: https://doi.org/10.1016/B0-12-369397-7/00401-5

Hartmann M, Six J (2023) Soil structure and microbiome functions in agroecosystems. Nat Rev Earth Environ 4(1):4–18

Hossain A, Farooq M, El Sabagh A, Hasanuzzaman M, Erman M, Islam T (2020) Morphological, physio-biochemical and molecular adaptability of legumes of Fabaceae to drought stress, with special reference to Medicago sativa L. In: Hasanuzzaman M, Araújo S, Gill SS (eds) The plant family Fabaceae: biology and physiological responses to environmental stresses. Springer, pp 289–317

Imadi SR, Shah SW, Kazi AG, Azooz MM, Ahmad P (2016) Phytoremediation of saline soils for sustainable agricultural productivity. Plant Metal Interact:455–468

Johan PD, Ahmed OH, Omar L, Hasbullah NA (2021) Phosphorus transformation in soils following co-application of charcoal and wood ash. Agronomy 11:2010. https://doi.org/10.3390/agronomy11102010

Kalayu G (2019) Phosphate solubilizing microorganisms: promising approach as biofertilizers. Int J Agron 2019:1–7. https://doi.org/10.1155/2019/4917256

Kang S-M, Khan AL, Waqas M, You Y-H, Hamayun M, Joo G-J, Shahzad R, Choi K-S, Lee I-J (2015) Gibberellin-producing Serratia nematodiphila PEJ1011 ameliorates low temperature stress in Capsicum annuum L. Eur J Soil Biol 68:85–93

Kanwar MK, Yu J, Zhou J (2018) Phytomelatonin: recent advances and future prospects. J Pineal Res 65(4):e12526

Kelley HW (1983) Keeping the land alive: soil erosion—its causes and cures, vol 50. Food & Agriculture Org

Lal R (2005) Influence of soil erosion on carbon dynamics in the world. In: Roose EJ et al (eds) Soil erosion and carbon dynamics. CRC Press, pp 23–35

Leventhal GE, Ackermann M, Schiessl KT (2019) Why microbes secrete molecules to modify their environment: the case of iron-chelating siderophores. J R Soc Interface 16:20180674. https://doi.org/10.1098/rsif.2018.0674

Li Z, Alami MM, Tang H, Zhao J, Nie Z, Hu J, Shu S, Zhu D, Yang T (2022) Applications of Streptomyces jingyangensis T. and Bacillus mucilaginosus A. improve soil health and mitigate the continuous cropping obstacles for Pinellia ternata (Thunb.) Breit. Industr Crops Products 180:114691. https://doi.org/10.1016/j.indcrop.2022.114691

Martínez-Mena M, Perez M, Almagro M, García-Franco N, Díaz-Pereira E (2021) Long-term effects of sustainable management practices on soil properties and crop yields in rainfed Mediterranean almond agroecosystems. Eur J Agron 123:126207

Mishra J, Dutta V, Arora NK (2020) Biopesticides in India: technology and sustainability linkages. 3 Biotech 10:210. https://doi.org/10.1007/s13205-020-02192-7

Mittal U, Kumar V, Kukreja S, Singh B, Pandey NK, Goutam U (2023 Jun) Role of beneficial elements in developing resilience to abiotic and biotic stresses in plants: present status and future prospects. J Plant Growth Regul 42(6):3789–3813

Mukherjee A, Gaurav AK, Singh S, Yadav S, Bhowmick S, Abeysinghe S, Verma JP (2022) The bioactive potential of phytohormones: a review. Biotechnol Rep 35:e00748. https://doi.org/10.1016/j.btre.2022.e00748

Muleta D, Assefa F, Börjesson E, Granhall U (2013) Phosphate-solubilising rhizobacteria associated with Coffea arabica L. in natural coffee forests of southwestern Ethiopia. J Saudi Soc Agric Sci 12:73–84. https://doi.org/10.1016/j.jssas.2012.07.002

Müller M, Munné-Bosch S (2021) Hormonal impact on photosynthesis and photoprotection in plants. Plant Physiol 185:1500–1522. https://doi.org/10.1093/plphys/kiaa119

Odoh CK, Eze CN, Obi CJ, Anyah F, Egbe K, Unah UV, Akpi UK, Adobu US (2020) Fungal biofertilizers for sustainable agricultural productivity. In: Yadav AN, Mishra S, Kour D, Yadav N, Kumar A (eds) Agriculturally important fungi for sustainable agriculture, fungal biology. Springer International Publishing, Cham, pp 199–225. https://doi.org/10.1007/978-3-030-45971-0_9

Pathania P, Rajta A, Singh PC, Bhatia R (2020) Role of plant growth-promoting bacteria in sustainable agriculture. Biocatal Agric Biotechnol 30:101842. https://doi.org/10.1016/j.bcab.2020.101842

Pullaiah T, Rao MS, Sreedevi E (2017) Plant tissue culture: theory & practicals, 2nd edn. Scientific Publishers

Raklami A, Bechtaoui N, Tahiri A, Anli M, Meddich A, Oufdou K (2019) Use of rhizobacteria and mycorrhizae consortium in the open field as a strategy for improving crop nutrition, productivity and soil fertility. Front. Microbiol 10:1106. https://doi.org/10.3389/fmicb.2019.01106

Raza A, Mehmood SS, Tabassum J, Batool R (2019) Targeting plant hormones to develop abiotic stress resistance in wheat. In Wheat production in changing environments: responses, adaptation and tolerance, pp 557–577

Reed KM, Bargmann BOR (2021) Protoplast regeneration and its use in new plant breeding technologies. Front Genome Ed 3:734951. https://doi.org/10.3389/fgeed.2021.734951

Sabagh AE, Mbarki S, Hossain A, Iqbal MA, Islam MS, Raza A et al (2021) Potential role of plant growth regulators in administering crucial processes against abiotic stresses. Front Agron 3:648694

Saeed Q, Xiukang W, Haider FU, Kučerik J, Mumtaz MZ, Holatko J, Naseem M, Kintl A, Ejaz M, Naveed M, Brtnicky M, Mustafa A (2021) Rhizosphere bacteria in plant growth promotion,

biocontrol, and bioremediation of contaminated sites: a comprehensive review of effects and mechanisms. IJMS 22:10,529. https://doi.org/10.3390/ijms221910529
Sagar A, Rathore P, Ramteke PW, Ramakrishna W, Reddy MS, Pecoraro L (2021) Plant growth promoting rhizobacteria, arbuscular mycorrhizal fungi and their synergistic interactions to counteract the negative effects of saline soil on agriculture: key macromolecules and mechanisms. Microorganisms 9:1491. https://doi.org/10.3390/microorganisms9071491
Sah SK, Reddy KR, Li J (2016) Abscisic acid and abiotic stress tolerance in crop plants. Front Plant Sci 7. https://doi.org/10.3389/fpls.2016.00571
Shabbir R, Javed T, Afzal I, Sabagh AE, Ali A, Vicente O, Chen P (2021) Modern biotechnologies: innovative and sustainable approaches for the improvement of sugarcane tolerance to environmental stresses. Agronomy 11(6):1042
Sharma L, Priya M, Kaushal N, Bhandhari K, Chaudhary S, Dhankher OP et al (2020) Plant growth-regulating molecules as thermoprotectants: functional relevance and prospects for improving heat tolerance in food crops. J Exp Bot 71(2):569–594
Sheng TC (1989) Soil conservation for small farmers in the humid tropics (No. 60). Food & Agriculture Org
Sonnenschein EC, Stierhof M, Goralczyk S, Vabre FM, Pellissier L, Hanssen KØ, De La Cruz M, Díaz C, De Witte P, Copmans D, Andersen JH, Hansen E, Kristoffersen V, Tormo JR, Ebel R, Milne BF, Deng H, Gram L, Jaspars M, Tabudravu JN (2017) Pseudochelin A, a siderophore of Pseudoalteromonas piscicidaS2040. Tetrahedron 73:2633–2637. https://doi.org/10.1016/j.tet.2017.03.051
Tang J, Li Y, Zhang L, Mu J, Jiang Y, Fu H, Zhang Y, Cui H, Yu X, Ye Z (2023) Biosynthetic pathways and functions of Indole-3-acetic acid in microorganisms. Microorganisms 11:2077. https://doi.org/10.3390/microorganisms11082077
Timofeeva A, Galyamova M, Sedykh S (2022) Prospects for using phosphate-solubilizing microorganisms as natural fertilizers in agriculture. Plan Theory 11:2119. https://doi.org/10.3390/plants11162119
Wahab A, Muhammad M, Munir A, Abdi G, Zaman W, Ayaz A, Khizar C, Reddy SPP (2023) Role of arbuscular mycorrhizal fungi in regulating growth, enhancing productivity, and potentially influencing ecosystems under abiotic and biotic stresses. Plan Theory 12:3102. https://doi.org/10.3390/plants12173102
Wang X, Mavrodi DV, Ke L, Mavrodi OV, Yang M, Thomashow LS, Zheng N, Weller DM, Zhang J (2015) Biocontrol and plant growth-promoting activity of rhizobacteria from Chinese fields with contaminated soils. Microb Biotechnol 8:404–418. https://doi.org/10.1111/1751-7915.12158
Wang L, Rengel Z, Zhang K, Jin K, Lyu Y, Zhang L, Cheng L, Zhang F, Shen J (2022) Ensuring future food security and resource sustainability: insights into the rhizosphere. iScience 25:104168. https://doi.org/10.1016/j.isci.2022.104168
Xuan L, Li J, Wang X, Wang C (2020) Crosstalk between hydrogen sulfide and other signal molecules regulates plant growth and development. Int J Mol Sci 21(13):4593
Yan H, Yang Z, Chen S, Wu J (2023) Exploration and development of artificially synthesized plant growth regulators. Adv Agrochem:S2773237123000552. https://doi.org/10.1016/j.aac.2023.07.008
Yu Y, Gui Y, Li Z, Jiang C, Guo J, Niu D (2022) Induced systemic resistance for improving plant immunity by beneficial microbes. Plan Theory 11:386. https://doi.org/10.3390/plants11030386
Zhang S, Deng Z, Borham A, Ma Y, Wang Y, Hu J, Wang J, Bohu T (2023) Significance of soil siderophore-producing bacteria in evaluation and elevation of crop yield. Horticulturae 9:370. https://doi.org/10.3390/horticulturae9030370

Integration of Plant Hormones in the Biological System as an Opportunity for Sustainable Crop Production

13

Fadime Karabulut, Mohammad Faizan, Danyal Rasheed, Zahoor Ahmad, Gafoor Unnisa, and Ahmad Faraz

Abstract

Plant growth and yield are influenced by a variety of physiological and biochemical processes, which are primarily controlled by phytohormones (PHs) in reaction to external stimuli. Hormone regulation sets off a chain of events that optimise physiological processes within the architecture of plants and enable them to thrive in less-than-ideal environments. PHs like jasmonates, ethylene, salicylic acid, and abscisic acid have been shown in numerous studies to play a part in how plants react to environmental stressors. It is common knowledge that environmental influences affect the functions of auxins, cytokinins, gibberellins, and relatively new phytohormones (PHs) like brassinosteroids and strigolactones in plant growth and development. Plant growth, nutrient availability in soil, nutrient uptake assistance, and stress tolerance can all be enhanced by the exogenous application of PHs. Furthermore, the use of microbes and hormones has been suggested as potential strategies for managing plant stress because of the roles that hormones play and the ability of microbes to produce hormones. This

F. Karabulut (✉)
Department of Biology, Faculty of Science, Firat University, Elazıg, Turkey

M. Faizan · G. Unnisa
Department of Botany, School of Sciences, Maulana Azad National Urdu University, Hyderabad, India

D. Rasheed
Department of Botany, Bacha Khan University, Charsadda, Pakistan

Z. Ahmad
Department of Botany, Constituent Punjab College, University of Central Punjab, Bahawalpur, Pakistan

A. Faraz
Department of Biotechnology, School of Life Sciences, Glocal University, Saharanpur, India

© The Author(s), under exclusive license to Springer Nature Singapore Pte Ltd. 2024

M. Faizan, S. Hayat (eds.), *Plant Growth Regulators: Resilience for Sustainable Agriculture*, https://doi.org/10.1007/978-981-97-2918-0_13

chapter suggests that the integration of plant hormones into the biological system can contribute to the main objectives of sustainable agriculture: guaranteeing that everyone has access to wholesome, safe food, changing to a sustainable consumption pattern, expanding the scale of nature-compatible production, producing equitable livelihoods, and strengthening resistance to stresses and vulnerabilities. Thus, quality and yield in crop production can be increased, resistance to pests can be developed, and tolerance to environmental stress factors can be achieved.

Keywords

Phytohormones · Plant physiological processes · Sustainable production

13.1 Introduction

Phytohormones are endogenous signalling molecules essential for full-fledged defense and growth of plants (Jang et al. 2020; Küpers et al. 2020). Signaling pathways are connected by a sophisticated structure that controls physiological functions to quickly adjust to surrounding stressors. The majority of phytohormone compounds' physiological functions are now understood thanks to substantial advancements in the molecular mechanisms underlying hormone biosynthesis, response, and transport. The majority of hormones have either well-established or recently developed biosynthesis pathways (Raza 2021, 2022). As a result, hormone levels seem to be highly controlled and responsive to environmental changes. Many aspects of plant growth and development depend on the local and long-distance transport of auxin, but plant defence also depends on the transport of volatile compounds like ethylene and methyl jasmonate (Santner et al. 2009).

Phytohormones are also affected by environmental stresses. Some environmental changes that interfere with plant development and optimal growth include stressful environments (Raza et al. 2020; Hossain et al. 2021). Consequently, grain yields and biomass are significantly reduced (Gururani et al. 2015a, b). To survive under pressure, plants have evolved efficient sensing, signalling, and reaction systems. Phytohormones, also known as cellular signalling molecules, are one of these response mechanisms used by plants. Low-concentration chemical messengers are essential for regulating plant responses (Davies 2010; Williams 2011; Raza 2021, 2022). Plants naturally produce small organic molecules known as phytohormones, which are referred to as plant hormones as well. These molecules act as signalling agents, controlling nearly every biological process that is known to occur in plants, including development, growth, and responses to different environmental stressors. Plant hormones are classified into several groups, including auxins (AUXs), strigolactones (SLs), gibberellins (GAs), salicylic acid (SA), cytokinins (CKs), ethylene (ET), brassinosteroids (BRs), and jasmonates (JAs) (Santner et al. 2009; Waters et al. 2017). Despite having some physiological similarities, these plant hormones typically serve distinct purposes (Nemhauser et al. 2006). It is demonstrated in this chapter that plants use and depend on these hormones to regulate their biological functions so they can adapt to changes in their surroundings.

13.2 Phytohormones

Plant growth regulators are another name for plant hormones. But not all plant growth regulators are true hormones.

13.2.1 Auxin

The development and relative growth of roots and shoots are mainly regulated by auxins, which are endogenous plant growth regulators (Sachs 2005). Auxins and cytokinins have been shown to cooperate in certain cellular processes and physiological processes, including the development of the leaf during seed maturation, apical dominance, cell expansion, and cell cycle progression (Tromas et al. 2009; Jurado et al. 2010). In addition, auxin is an important and essential hormone synthesised by plants and microorganisms (Sethia et al. 2015). Indole-3-acetic acid (IAA) is an endogenous auxin whose effects have been extensively studied, even at nanomolar concentrations. This auxin is responsible for the majority of naturally occurring bioactive auxin-like effects (Woodward and Bartel 2005). When plants are stressed by their environment, auxins affect their growth. These environmental changes can also result in altered morphogenesis, which can distort plant growths, as well as subsequent changes in auxin homeostasis (Tognetti et al. 2012). The following are additional conjugates of IAA: phenylacetic acid, indole-3-acetaldehyde, 4-chloro-indoleacetic acid, indole-3-butyric acid (IBA), and indole-3-acetyl aspartate. The phloem transports auxin from one epidermal cell to another via vascular tissues (cambium and procambium) to the roots (Davies 2004). IAA and cytokinin work together to stimulate cell division in the cambium. IAA induces vascular tissue (xylem and phloem) differentiation, cell proliferation, root initiation in root cuttings, and root differentiation in tissue culture. Auxin exhibits apical dominance, fruit abscission, leaf senescence, and a tropistic response, which involves roots and shoots bending in reaction to gravity and light. It starts fruit development and growth and ripening, and in dioecious flowers, it promotes flowering and feminisation (Hauvermale et al. 2012). Because auxin and cytokinin concentrations in plants are inversely correlated, treating a plant with the former can quickly inhibit the production of cytokinin, which is used to induce shoot and root development in in vitro (Nordstrom et al. 2004). Exogenous auxin exposure boosts the synthesis of GA and ethylene while inhibiting the transport of ethylene auxin (Woodward and Bartel 2005). Plant steroids in maize respond favourably to exogenous application of brassinosteroids (Kim et al. 2000). Furthermore, IAA levels are reduced when this plant is exposed to ABA and Melon forms esterified IAA conjugates (Dunlap and Robacker 1990).

In the cellular environment, auxin homeostasis, distribution, and metabolism are correlated with a variety of abiotic stressors. Such changes in the distribution of auxin during stress conditions have been attributed to two molecular mechanisms: modifications in the expression of the PIN gene, which is required for the transport of polar auxin, and inhibition of that transport brought on by phenolic compounds accumulated during exposure to stress (Potters et al. 2009). Additionally, peroxidases catalyse the breakdown of IAA, which controls auxin metabolism induced by stress (Jain and Khurana 2009). Auxins have many functions in plants, but one of the most significant is lateral root development. As a result, auxins may be thought of as stress hormones that modify the way that some stress-responsive genes are expressed, mediating stress in one way or another. Lateral roots are essential to the growth of plants because they control the structure of the root system, maintain the stability of the plant, and ensure that the entire organism efficiently absorbs water and nutrients.

13.2.2 Gibberellic Acids or Gibberellins

Glyceraldehyde-3-phosphate is converted to galacto-saccharides (GAs) in developing seeds and early shoot tissues. Starting in chloroplasts, the site of synthesis is subsequently transported through vascular tissues, cytoplasmic membranes, and other mechanisms (Davies 2004). Deactivation, which is accomplished by adding hydroxyl, methyl, or other functional groups, controls the synthesis of GA. GA promotes stem hyper-elongation, cell elongation, and cell division, all of which increase plant height (Hauvermale et al. 2012). During the germination process, GA also initiates cold- and light-induced seed germination. Additionally, it causes virility in dioecious flowers and stimulates different enzymes, including amylase, in cereal grains that are germinating. Fruit set and growth are triggered by the external application of GA. Among the most well-known and reputable groups of regulatory hormones are GAs. They are involved in several stages of plant development, including fruit development, inter-nodal elongation, seed germination, and flowering stimulation (Islam et al. 2021). GA is an expert in controlling plant growth, particularly in the face of biotic (Verma et al. 2016) and abiotic stress (Gururani et al. 2015a, b), according to numerous studies conducted by physiologists and molecular geneticists. Just a tiny portion of the more than 250 GAs identified as this group's members have various roles in plant development. According to some, there is a complicated cellular regulation of GAs. For instance, small multigene families of enzymes control the steps in the metabolic pathway of GAs. Each of these performers has a distinct style of expression. GA3 oxidases (GA3ox) and GA20 oxidases (GA20ox), two families of genes that encode GAs and are closely related to GA biosynthesis, catalyse the last processes involved in the production of bioactive GAs (Hedden and Thomas 2012). GA cleaves the repressor protein to provide a signal that triggers a cascade of physiological and biochemical events. Plants develop stress tolerance due to GA accumulation in their tissues, which can occur from exogenous application or endogenous biosynthesis. For instance, ectopic

GA2ox6 expression enhanced rice seed yield and drought stress tolerance when there was a water deficit (Lo et al. 2017). Plants can adjust to less-than-ideal growth conditions because GA plays a role in maintaining redox balance. GRAS transcription factors are important for signal transduction pathways and GA biosynthesis in plant development and signalling.

13.2.3 Cytokinins

Members of the plant hormone class, cytokinins are essential to the cell cycle and development of plants. Cytokinins are important PHs that are frequently thought of as auxin antagonists/synergists, and ABA antagonists respond to different stresses in a range of plant development processes (Danilova et al. 2016). Auxin-cytokinin is essential for the morphogenesis of plants in vitro (Werner et al. 2001). These derivatives of adenine have the same chemical structure as zeatin and other N6-substituted purine derivatives found in developing seeds and root tips as ribosides and ribotides. Higher plants naturally contain the cytokinins isopentenyl adenine, zeatin, and dihydro zeatin. Through the xylem, from the roots to the shoots, the hormone is transferred (Davies 2004). Exogenous cytokinin application, together with auxin in tissue culture, induces cell division, resulting in morphogenesis and shoot initiation. Cytokinins also regulate photosynthesis by affecting gene expression in chloroplasts. Reduced leaf senescence, increased stomatal opening, and delayed chlorophyll breakdown are all effects of cytokinin. It helps chlorophyll accumulation and increases photosynthetic activity by promoting the conversion of etioplasts to chloroplasts (Kulaeva and Prokoptseva 2004). Research conducted by some researchers demonstrates that plants exhibit two distinct forms of cytokinin activity: signalling activity that governs functional changes over extended distances from root to shoot (endocrine) and local activity that controls cell division (paracrine/autocrine) (Kudo et al. 2010; Heyl et al. 2006).

Adenine derivatives containing aromatic or isoprenoid side chains, known as plant endogenous cytokinins, are found in smaller amounts in plants (Sakakibara 2006). Depending on how the side chain is hydroxylated and reduced, isoprenoid cytokinins can be classified as trans-zeatin (tZ), cis zeatin (cZ), isopentenyl adenine (iP), or dihydro zeatin-type derivatives. The biosynthesis rate of isoprenoid cytokinins is catalysed by isopentenyl transferase (IPT). It seems that during osmotic stress, cytokinins promote normal growth and development.

13.2.4 Jasmonic Acid

Plants contain large amounts of jasmonates, which include methyl jasmonate, JA, and jasmonoyl-isoleucine. These compounds regulate plant growth. Their cyclopentanone ring structure is a result of lipoxygenase (LOX) oxygenating polyunsaturated fatty acids (PUFAs). The synthesis of JA occurs through the octadecanoid pathway (Pirbalouti et al. 2014). Lipid-mediated substances are translocated from chloroplast membranes to the cytoplasm and eventually to peroxisomes by this pathway, which is involved in the metabolism of membrane fatty acids (Wani et al. 2016; Per et al. 2018). These substances have an impact on defence responses, secondary metabolism, senescence, reproduction, and other processes (Fahad et al. 2015; Per et al. 2018). They are crucial for seed seedling flowering and growth, germination, fruit ripening, growth inhibition, and senescence (Per et al. 2018). In addition to the developmental processes, JA causes plants to activate their defence mechanisms against pathogens and other abiotic stresses (Kumudini et al. 2018). When plants are stressed, JA interacts with other phytohormones to form a complex network (Wasternack and Strnad 2016).

Jasmonates are a broad class of compounds that include amino acid conjugates, octadecanoid cis(+) 12 oxophytodienoic acid (OPDA), 12-OH-JA and 11-OH-JA, the precursor of JAs, jasmonic acid methyl ester (JAME), and many more. When plants respond to different stresses, these substances are usually involved. The entire body of the plant contains jasmonates. But sensitive growth regions exhibit extremely high concentrations, including immature fruits, young leaves, shoot tips, and root tips (Wasternack 2007). The principal subcellular sites of JA biosynthesis are peroxisomes and chloroplasts, whereas the primary plant organs that biosynthesise jasmonates are roots and leaves (Cheong and Choi 2003). According to studies, JAs mediate a variety of biological processes, including reproductive organs and the development (for the embryo, sex determination, fruit ripening, fruit ripening, root growth, leaf senescence and movement, trichome formation, gravitropism, and tuberisation) (Wasternack and Hause 2013; Wasternack 2014). Additionally, jasmonates mediate signals linked to insect defence mechanisms or injuries caused by pathogens. Because these compounds can significantly protect plants against stress, research on the role of JAs in abiotic stress tolerance has focused on these compounds (Takeuchi et al. 2011). In response to various stresses, for instance, there has been evidence of JA-mediated plant reactions (Sasaki-Sekimoto et al. 2005). JA-mediated secondary metabolite production has also been linked to plant movements associated with adaptation to seasonal and circadian rhythms (Chen et al. 2006). Jasmonates are reported to control symbiotic associations with microorganisms (Wasternack and Hause 2013). Numerous studies using different environmental stressors show changes in JA profiles in response to different abiotic stresses.

13.2.5 Salicylic Acid

This well-known phenolic acid is naturally produced by plants during defence and controls pigment accumulation, stomatal closure, Rubisco activity, and chloroplast structure, among other aspects of fruit ripening and plant growth (Wani et al. 2016). Two pathways are used to synthesise SA, the phenylalanine pathway and the isochorismate pathway, with the majority of plants having a higher abundance of the former (Wani et al. 2016). At low concentrations, SA raises plant antioxidant levels; at higher concentrations, it kills cells. Heat shock protein, antioxidant, chaperone, and secondary metabolite genes are activated by SA (Jumali et al. 2011).

Almost all plant species, regardless of group, contain salicylic acid, one of the most significant phenolic endogenous growth regulators. Fresh biomass of barley, soybean, and rice plants has been reported to contain 1 $\mu g\ g^{-1}$ of SA. In plants, it controls several physiological and biochemical functions. It has undergone extensive investigation in plants to see if it can encourage systemic acquired resistance to a range of pathogens (Misra and Saxena 2009). However, the function of salicylic acid (SA) in different plant species, depending on the degree and duration of osmotic stress, has not been fully elucidated. At the same time, the exogenous application of SA mitigated salinity's detrimental effects on plant development (Horváth et al. 2007).

13.2.6 Brassinosteroids

Numerous physiological processes are attributed to this set of 40 plant steroid hormones. Furthermore, it has been established that a multitude of cellular reactions are set off in the processes of pollen tube growth, photomorphogenesis, xylem differentiation, stem elongation, leaf epinasty, and root inhibition. The proton pump is activated, ethylene biosynthesis is induced, gene expression linked to photosynthesis is controlled, and among the processes started are adaptive reactions to environmental stress. Under different conditions, they control senescence, stomatal formation, and plant structure (Mandava 1988). Based on their structural carbon number, brassinosteroids are categorised (Vardhini 2014). 24-Epibrassinolide and 28-homobrassinolide are the two brassinosteroids that are most commonly used (Bucker-Neto et al. 2017). Polyhydroxylated steroids, or PHs, are called brassinosteroids (BRs). Plant growth, reproduction, xylem differentiation, photomorphogenesis, cell elongation, and division are among the biochemical and physiological processes they regulate (Nolan et al. 2020). BRs can be conjugated or free (Bajguz 2011). The biological activity and nature of BRs vary. The most active BR known to science is brassinolide (BL), extracted and refined from *Brassica napus* pollen (Grove et al. 1979). Auxins and BRs are closely associated, because they both regulate their transport, coordinate plant organ tropic responses (Li et al. 2005), and stimulate the initiation of lateral root primordial during lateral root development (Bao et al. 2004). The synthesis of BRs was probably carried out in the endoplasmic reticulum. Only in plants has it been proposed that a protein complex of enzymes, called metabolites, could effectively direct the substrate to the designated enzymes in a single biosynthetic pathway. BRs are biosynthesised through the cytochrome P450 (CYP) triterpenoid pathway (Choe 2007; Chung et al. 2011). The BR synthetic pathway is started by the hydroxylation of campesterol derived from membranes (Choe et al. 1998). The hydroxyl group is added to the side chain of steroid rings (Vukašinović and Russinova 2018). BR metabolism includes some steps that are necessary to keep the proper amounts of bioactive BRs in cells, such as acylation, glycosylation, and sulfonation (Saini et al. 2015). Many studies using exogenous BRs have shown that plants can withstand abiotic stress (Divi et al. 2016). But just like other PHs, BRs are only slightly necessary. As a result, concentration affects how plants react when exogenous BR is applied. While the opposite is seen at low concentrations, plant growth appears to be inhibited by high BR application rates (Belda-Palazon et al. 2018). Oxidative stress is brought on by abiotic stressors that increase the production of ROS; however, BRs help regulate the amount of ROS in cells under stressful conditions. For instance, plants of *Brassica juncea* L. exposed to combined variety of stress showed increased enzymatic antioxidant activities and ROS homeostasis upon treatment with 28-homobrassinolide (Kaur et al. 2018). Comparable outcomes were noted in *Lycopersicum esculentum*, where BR treatment reduced photosynthesis inhibition brought on by below-optimal temperature and raised carboxylation and antioxidant system activities (Ogweno et al. 2007). BRs may control certain important agronomic characteristics in plants, including leaf angle, plant height, and inflorescence structure (Yang et al. 2018). All of these studies support the significant role that BRs play and show that BRs (as anti-stress agents) are powerful PHs because of their varied functions.

13.2.7 Strigolactones

As chemicals with biological activity that facilitate information transfer between distinct organisms, they are members of the semiochemical family (Zwanenburg and Pospíšil 2013). The composition of strigol, which was initially identified and named after *Gossypium hirsutum* L. root exudates in early 1966 (Zwanenburg et al. 2016), was clarified in 1972. Later in 1992, Hauck et al. isolated sorgolactone, a chemical that shares structural similarities with strigol, from the root exudates of *Sorghum* and *Striga* (Hauck et al. 1992). Three fused rings make up the ABC skeleton of strigolactones. The majority of terrestrial plants, such as higher, non-vascular plants like mosses, liverworts, and stoneworts, produce strigolactones with distinct structural variations (Screpanti et al. 2016). Eighty percent of plants secrete strigolactones (SLs); PHs are derived from carotenoid sources that indicate a symbiotic relationship with arbuscular soil mycorrhizae (Akiyama and Hayashi 2006). The first explanation for these PHs was that they were an "ecological signal" for the germination of a symbiotic relationship and parasitic seeds between plants and helpful microbes. *Striga lutea* was found to benefit from the first natural SL, known as strigol, which stimulates germination. For germination and early development, this obligatory hemiparasitic plant needs a living host; subsequently, these substances were dubbed strigolactones as a group (Cook et al. 1972). The functional roles of strigolactone analogues in various developmental processes including the regulation of plant organ architecture have been revealed by subsequent characterisation (Kapulnik et al. 2011; Kohlen et al. 2011). Similar to additional plant hormones, SLs primarily function in development, and in SL-regulated developmental processes, their interaction with auxin is predominant (Hayward et al. 2009). Carotenoid cleavage dioxygenase (CCD) enzymes are responsible for the apocarotenoids known as SLs (Booker et al. 2005).

SLs support the symbiotic relationship between plants and microbes in the rhizosphere. SLs also affect the quantitative growth of symbiotic nitrogen-fixing bacteria in root nodule symbiosis (Rochange et al. 2019). Plants produce SLs to a tightly controlled extent, which is determined by the stresses they encounter during different phases of their development (Marzec et al. 2013). The exogenous application of GR24 improved rice seedling growth, net photosynthetic rate, stomatal conductance, and intercellular CO_2 concentrations (Ling et al. 2020). It also increased POD and SOD activity. According to Umehara et al. (2008), the plant produces a lot of SLs, which inhibit shoot branching and promote symbiosis in the event of nutrient shortage. Moreover, it has been demonstrated that SLs are crucial in nitrogen and phosphorus deficiencies, altering the architecture of roots and shoots and encouraging rhizobial bacteria and AM fungi to form symbiotic relationships (Marzec 2016).

13.2.8 Melatonin

N-Acetyl-5-methoxytryptamine, another name for melatonin, was initially identified by scientists in 1995 in vascular plants. Melatonin is a pleiotropic substance that affects plants in a variety of ways (Wang et al. 2017). Plant melatonin supports a variety of physiological processes in response to various stressors (Arnao and Hernández-Ruiz 2015). The diverse effects of melatonin have led to the proposal of the plant melatonin receptor regulator molecule as both a master regulator and a novel plant hormone. Numerous cellular and physiological effects of melatonin are possible, including modifications to the permeability of membranes mediated by intracellular Ca^{2+} and ion transporters, alterations in stomata opening and closing, and modifications to the metabolism of carbohydrates, lipids, nitrogen, and metabolites that act as osmoprotectants. The processes that are affected by melatonin include photosynthesis, productivity optimisation, rooting induction, tropism, seed germination promotion, growth promotion, and leaf water/CO_2 exchange (Sharif et al. 2018). In addition, it controls parthenocarpy, internal biological clocks, and other processes like senescence and maturation. In addition, melatonin protects plants from biotic and abiotic stressors by acting as an endogenous plant biostimulant (Sun et al. 2014). A receptor that interacts with one of the essential elements of any hormonal function in an animal or plant is the hormone, which starts downstream signalling chain elements. The lack of a recognised receptor is one of the most significant obstacles to melatonin research in plants. The assessment of melatonin as a plant hormone has also been hampered by this. However, the receptor was not discovered until numerous biochemical and physiological data on melatonin were gathered, much like with other plant hormones (Zhang and Chang 2016).

13.2.9 Polyamines

The result of cellular metabolism is the small aliphatic nitrogenous bases known as polyamines (PAs), though not designed to function as hormones, because of their ability to regulate a variety of plant development and growth processes, as well as their reactions to abiotic stress. A novel class of plant growth regulators called PAs has been proposed (Chen et al. 2019). PAs were originally regarded as polycations with distinct structural roles because of the belief that their biological activities were connected to their capacity to bind with anionic macromolecules. But later research has demonstrated that PAs regulate basic cellular functions like protein synthesis and gene expression (Childs et al. 2017).

Research indicates that PAs provide defence against environmental stressors (Chen et al. 2019). Plants primarily use spermidine (Spd), spermine (Spm), and putrescine (Put) as PAs; however, less focus has been placed on cadaverine (Cad) and diamino propane (Dap). Both conjugated (covalently and non-covalently bound) and free-form states are known to contain PAs (Gholami et al. 2013). Put, a precursor to Spm and Spd, is the primary by-product of the biosynthesis pathway of PAs (Xu et al. 2009). The three distinct pathways involved in the biosynthesis of put in plants are the most common ones, which are arginine by arginine decarboxylase (ADC), ornithine by ornithine decarboxylase (ODC) (Orn), and citrulline by citrulline decarboxylase (CDC) (Chen et al. 2019). Diamine oxidase is necessary for the breakdown of PAs, and PA oxidase and PA metabolism are closely linked to numerous other plant metabolic pathways. The H_2O_2 generated by PA oxidation is involved in signal transduction and plant responses to biotic and environmental stress points (Sariyev et al. 2020). Due to their shared precursor (S-adenosyl methionine), the competing ethylene synthesis and the PA biosynthesis pathways are connected. Furthermore, nitric oxide production, which initiates a process of signal transduction associated with plant growth, is closely linked to PA metabolism (Pál et al. 2015). Plant organ and tissue specificity governs the distribution of PAs. Put is the most prevalent PA present in leaves, whereas Spd levels are higher in other plant organs (Takahashi et al. 2017). In reaction to various abiotic stressors, polyamines form. The protective roles of polyamines under stress have generally been demonstrated by transgenic plants that overexpress the enzymes involved in PA biosynthesis. Moreover, plants are believed to benefit from the exogenous application of PAs because some plant species have heightened resistance to stress (Qi et al. 2010; Minocha et al. 2014).

13.2.10 Abscisic Acid

The hormone is crucial for plant adaptation and is generated by plants in reaction to abiotic stress. Because of its function in the abscission of leaves, ABA was formerly known as "abscisin" (Wani et al. 2016). The phytohormone abscisic acid was identified as the cause of seed dormancy in the early 1960s (Cornforth et al. 1965). Mature leaves and roots respond to drought or water stress by producing ABA from glyceraldehyde-3-phosphate via carotenoids and isopentenyl diphosphate (Normanly et al. 2004). Senescence, stomatal closure, embryo maturation, and seed resting are among the stages of plant development that are regulated by abnormal basal stem temperature. It causes the synthesis of storage proteins, which helps seeds develop their ability to withstand stress (Gururani et al. 2015a, b). Together with controlling photosynthesis, the photosynthate pool's migration between the sink and source tissues is likewise regulated by ABA (Pinheiro et al. 2011; Barickman et al. 2014). There is evidence that a sudden rise in ABA levels can change gene expression and activate signalling pathways (Wani et al. 2016). ABA contributes to root growth and architectural alterations in situations where there is a water deficit. It also controls antioxidant enzyme synthesis, osmoprotectant synthesis, and the preservation of cell turgor (Chaves et al. 2003).

Plant development and stress-adaptive reactions have been found to involve ABA (Iqbal 2015). For example, ABA contributes to a plant's ability to adapt to environmental stressors, to control the opening and closing of stomata to regulate water levels in the plant stem, and to facilitate seed dormancy and maturation. In situations where there is a water shortage, the leaves are where ABA is more concentrated after it is transported there via the xylem after being biosynthesised in the roots. This increase in ABA in guard cells acts as a signal to initiate a signalling cascade that regulates cell turgor (Awan et al. 2017). As a result of stomata closing in response to environmental stress, which slows down cell growth by decreasing the rate of transpiration, ABA is distributed and produced (Peleg and Blumwald 2011). Phosphatases and kinases are involved in ABA signalling to facilitate rapid responses to a range of abiotic stresses (Arkhipova et al. 2020). ABA's function in plant-water relations makes it significant (Dar et al. 2017). Carotenoids provide the building blocks for the ABA biosynthetic pathway. After b-carotene is oxidatively cleaved by the enzyme 9-cisepoxy-carotenoid di-oxygenase (NCED), stress stimuli quickly start the biosynthesis of ABA. The subsequent oxidation produces ABA (Xu et al. 2013; Llanes et al. 2014). In plants under abiotic stress, ABA not only interacts with other PHs to regulate physiological responses, but it also mediates physiological responses through its own signals (He et al. 2018). Heat shock proteins, ROS-detoxifying enzymes, proline biosynthesis, unsaturated fatty acids, and antioxidants are just a few of the biochemical defences that are triggered and activated by ABA (Yin et al. 2018; Singhal et al. 2021).

13.2.11 Ethylene

Ethylene is a gas that is essential to numerous morphophysiological processes, including the triple response, which occurs when seeds germinate, blossom, and ripen and plants react to external stimuli (Awan et al. 2017). For example, for a temperature stress capacity to be below ideal, ET homeostasis regulation is necessary (Yang et al. 2017). Fruit ripening, seedling senescence, and leaf abscission are all influenced by this important metabolite, which is made from methionine in a variety of tissues under stress. 1-Aminocyclopropane-1-carboxylic acid (ACC) is produced by ACC synthetase, while SAM synthetase converts ethylene to S-adenosylmethionine (SAM). In the presence of oxygen, ACC oxidase breaks down ACC to produce ethylene, carbon dioxide, and cyanide (Yang and Hoffman 1984; Maheshwari et al. 2015). Even at the peak of ethylene biosynthesis, L-methionine levels remain constant because of this recycling. The extremely labile enzyme ACC synthase, which tends to limit the rate of biosynthesis, has an impact on the ET biosynthetic pathway (Kendrick and Chang 2008). When ethylene is produced in the senescing or maturing tissues of plants, it shields the apical hook of the seedling. It also affects the formation and differentiation of incidental roots as well as the growth of roots and shoots. In dioecious flowers, ethylene also promotes feminisation, flowering, fruit ripening, and leaf and fruit abscission (Wang and Irving 2011). The amount of ET biosynthesis in various plant tissues has been measured, including ripening fruits and wilting flowers in abiotically stressed plants (Kendrick and Chang 2008). In stressful environments, it suppresses shoot/leaf expansion, photosynthesis, and root development (Pierik et al. 2006). Ethylene counteracts drought and ABA-induced stomatal closure, according to similar research in tomato seedlings (Vysotskaya et al. 2011; Wilkinson and Davies 2009).

13.3 The Relationships Between the Hormones

Many interactions exist between plant hormones and other hormones, as demonstrated by physiology data. Consequently, modifications in hormone levels or responses may be a part of hormone interaction. By activating ACC synthase, auxin can also affect the biosynthesis of ethylene (Tsuchisaka and Theologis 2004), which puts ethylene transcriptionally controlling a number of genes needed for auxin biosynthesis (Stepanova et al. 2008). The level of hormone distribution is also where hormone interactions take place. For example, during the initiation of lateral roots, auxin and cytokinin exhibit opposing actions. For plant growth and morphological shaping, the processes of polar auxin transport and auxin gradient formation are crucial (Feraru and Friml 2008). During root development, auxin stimulates the initiation of lateral roots, whereas cytokinin has the opposite effect. By impeding the founder cells of lateral roots from creating a local auxin gradient, cytokinin also suppresses the initiation of lateral roots (Laplaze et al. 2007). Brassinosteroids can induce genes induced by auxin, and they can also induce common target genes repressed by auxin (Mouchel et al. 2006). When roots grow, ethylene and ABA may act in opposition to GA (Weiss and Ori 2007).

Plants use de novo synthesis, degradation, and inactivation to systematically control the amounts of plant hormones. This guarantees that phytohormones function as best they can react to both the internal and external environments. It appears that plant hormones are found throughout the entire plant, but different tissues and differentiated cell types have different concentrations at different times. Certain tissues contain plant hormones responsible for development, such as BR, GA, CK, and IAA (Novák et al. 2017). Furthermore, it has been demonstrated that every plant hormone affects photosynthesis, which is essential for plant growth, either positively or negatively (Gururani et al. 2015a, b). Plant hormone levels can therefore be measured throughout different tissues and developmental stages to learn more about how different tissues function. Plant hormone biosynthesis and degradation pathways have been extensively studied, with the identification of genes accountable for crucial stages in these metabolic and catabolic processes (Mostofa et al. 2018; Wei and Li 2020; Pattyn et al. 2021; Vukašinović et al. 2021). Cytokinin oxidase (CKX) or glucosylation can inactivate CKs irreversibly (Bajguz and Piotrowska 2009) or reversibly (Werner et al. 2006). In plants, in response to environmental stress, ET, ABA, SA, and JA are functional (Dar et al. 2015; Kazan 2015). To control the functions of plant hormones as signalling molecules in the control of biosynthesis and degradation pathways, plants have evolved complex systems (Park et al. 2017). Studies suggest that some transporters control the spatial distribution of other hormones found in plants. In plants, CK transporters come in various forms. Some regulate intracellular CK trafficking, while others are engaged in long-distance or intercellular translocation (Kang et al. 2017; Liu et al. 2019). Different transporters move ABA between cells or tissues to control physiological reactions like stomatal movement, distal organ stress responses, and seed germination (Kang et al. 2015; Kuromori et al. 2018). Amino acid transporters carry the ET precursor ACC (Choi et al. 2019). The spatial distribution of these hormones is regulated by the transport of JAs (Kanno et al. 2016; Tal et al. 2016). Additionally, 12-oxophytodienoic acid, JA/JA-Ile transporters, and JA precursors control the intracellular or tissue distribution of JA (Footitt et al. 2007; Li et al. 2017). It has been demonstrated that the release of this hormone into rhizophores and its tissue-specific distribution are regulated by an ATP binding cassette (ABC) transporter (Sasse et al. 2015). The entire plant's physiological states and developmental processes are controlled by the active, spatial, and temporal regulation of plant hormones. Once plant hormones have conjugated with different metabolites, like sugars and amino acids, they may be kept in dormant state. Hormones are activated when required and released from their conjugated form. The vacuole-stored ABA-glucose conjugates serve as an example of this (Xu et al. 2012). It should be mentioned that in the instance of JA signalling, the JA receptor complex recognises JA-Ile as an endogenous signalling molecule (Fonseca et al. 2009). Multiple spatially organised plant hormones can be monitored simultaneously using a technique called hormonemy, which offers a fantastic chance to obtain vital information regarding the physiological and developmental circumstances of plant tissues, organs, and cells. This method can also aid in the discovery of helpful biomarkers that can track a plant's physiological and developmental state in a variety of environmental circumstances. Numerous stress tolerances are improved by phytohormones and their roles in various physiological processes. Plant growth, development, and adaptation/acclimatisation are regulated by these phytohormones in concert (Fig. 13.1).

13.4 The Significance of Concurrently Characterising Multiple Plant Hormones

Based on numerous physiological and molecular studies, each phytohormone has a specific physiological role. It is also established that the majority of the biological processes that have been proposed are not regulated by distinct hormones. Plant hormones that control biological processes in plants also have various and intricate factors to take into account (Ross et al. 2011). Also, plant hormones can act in a unidirectional or bidirectional manner, having both antagonistic and synergistic

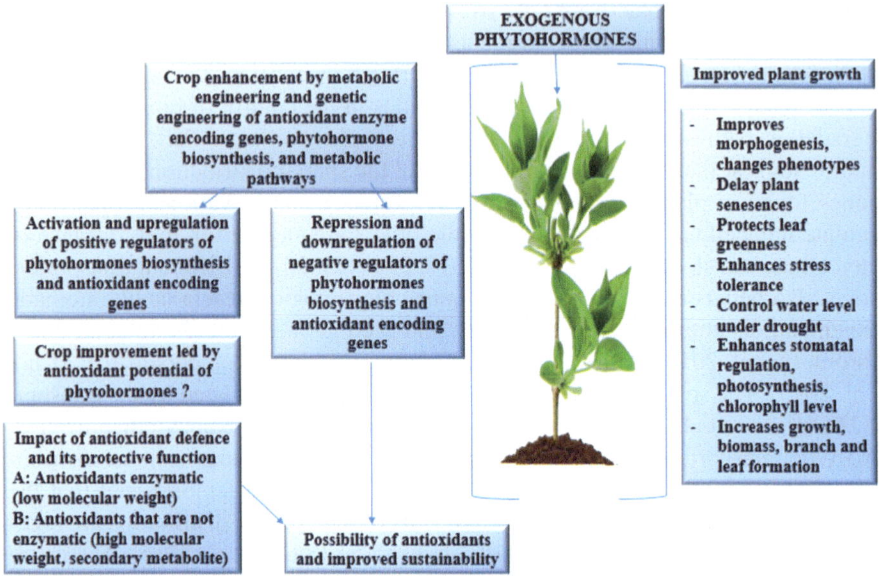

Fig. 13.1 Several biological processes that phytohormones are involved in improve a variety of stress tolerances

effects. Furthermore, they have a multilevel impact on output processes, signalling, biosynthesis, gene expression, and other processes (Altmann et al. 2020; Aerts et al. 2021). This includes how plant hormones interact with one another to help plants defend themselves against necrotrophic and biotrophic pathogens (Vlot et al. 2009). Plants are now better able to balance energetic processes like growth and defence responses to increase their environmental fitness thanks to extensive hormonal studies (Shigenaga et al. 2017; Van Butselaar and Van den Ackerveken 2020). Moreover, hormones may be involved in many biological processes in plants, including growth (El-Showk et al. 2013) and reactions to abiotic stress (Yu et al. 2020). It might aid in the optimisation of conflicting physiological processes carried out by antagonistic hormones in plants. For clarification, hormone analysis, or the simultaneous profiling of plant hormones, is essential.

Numerous peptides that resemble hormones and small peptides released into extracellular spaces are crucial signalling molecules that play a role in regulating plant development, stress responses, and cell differentiation in the shoot or root meristem. Additionally, signalling peptides play a role in controlling the processes of cell growth and differentiation, including stomata (Kim et al. 2021). The most well-known instance of this is seen in *Arabidopsis*, where the shoot apical meristem is maintained in part by CLAVATA3 (CLV3). A putative signalling peptide that is secreted into the extracellular space is found in the 96-aa peptide CLV3 (Fletcher et al. 1999). Moreover, plants' systemic reactions to environmental factors that cause pathogen invasion, salinity stress, drought stress, and nutrient shortages are also mediated by hormone-like peptides. Dehydration signalling is transferred from

Table 13.1 Hormone localisation and precursors of plant hormone production

Phytohormones	Localisations of hormone occurrence in plants	Precursors of plant hormones
Auxins	Forming seeds, immature leaves, and leaf primordia	Tryptophan
Abscisic acid	Roots, seeds, and mature leaves, especially when they are stressed by water	Isopentenyl pyrophosphate
Salicylic acid	Leaves, especially when attacked by pathogens	Phenylalanine
Gibberellins	Seeds and early shoot tissues in development	Isopentenyl pyrophosphate
Strigolactones	Shoots and roots	Isopentenyl pyrophosphate
Cytokinins	Root tips, developing seeds, leaves, stem, flowers, siliques, fruits, and shoot meristem	Isopentenyl pyrophosphate
Jasmonates	Roots and leaves	α-Linolenic acid
Melatonin	Leaves, stems, seeds, roots, and fruits	Tryptophan
Polyamines	The majority of tissues, especially those going through ripening or senescence in reaction to stress	Arginine
Brassinosteroids	Leaves, shoots, roots, fruits, seeds, and pollen	Isopentenyl pyrophosphate
Ethylene	Tissues that are ripening (fruits), roots, and shoots, especially while under stress	Methionine

Arabidopsis roots to shoots during stressful situations. Using a cellular signalling pathway and a minimum number of meristem receptors, this activates the genes responsible for synthesising ABA (Takahashi et al. 2018). Comprehensive peptide analyses of putative signalling peptides or genes uncover a large number of secretory hormone-like peptides that are unknown in plants (Wang et al. 2020). The diverse biological functions and biosynthesis of phytohormones are substantially influenced by their chemical structures (Table 13.1).

13.5 Conclusion and Future Perspectives

In response to growing demand and diminishing crop yields. Plant science is expected to provide solutions due to global climate change. Thus, it ought to offer methods for raising crop output. Moreover, this area has been shown to boost crop yields while requiring fewer inputs. Another expected benefit is enhanced resistance to abiotic and biotic stresses. Through their extensive research, scientists have gathered a great deal of knowledge about the physiological, biochemical, and functional properties of genes to better understand the mechanisms. Hormonomy offers further information about the physiological and developmental state of plants or crops in addition to other molecular or biochemical techniques. A variety of sensors have been created to track the dynamics of plant hormones in real time, thereby addressing the issues related to the physiological function of these hormones. Plant hormone responses are dynamically and spatially regulated, as evidenced by the identification of plant hormones and their precursors in vascular flow as well as the application of direct plant hormone sensors. Not only are direct sensors for known plant hormones expected to exist, but new and enhanced ones are also expected. It will allow to build a more thorough spatial and dynamic picture of how plant hormone function within organisms.

References

Aerts N, Pereira Mendes M, Van Wees SCM (2021) Multiple levels of crosstalk in hormone networks regulating plant defense. Plant J 105:489–504

Akiyama K, Hayashi H (2006) Strigolactones: chemical signals for fungal symbionts and parasitic weeds in plant roots. Ann Bot 97:925–931. https://doi.org/10.1093/aob/mcl063

Altmann M, Altmann S, Rodriguez PA, Weller B, Elorduy Vergara L, Palme J et al (2020) Extensive signal integration by the phytohormone protein network. Nature 583:271–276

Arkhipova T, Martynenko E, Sharipova G, Kuzmina L, Ivanov I, Garipova M et al (2020) Effects of plant growth promoting rhizobacteria on the content of abscisic acid and salt resistance of wheat plants. Plants 9:1429. https://doi.org/10.3390/plants9111429

Arnao MB, Hernández-Ruiz J (2015) Melatonin: synthesis from tryptophan and its role in higher plants. In: D'Mello JPF (ed) Amino acids in higher plants. CAB International, Boston, MA, pp 390–435. https://doi.org/10.1079/9781780642635.0390

Awan FK, Khurshid MY, Mehmood AJIJIRB (2017) Plant growth regulators and their role in abiotic stress management. Int J Innov Res Biosci 1:9–21

Bajguz A (2011) Brassinosteroids – occurence and chemical structures in plants. In: Hayat S, Ahmad A (eds) Brassinosteroids: a class of plant hormone. Springer, Dordrecht, pp 1–27. https://doi.org/10.1007/978-94-007-0189-2_1
Bajguz A, Piotrowska A (2009) Conjugates of auxin and cytokinin. Phytochemistry 70:957–969
Bao L, Locovei S, Dahl GP (2004) Annexin membrane channels are mechanosensitive conduits for ATP. FEBS Lett 572:65–68. https://doi.org/10.1016/j.febslet.2004.07.009
Barickman TC, Kopsell DA, Sams CE (2014) Abscisic acid increases carotenoid and chlorophyll concentrations in leaves and fruit of two tomato genotypes. J Am Soc Hortic Sci 139:261–266
Belda-Palazon B, Gonzalez-Garcia MP, Lozano-Juste J, Coego A, Antoni R, Julian J et al (2018) PYL8 mediates ABA perception in the root through noncell-autonomous and ligand-stabilization-based mechanisms. Proc Natl Acad Sci USA 115:E11857–E11863. https://doi.org/10.1073/pnas.1815410115
Booker J, Sieberer T, Wright W, Williamson L, Willett B, Stirnberg P et al (2005) MAX1 encodes a cytochrome P450 family member that acts downstream of MAX3/4 to produce a carotenoid-derived branch-inhibiting hormone. Dev Cell 8:443–449. https://doi.org/10.1016/j.devcel.2005.01.009
Bucker-Neto L, Paiva ALS, Machado RD et al (2017) Interactions between plant hormones and heavy metals responses. Genet Mol Biol 40:373–386
Chaves MM, Maroco JP, Pereira JS (2003) Understanding plant responses to drought: from genes to the whole plant. Funct Plant Biol 30:239–264
Chen Y-H, Aires F, Francis JA, Miller JR (2006) Observed relationships between Arctic longwave cloud forcing and cloud parameters using a neural network. J Clim 19:4087–4104. https://doi.org/10.1175/JCLI3839.1
Chen D, Shao Q, Yin L, Younis A, Zheng B (2019) Polyamine function in plants: development, and roles in abiotic stress responses. Front Plant Sci 9:1945. https://doi.org/10.3389/fpls.2018.01945
Cheong JJ, Choi YD (2003) Methyl jasmonate as a vital substance. Trends Genet 19:409–413. https://doi.org/10.1016/S0168-9525(03)00138-0
Childs C, Holdsworth RE, Jackson CA-L, Manzocchi T, Walsh JJ, Yielding G (2017) Introduction to the geometry and growth of normal faults. Geol Soc 439:1–9. https://doi.org/10.1144/SP439.24
Choe S (2007) Brassinosteroid biosynthesis and metabolism. In: Davies PJ (ed) Plant hormones: biosynthesis, signal transduction, action. Kluwer Academic Publishers, Dordrecht, pp 156–178. https://doi.org/10.1007/978-1-4020-2686-7_8
Choe S, Dilkes BP, Fujioka S, Takatsuto S, Sakurai A, Feldmann KA (1998) The DWF4 gene of Arabidopsis encodes a cytochrome P450 mediates multiple 22að*-hydroxylation steps in brassinosteroid biosynthesis. Plant Cell 10:231–243. https://doi.org/10.1105/tpc.10.2.231
Choi J, Eom S, Shin K, Lee R-A, Choi S, Lee J-H et al (2019) Identification of lysine histidine transporter 2 as an 1-aminocyclopropane carboxylic acid transporter in *Arabidopsis thaliana* by transgenic complementation approach. Front Plant Sci 10:1092
Chung KY, Rasmussen SG, Liu T, Li S, DeVree BT, Chae PS et al (2011) Conformational changes in the G protein Gs induced by the b2 adrenergic receptor. Nature 477:611–615. https://doi.org/10.1038/nature10488
Cook CE, Whichard LP, Wall ME, Egley GH, Coggon P, Luhan PA et al (1972) Germination stimulants. II. The structure of strigol-a potent seed germination stimulant for witchweed (Striga lutea Lour.). J Am Chem Soc 94:6198–6199. https://doi.org/10.1021/ja00772a048
Cornforth JW, Milborow BV, Ryback G, Wareing PF (1965) Identity of sycamore 'dormin' with abscisin II. Nature 205:1269–1270. https://doi.org/10.1038/2051269b0
Danilova MN, Kudryakova NV, Doroshenko AS, Zabrodin DA, Vinogradov NS, Kuznetsov VV (2016) Molecular and physiological responses of Arabidopsis thaliana plants deficient in the genes responsible for ABA and cytokinin reception and metabolism to heat shock. Russ J Plant Physiol 63:308–318. https://doi.org/10.1134/S1021443716030043
Dar TA, Uddin M, Khan MMA, Hakeem KR, Jaleel H (2015) Jasmonates counter plant stress: a review. Environ Exp Bot 115:49–57

Dar NA, Amin I, Wani W, Wani SA, Shikari AB, Wani SH et al (2017) Abscisic acid: a key regulator of abiotic stress tolerance in plants. Plant Gene 11:106–111. https://doi.org/10.1016/j.plgene.2017.07.003
Davies PJ (2004) The plant hormones: their nature, occurrence, and functions. In: Davies PJ (ed) Plant hormones biosynthesis, signal transduction, action! Kluwer Academic Publishers, pp 1–35
Davies PJ (2010) The plant hormones: their nature, occurrence, and functions. In: Davies PJ (ed) Plant hormones. Springer, Dordrecht, pp 1–15. https://doi.org/10.1007/978-1-4020-2686-7
Divi UK, Rahman T, Krishna P (2016) Gene expression and functional analyses in brassinosteroid-mediated stress tolerance. Plant Biotechnol J 14:419–432. https://doi.org/10.1111/pbi.12396
Dunlap JR, Robacker KM (1990) Abscisic acid alters the metabolism of indole-3-acetic acid in senescing flowers of Cucumis melo L. Plant Physiol 94:870–874
El-Showk S, Ruonala R, Helariutta Y (2013) Crossing paths: cytokinin signalling and crosstalk. Development 140:1373–1383
Fahad SS, Hussain A, Matloob FA et al (2015) Phytohormones and plant responses to salinity stress: a review. Plant Growth Regul 75:391–404
Feraru E, Friml J (2008) PIN polar targeting. Plant Physiol 147:1553–1559
Fletcher JC, Brand U, Running MP, Simon R, Meyerowitz EM (1999) Signaling of cell fate decisions by CLAVATA3 in Arabidopsis shoot meristems. Science 283:1911–1914
Fonseca S, Chini A, Hamberg M, Adie B, Porzel A, Kramell R et al (2009) (+)-7-iso-Jasmonoyl-L-isoleucine is the endogenous bioactive jasmonate. Nat Chem Biol 5:344–350
Footitt S, Dietrich D, Fait A, Fernie AR, Holdsworth MJ, Baker A et al (2007) The COMATOSE ATP-binding cassette transporter is required for full fertility in Arabidopsis. Plant Physiol 144:1467–1480
Gholami M, Fakhari AR, Ghanati F (2013) Selective regulation of nicotine and polyamines biosynthesis in tobacco cells by enantiomers of ornithine. Chirality 25:22–27. https://doi.org/10.1002/chir.22107
Grove MD, Spencer GF, Rohwedder WK, Mandava N, Worley JF, Warthen JD Jr et al (1979) Brassinolide, a plant growth promoting steroid isolated from Brassica napus pollen. Nature 281:216–217. https://doi.org/10.1038/281216a0
Gururani MA, Mohanta TK, Bae H (2015a) Current understanding of the interplay between phytohormones and photosynthesis under environmental stress. Int J Mol Sci 16:19,055–19,085
Gururani MA, Venkatesh J, Tran L-S (2015b) Regulation of photosynthesis during abiotic stress-induced photoinhibition. Mol Plant 8:1304–1320. https://doi.org/10.1016/j.molp.2015.05.005
Hauck C, Müller S, Schildknecht H (1992) A germination stimulant produced for parasitic flowering plants from Sorghum bicolor, a genuine host plant. J Plant Physiol 139:474–478
Hauvermale AL, Ariizumi T, Steber CM (2012) Gibberellin signalling: a theme and variations on DELLA repression. Plant Physiol 160:83–92
Hayward A, Stirnberg P, Beveridge C, Leyser O (2009) Interactions between auxin and strigolactone in shoot branching control. Plant Physiol 151:400–412. https://doi.org/10.1104/pp.109.137646
He M, He C-Q, Ding N-Z (2018) Abiotic stresses: general defenses of land plants and chances for engineering multistress tolerance. Front Plant Sci 9:1771. https://doi.org/10.3389/fpls.2018.01771
Hedden P, Thomas SG (2012) Gibberellin biosynthesis and its regulation. Biochem J 444:11–25. https://doi.org/10.1042/BJ20120245
Heyl A, Werner T, Schmulling T (2006) Cytokinin metabolism and signal transduction. In: Hedden P, Thomas SG (eds) Plant hormone signaling. Blackwell Publishing Ltd, Oxford, pp 93–124
Horváth E, Pál M, Szalai G, Páldi E, Janda T (2007) Exogenous 4-hydroxybenzoic acid and salicylic acid modulate the effect of short-term drought and freezing stress on wheat plants. Biol Plantarum 51:480–487. https://doi.org/10.1007/s10535-007-0101-1
Hossain A, Raza A, Maitra S, Asaduzzaman M et al (2021) Strigolactones: a novel carotenoid-derived phytohormone—biosynthesis, transporters, signalling, and mechanisms in abiotic stress. In: Aftab T, Hakeem KR (eds) Plant growth regulators. Springer, Cham, pp 275–303. https://doi.org/10.1007/978-3-030-61153-8_13

Iqbal MA (2015) Improving the growth and yield of canola (Brassica napus L.) with seed treatment and foliar sprays of brassica (Brassica napus L.) and moringa (Moringa oleifera L.) leaf extracts. Am Eur J Agric Environ Sci 14:1067–1073. https://doi.org/10.5829/idosi.aejaes.2014.14.10.12429

Islam MS, Hasan MK, Islam B, Renu NA, Hakim MA, Islam MR et al (2021) Responses of water and pigments status, dry matter partitioning, seed production, and traits of yield and quality to foliar application of GA3 in mung bean (Vigna radiata L.). Front Agron 2:596850. https://doi.org/10.3389/fagro.2020.596850

Jain M, Khurana JP (2009) Transcript profiling reveals diverse roles of auxin-responsive genes during reproductive development and abiotic stress in rice. FEBS J 276:3148–3162. https://doi.org/10.1111/j.1742-4658.2009.07033.x

Jang G, Yoon Y, Choi YD (2020) Crosstalk with jasmonic acid integrates multiple responses in plant development. Int J Mol Sci 21:305. https://doi.org/10.3390/ijms21010305

Jumali SS, Said IM, Ismail I, Zainal Z (2011) Genes induced by high concentration of salicylic acid in Mitragyna speciosa. Aust J Crop Sci 5:296

Jurado S, Abraham Z, Manzano C, López-Torrejón G, Pacios LF, Del Pozo JC (2010) The Arabidopsis cell cycle F-box protein SKP2A binds to auxin. Plant Cell 22:3891–3904. https://doi.org/10.1105/tpc.110.078972

Kang J, Yim S, Choi H, Kim A, Lee KP, Lopez-Molina L et al (2015) Abscisic acid transporters cooperate to control seed germination. Nat Commun 6:8113

Kang J, Lee Y, Sakakibara H, Martinoia E (2017) Cytokinin transporters: go and stop in signaling. Trends Plant Sci 22:455–461

Kanno Y, Oikawa T, Chiba Y, Ishimaru Y, Shimizu T, Sano N et al (2016) AtSWEET13 and AtSWEET14 regulate gibberellin-mediated physiological processes. Nat Commun 7:13,245

Kapulnik Y, Resnick N, Mayzlish-Gati E, Kaplan Y, Wininger S, Hershenhorn J et al (2011) Strigolactones interact with ethylene and auxin in regulating root-hair elongation in Arabidopsis. J Exp Bot 62:2915–2924. https://doi.org/10.1093/jxb/erq464

Kaur H, Sirhindi G, Bhardwaj R, Alyemeni MN, Siddique KHM, Ahmad P (2018) 28-homobrassinolide regulates antioxidant enzyme activities and gene expression in response to salt and temperature-induced oxidative stress in Brassica juncea. Sci Rep 28:8735. https://doi.org/10.1038/s41598-018-27032-w

Kazan K (2015) Diverse roles of jasmonates and ethylene in abiotic stress tolerance. Trends Plant Sci 20:219–229

Kendrick MD, Chang C (2008) Ethylene signaling: new levels of complexity and regulation. Curr Opin Plant Biol 11:479–485. https://doi.org/10.1016/j.pbi.2008.06.011

Kim SK, Chang SC, Lee EJ et al (2000) Involvement of brassinosteroids in the gravitropic response of primary root of maize. Plant Physiol 123:997–1004

Kim M-J, Jeon BW, Oh E, Seo PJ, Kim J (2021) Peptide signaling during plant reproduction. Trends Plant Sci 26:822–835

Kohlen W, Charnikhova T, Liu Q, Bours R, Domagalska MA, Beguerie S et al (2011) Strigolactones are transported through the xylem and play a key role in shoot architectural response to phosphate deficiency in nonarbuscular mycorrhizal host Arabidopsis. Plant Physiol 155:974–987. https://doi.org/10.1104/pp.110.164640

Kudo T, Kiba T, Sakakibara H (2010) Metabolism and long-distance translocation of cytokinins. J Integr Plant Biol 52:53–60

Kulaeva ON, Prokoptseva OS (2004) Recent advances in the study of mechanisms of action of phytohormones. Biochemistry (Mosc) 69:233–247

Kumudini BS, Jayamohan NS, Patil SV, Govardhana M (2018) Primary plant metabolism during plant-pathogen interactions and its role in defense. In: Ahmad P, Ahanger MA, Singh VP et al (eds) Plant metabolites and regulation under environmental stress. Academic Press, Elsevier, pp 215–229

Küpers JJ, Oskam L, Pierik R (2020) Photoreceptors regulate plant developmental plasticity through auxin. Plan Theory 9:940. https://doi.org/10.3390/plants9080940

Kuromori T, Seo M, Shinozaki K (2018) ABA transport and plant water stress responses. Trends Plant Sci 23:513–522
Laplaze L et al (2007) Cytokinins act directly on lateral root founder cells to inhibit root initiation. Plant Cell 19:3889–3900
Li L, Xu J, Xu ZH, Xue HW (2005) Brassinosteroids stimulate plant tropisms through modulation of polar auxin transport in Brassica and Arabidopsis. Plant Cell 17:2738–2753. https://doi.org/10.1105/tpc.105.034397
Li Q, Zheng J, Li S, Huang G, Skilling SJ, Wang L et al (2017) Transporter-mediated nuclear entry of jasmonoyl-isoleucine is essential for jasmonate signaling. Mol Plant 10:695–708
Ling F, Su Q, Jiang H et al (2020) Effects of strigolactone on photosynthetic and physiological characteristics in salt-stressed rice seedlings. Sci Rep 10:6183. https://doi.org/10.1038/s41598-020-63352-6
Liu C-J, Zhao Y, Zhang K (2019) Cytokinin transporters: multisite players in cytokinin homeostasis and signal distribution. Front Plant Sci 10:693
Llanes A, Masciarelli O, Ordoñez R, Isla MI, Luna V (2014) Differential growth responses to sodium salts involve different ABA catabolism and transport in the halophyte Prosopis strombulifera. Biol Plant 58:80–88. https://doi.org/10.1007/s10535-013-0365-6
Lo SF, Ho THD, Liu YL, Jiang MJ, Hsieh KT, Chen KT et al (2017) Ectopic expression of specific GA 2 oxidase mutants promotes yield and stress tolerance in rice. Plant Biotechnol J 15:850–864. https://doi.org/10.1111/pbi.12681
Maheshwari DK, Dheeman S, Agarwal M (2015) Phytohormone-producing PGPR for sustainable agriculture. In: Maheshwari DK (ed) Bacterial metabolites in sustainable agroecosystem. Springer, Cham, pp 159–182. https://doi.org/10.1007/978-3-319-24654-3_7
Mandava NB (1988) Plant growth promoting brassinosteroids. Ann Rev Plant Physiol Plant Mol Biol 39:23–52
Marzec M (2016) Perception and signaling of strigolactones. Front Plant Sci 7:1260. https://doi.org/10.3389/fpls.2016.01260
Marzec M, Muszynska A, Gruszka D (2013) The role of strigolactones in nutrient-stress responses in plants. Int J Mol Sci 14:9286–9304. https://doi.org/10.3390/ijms14059286
Minocha R, Majumdar R, Minocha SC (2014) Polyamines and abiotic stress in plants: a complex relationship. Front Plant Sci 5:175. https://doi.org/10.3389/fpls.2014.00175
Misra N, Saxena P (2009) Effect of salicylic acid on proline metabolism in lentil grown under salinity stress. Plant Sci 177:181–189. https://doi.org/10.1016/j.plantsci.2009.05.007
Mostofa MG, Li W, Nguyen KH, Fujita M, Tran L-SP (2018) Strigolactones in plant adaptation to abiotic stresses: an emerging avenue of plant research. Plant Cell Environ 41:2227–2243
Mouchel CF, Osmont KS, Hardtke CS (2006) BRX mediates feedback between brassinosteroid levels and auxin signalling in root growth. Nature 443:458–461
Nemhauser JL, Hong F, Chory J (2006) Different plant hormones regulate similar processes through largely nonoverlapping transcriptional responses. Cell 126:467–475
Nolan T, Vukasinovi N, Liu D, Russinova E, Yina Y (2020) Brassinosteroids: multidimensional regulators of plant growth, development, and stress responses. Plant Cell 32:295–318. https://doi.org/10.1105/tpc.19.00335
Nordstrom A, Tarkowski P, Tarkowska D et al (2004) Auxin regulation of cytokinin biosynthesis in Arabidopsis thaliana: a factor of potential importance for auxin cytokinin-regulated development. Proc Natl Acad Sci USA 101:8039–8044
Normanly J, Slovin JP, Cohen JD (2004) Hormone biosynthesis, metabolism and its regulation. In: Davies PJ (ed) Plant hormones biosynthesis, signal transduction, action! Kluwer Academic Publishers, pp 36–62
Novák O, Napier R, Ljung K (2017) Zooming in on plant hormone analysis: tissue- and cell-specific approaches. Annu Rev Plant Biol 68:323–348
Ogweno JO, Song XS, Shi K et al (2007) Brassinosteroids alleviate heat-induced inhibition of photosynthesis by increasing carboxylation efficiency and enhancing antioxidant systems in lycopersicon esculentum. J Plant Growth Regul 27:49–57. https://doi.org/10.1007/s00344-007-9030-7

Pál M, Szalai G, Janda T (2015) Speculation: polyamines are important in abiotic stress signaling. Plant Sci 237:16–23. https://doi.org/10.1016/j.plantsci.2015.05.003
Park J, Lee Y, Martinoia E, Geisler M (2017) Plant hormone transporters: what we know and what we would like to know. BMC Biol 15:93
Pattyn J, Vaughan-Hirsch J, Van de Poel B (2021) The regulation of ethylene biosynthesis: a complex multilevel control circuitry. New Phytol 229:770–782
Peleg Z, Blumwald E (2011) Hormone balance and abiotic stress tolerance in crop plants. Curr Opin Plant Biol 14:290–295. https://doi.org/10.1016/j.pbi.2011.02.001
Per TS, Khanb MIR, Anjuma NA et al (2018) Jasmonates in plants under abiotic stresses: crosstalk with other phytohormones matters. Environ Exp Bot 145:104–120
Pierik R, Tholen D, Poorter H et al (2006) The Janus face of ethylene: growth inhibition and stimulation. Trends Plant Sci 11:176–118
Pinheiro C, Antonio C, Ortuno MF et al (2011) Initial water deficit effects on Lupinus albus photosynthetic performance, carbon metabolism, and hormonal balance: metabolic reorganization prior to early stress responses. J Exp Bot 62:4965–4974
Pirbalouti GA, Sajjadi SE, Parang K (2014) A review (research and patents) on jasmonic acid and its derivatives. Arch Pharm 347:229–239
Potters G, Pasternak TP, Guisez Y, Jansen MA (2009) Different stresses, similar morphogenic responses: integrating a plethora of pathways. Plant Cell Environ 32:158–169. https://doi.org/10.1111/j.1365-3040.2008.01908.x
Qi YC, Wang FF, Zhang H, Liu WQ (2010) Overexpression of suaeda salsa S-adenosylmethionine synthetase gene promotes salt tolerance in transgenic tobacco. Acta Physiol Plant 32:263–269. https://doi.org/10.1007/s11738-009-0403-3
Raza A (2021) Eco-physiological and biochemical responses of rapeseed (Brassica napus L.) to abiotic stresses: consequences and mitigation strategies. J Plant Growth Regul 40:1368–1388. https://doi.org/10.1007/s00344-020-10231-z
Raza A (2022) Metabolomics: a systems biology approach for enhancing heat stress tolerance in plants. Plant Cell Rep 41:741–763. https://doi.org/10.1007/s00299-020-02635-8
Raza MA, van der Werf W, Ahmed M, Yang W (2020) Removing top leaves increases yield and nutrient uptake in maize plants. Nutr Cycl Agroecosyst 118:57–73. https://doi.org/10.1007/s10705-020-10082-w
Rochange S, Goormachtig S, Lopez-Raez JA, Gutjahr C (2019) The role of strigolactones in plant-microbe interactions. In: Koltai H, Prandi C (eds) Strigolactones-biology and applications. Springer, Cham, pp 121–142. https://doi.org/10.1007/978-3-030-12153-2_4
Ross JJ, Weston DE, Davidson SE, Reid JB (2011) Plant hormone interactions: how complex are they? Physiol Plant 141:299–309
Sachs T (2005) Auxins role as an example of the mechanisms of shoot/root relations. Plant Soil 268:13–19. https://doi.org/10.1007/s11104-004-0173-z
Saini S, Sharma I, Pati PK (2015) Versatile roles of brassinosteroid in plants in the context of its homoeostasis, signaling and crosstalks. Front Plant Sci 6:950. https://doi.org/10.3389/fpls.2015.00950
Sakakibara H (2006) Cytokinins: activity, biosynthesis, and translocation. Annu Rev Plant Biol 57:431–449. https://doi.org/10.1146/annurev.arplant.57.032905.105231
Santner A, Calderon-Villalobos LIA, Estelle M (2009) Plant hormones are versatile chemical regulators of plant growth. Nat Chem Biol 5(5):301–307
Sariyev A, Barutcular C, Acar M, Hossain A, E. L., and Sabagh, A. (2020) Sub-surface drip irrigation in associated with H2O2 improved the productivity of maize under clay-rich soil of Adana, Turkey. Phyton Int J Exp Bot 89:519–528. https://doi.org/10.32604/phyton.2020.09142
Sasaki-Sekimoto Y, Taki N, Obayashi T (2005) Coordinated activation of metabolic pathways for antioxidants and defence compounds by jasmonates and their roles in stress tolerance in Arabidopsis. Plant J 44:653–668. https://doi.org/10.1111/j.1365-313X.2005.02560.x
Sasse J, Simon S, Gübeli C, Liu G-W, Cheng X, Friml J et al (2015) Asymmetric localizations of the ABC transporter PaPDR1 trace paths of directional strigolactone transport. Curr Biol 25:647–655

Screpanti C, Fonné-Pfister R, Lumbroso A et al (2016) Strigolactone derivatives for potential crop enhancement applications. Bioorg Med Chem Lett 26:2392–2400
Sethia B, Mustafa M, Manohar S et al (2015) Indole acetic acid production by fluorescent Pseudomonas spp. from the rhizosphere of Plectranthus amboinicus (Lour.) Spreng. and their variation in extragenic repetitive DNA sequences. Indian J Exp Biol 53:342–349
Sharif R, Xie C, Zhang H, Arnao MB, Ali M, Ali Q et al (2018) Melatonin and its effects on plant systems. Molecules 23:2352. https://doi.org/10.3390/molecules23092352
Shigenaga AM, Berens ML, Tsuda K, Argueso CT (2017) Towards engineering of hormonal crosstalk in plant immunity. Curr Opin Plant Biol 38:164–172
Singhal RK, Saha D, Skalicky M, Mishra UN, Chauhan J, Behera LP et al (2021) Crucial cell signaling compounds crosstalk and integrative multi-omics techniques for salinity stress tolerance in plants. Front Plant Sci 12:670369. https://doi.org/10.3389/fpls.2021.670369
Stepanova AN et al (2008) TAA1-mediated auxin biosynthesis is essential for hormone crosstalk and plant development. Cell 133:177–191
Sun Q, Zhang N, Wang J, Zhang H, Li D, Shi J et al (2014) Melatonin promotes ripening and improves quality of tomato fruit during postharvest life. J Exp Bot 66:657–668. https://doi.org/10.1093/jxb/eru332
Takahashi Y, Tahara M, Yamada Y, Al E (2017) Characterization of the polyamine biosynthetic pathways and salt stress response in Brachypodium distachyon. J Plant Growth Regul 37:625–634. https://doi.org/10.1007/s00344-017-9761-z
Takahashi F, Suzuki T, Osakabe Y, Betsuyaku S, Kondo Y, Dohmae N et al (2018) A small peptide modulates stomatal control via abscisic acid in long-distance signalling. Nature 556:235–238
Takeuchi K, Gyohda A, Tominaga M, Kawakatsu M, Hatakeyama A, Ishii N et al (2011) RSOsPR10 expression in response to environmental stresses is regulated antagonistically by jasmonate/ethylene and salicylic acid signaling pathways in rice roots. Plant Cell Physiol 52:1686–1696. https://doi.org/10.1093/pcp/pcr105
Tal I, Zhang Y, Jørgensen ME, Pisanty O, Barbosa ICR, Zourelidou M et al (2016) The Arabidopsis NPF3 protein is a GA transporter. Nat Commun 7:11486
Tognetti VB, Müllenbock PER, Van Breusegem F (2012) Stress homeostasis-the redox and auxin perspective. Plant Cell Environ 35:321–333. https://doi.org/10.1111/j.1365-3040.2011.02324.x
Tromas A, Braun N, Muller P et al (2009) The auxin binding protein 1 is required for differential auxin responses mediating root growth. PLoS One 4:e6648. https://doi.org/10.1371/journal.pone.0006648
Tsuchisaka A, Theologis A (2004) Unique and overlapping expression patterns among the Arabidopsis 1-amino-cyclopropane-1-carboxylate synthase gene family members. Plant Physiol 136:2982–3000
Umehara M, Hanada A, Yoshida S, Akiyama K, Arite T, Takeda-Kamiya N et al (2008) Inhibition of shoot branching by new terpenoid plant hormones. Nature 455:195–200. https://doi.org/10.1038/nature07272
Van Butselaar T, Van den Ackerveken G (2020) Salicylic acid steers the growth–immunity tradeoff. Trends Plant Sci 25:566–576
Vardhini BV (2014) Brassinosteroids role for amino acids, peptides and amines modulation in stressed plants—a review. In: Anjum NA, Gill SS, Gill R (eds) Plant adaptation to environmental change: significance of amino acids and their derivatives. CAB International, Wallingford, pp 300–316
Verma V, Ravindran P, Kumar PP (2016) Plant hormone-mediated regulation of stress responses. BMC Plant Biol 16:86. https://doi.org/10.1186/s12870-016-0771-y
Vlot AC, Dempsey DA, Klessig DF (2009) Salicylic acid, a multifaceted hormone to combat disease. Annu Rev Phytopathol 47:177–206
Vukašinović N, Russinova E (2018) BRexit: possible brassinosteroid export and transport routes. Trends Plant Sci 23:285–292. https://doi.org/10.1016/j.tplants.2018.01.005
Vukašinović N, Wang Y, Vanhoutte I, Fendrych M, Guo B, Kvasnica M et al (2021) Local brassinosteroid biosynthesis enables optimal root growth. Nat Plants 7:619–632

Vysotskaya L, Wilkinson S, Davies WJ et al (2011) The effect of competition from neighbours on stomatal conductance in lettuce and tomato plants. Plant Cell Environ 34:729–737
Wang YH, Irving HR (2011) Developing a model of plant hormone interactions. Plant Signal Behav 6:494–500
Wang Y, Reiter RJ, Chan Z (2017) Phytomelatonin: a universal abiotic stress regulator. J Exp Bot 69:963–974. https://doi.org/10.1093/jxb/erx473
Wang S, Tian L, Liu H, Li X, Zhang J, Chen X et al (2020) Large-scale discovery of non-conventional peptides in maize and Arabidopsis through an integrated peptidogenomic pipeline. Mol Plant 13:1078–1093
Wani SH, Kumarb V, Shriramc V, Sah SK (2016) Phytohormones and their metabolic engineering for abiotic stress tolerance in crop plants. Crop J 4:162–176
Wasternack C (2007) Jasmonates: an update on biosynthesis, signal transduction and action in plant stress response, growth and development. Ann Bot 100:681–697. https://doi.org/10.1093/aob/mcm079
Wasternack C (2014) Action of jasmonates in plant stress responses and development-applied aspects. Biotechnol Adv 32:31–39. https://doi.org/10.1016/j.biotechadv.2013.09.009
Wasternack C, Hause B (2013) Jasmonates: biosynthesis, perception, signal transduction and action in plant stress response, growth and development. An update to the 2007 review in annals of botany. Ann Bot 111:1021–1058. https://doi.org/10.1093/aob/mct067
Wasternack C, Strnad M (2016) Jasmonate signaling in plant stress responses and development–active and inactive compounds. New Biotechnol 33:604–613
Waters MT, Gutjahr C, Bennett T, Nelson DC (2017) Strigolactone signaling and evolution. Annu Rev Plant Biol 68:291–322
Wei Z, Li J (2020) Regulation of brassinosteroid homeostasis in higher plants. Front Plant Sci 11:583622
Weiss D, Ori N (2007) Mechanisms of cross talk between gibberellin and other hormones. Plant Physiol 144:1240–1246
Werner T, Motyka V, Strnad M, Schmulling T (2001) Regulation of plant growth by cytokinin. Proc Natl Acad Sci USA 98:10487–10492
Werner T, Köllmer I, Bartrina I, Holst K, Schmülling T (2006) New insights into the biology of cytokinin degradation. Plant Biol 8:371–381
Wilkinson S, Davies WJ (2009) Ozone suppresses soil drying- and abscisic acid (ABA)-induced stomatal closure via an ethylene-dependent mechanism. Plant Cell Environ 32:949–959
Williams ME (2011) Introduction to phytohormones. Teaching tools in plant biology: lecture notes. Plant Cell 22:1. https://doi.org/10.1105/tpc.118.tt0718
Woodward AW, Bartel B (2005) Auxin: regulation, action, and interaction. Ann Bot 95:707–735
Xu C, Wu X, Zhang H (2009) Impact of D-Arg on drought resistance and endogenous polyamines in mycorrhizal Pinus massoniana. J Nanjing Forestry Univ 33:19–23
Xu Z-Y, Lee KH, Dong T, Jeong JC, Jin JB, Kanno Y et al (2012) A vacuolar β-glucosidase homolog that possesses glucose-conjugated abscisic acid hydrolyzing activity plays an important role in osmotic stress responses in Arabidopsis. Plant Cell 24:2184–2199
Xu Z-Y, Kim SY, Hyeon DY, Kim DH, Dong T, Park Y et al (2013) The arabidopsis NAC transcription factor ANAC096 cooperates with bZIP-type transcription factors in dehydration and osmotic stress responses. Plant Cell 25:4708–4724. https://doi.org/10.1105/tpc.113.119099
Yang SF, Hoffman NE (1984) Ethylene biosynthesis and its regulation in higher plants. Annu Rev Plant Physiol 35:155–189
Yang C, Li W, Cao J, Meng F, Yu Y, Huang J et al (2017) Activation of ethylene signaling pathways enhances disease resistance by regulating ROS and phytoalexin production in rice. Plant J 89:338–3531. https://doi.org/10.1111/tpj.13388
Yang J, Thames S, Best NB, Jiang H, Huang P, Dilkes BP, Eveland AL (2018) Brassinosteroids modulate meristem fate and differentiation of unique inflorescence morphology in Setaria viridis. Plant Cell 30:48–66. https://doi.org/10.1105/tpc.17.00816
Yin Y, Jiang X, Ren M, Xue M, Nan D, Wang Z et al (2018) AmDREB2C, from Ammopiptanthus mongolicus, enhances abiotic stress tolerance and regulates fatty acid composition in

transgenic Arabidopsis. Plant Physiol Biochem 130:517–528. https://doi.org/10.1016/j.plaphy.2018.08.002
Yu Z, Duan X, Luo L, Dai S, Ding Z, Xia G (2020) How plant hormones mediate salt stress responses. Trends Plant Sci 25:1117–1130
Zhang X, Chang X (2016) Conclusion. In: Zhang X, Chang X (eds) The logic of economic reform in China. China insights. Springer, Berlin; Heidelberg, pp 197–200. https://doi.org/10.1007/978-3-662-47404-4_11
Zwanenburg B, Pospíšil T (2013) Structure and activity of strigolactones: new plant hormones with a rich future. Mol Plant 6:38–62
Zwanenburg B, Pospíšil T, Ćavar Zeljković S (2016) Strigolactones: new plant hormones in action. Planta 243:1311–1326

Phytohormones for Combating Global Challenges: An Ecofriendly Approach

14

Saima Shaukat, Ghazala Mustafa, and Kejun Cheng

Abstract

Adverse climatic conditions cause significant harm to the photosynthetic system in plants. Prior research has indicated that phytohormones are essential not only for the plant development and growth but also for their regulation of responses to various climatic changes and abiotic and biotic stress conditions. Phytohormones are small compounds that have crucial functions in controlling plant growth and development, as well as enhancing stress resistance to ensure survival and adaptability in different conditions. Global agricultural losses are primarily caused by abiotic stress each year. Phytohormones, particularly cytokinins, present in fertilisers enhance plant growth by accelerating progression through the several stages of the plant cell cycle. In this study, we examine the possible influence of various phytohormones on global challenges. Additionally, we provide a summary of the current research advancements in understanding how plants respond to various stresses at both physiological and molecular levels. We focused on the regulation pathways of abscisic acid, indole acetic acid, cytokinins, gibberellic acid, brassinosteroids, jasmonates, and ethylene in response to different stressors. The global climate issue is a significant factor that causes abiotic stress, leading to a permanent reduction in crop yields and posing a threat to environmentally friendly farming. Moreover, external applications of plant hormones

S. Shaukat
Department of Plant Sciences, Quaid-i-Azam University, Islamabad, Pakistan

G. Mustafa (✉)
Department of Plant Sciences, Quaid-i-Azam University, Islamabad, Pakistan

Chemical Biology Center, Lishui Institute of Agriculture and Forestry Sciences, Lishui, China
e-mail: mghazala@qau.edu.pk

K. Cheng (✉)
Chemical Biology Center, Lishui Institute of Agriculture and Forestry Sciences, Lishui, China

© The Author(s), under exclusive license to Springer Nature Singapore Pte Ltd. 2024

M. Faizan, S. Hayat (eds.), *Plant Growth Regulators: Resilience for Sustainable Agriculture*, https://doi.org/10.1007/978-981-97-2918-0_14

have been demonstrated to increase the levels of naturally occurring plant hormones. In this chapter, we examine the existing knowledge on the function of phytohormones in plants during different stress situations and explore potential directions for future investigation.

Keywords

Phytohormones · Abiotic stress · Microbes · Metabolic engineering · Biofertilizer

14.1 Introduction

Changes in the environment brought about by human activities or dramatic seasonal variations in the climate are two primary barriers to plant survival. As sessile creatures, plants have to withstand these challenging environmental circumstances. Many environmental factors, such as flooding, severe weather, high salt levels, extremes in temperature, toxic substances, and direct sunlight, could interfere with a plant's ability to grow and develop to its maximum potential (Sabagh et al. 2022). A plant's morphology is determined by the combination of multiple endogenous and extrinsic inputs during growth and development, in addition to its fundamental genomic sequence. This process requires a class of growth regulators called phytohormones or plant hormones. This class of chemicals includes jasmonic acid (JA), ethylene, cytokinin, abscisic acid (ABA), auxin, brassinosteroids (BRs), and gibberellins (GAs), which all have low-concentration effects on various aspects of a plant's development as well as growth (Gray 2004).

Plants are stationary creatures that face a variety of biotic (such as pathogen infection and grazing) and abiotic (such as temperature fluctuations, light intensity, and availability of moisture and minerals) environmental stresses. In order to address these problems, plants utilise adaptive methods in their growth and reproduction, employing flexible systems that have been gained through vast evolutionary processes (Parsons et al. 2020). Hormones are essential chemical compounds that have a vital role in the growth and development of plants, as a result of intricate interactions among many factors (Brackmann et al. 2018). Phytohormones, or plant hormones, are distinct substances with their own inherent metabolisms and capacities. At low concentrations, they have the ability to impact several physiological processes in plants, such as growth, cell division, and maturation. Animal hormones and their characteristics differ significantly from those of plant hormones (Brackmann et al. 2018). Charles and Francis Darwin did research on *Phalaris canariensis*, notably observing the plant's responsiveness to both sunlight and gravity in its roots and young shoots. Moreover, research has demonstrated that these stimuli are sent to many areas of the plant during its growth and maturation stages (Darwin 1880; Morffy and Strader 2020).

Subsequently, Nicolai and Fritz conducted a laboratory experiment to examine the impact of *Avena sativa* coleoptile fragments. They observed that the segments exhibited curvature even in the absence of light from one side (Cholodny 1928; Went 1928; Du et al. 2020). The compound responsible for stimulating coleoptile growth was designated as 'auxin', derived from the Greek word 'auxein' which signifies 'to grow'. From the available data, it was concluded that auxin has a role in controlling the curvature of the coleoptile in response to light. It is worth mentioning that moving the stem portions to a darkened region led to faster growth (Shu et al. 2019; Du et al. 2020). The discovery of gibberellin (GA) stemmed from the finding of this plant hormone in the fungal pathogen *Gibberella fujikuroi*, which is notorious for causing a rice disease called 'foolish-seedling'. The excessive production of GA in plants leads to elongation and diminished structural integrity, leading in chlorosis and partial infertility (Binenbaum et al. 2018).

Gibberellins (GAs) are present in all vascular plants and are involved in the development of reproductive structures in lower plants such as ferns and lycophytes (Tanaka et al. 2014; Miyazaki et al. 2018). In more advanced plants, GA has a wider range of functions, including promoting cell proliferation, division, and seed germination, as well as initiating blooming in many species and enhancing organ development (Sponsel 2016). Cytokinins were identified in the 1950s (Miller et al. 1955) due to their capacity to stimulate cell division in plants. The auxin-cytokinin theory of plant growth was promptly proposed by Miller and Skoog upon their discovery. According to the concept, cytokinin and auxin play a vital role in plant development, exerting a substantial influence on the growth of shoots and roots (Werner et al. 2001). The shoot apical meristem (SAM) growth has been shown to rely heavily on cytokinin signaling. Cytokinins have a crucial role in regulating shoot growth and are also engaged in several root development processes (Wybouw and De Rybel 2019). According to Dubois, ethylene inhibits the growth of leaves in plants and ethylene slows down the plant growth. It appears that ethylene disrupts the cell cycle. Ethylene regulates the development and growth in plants. The most common association between ethylene and growth is the control of cell size; it can control cell division in addition to frequently limiting cell elongation. When it comes to development, ethylene is usually thought of as an 'aging' hormone, since it speeds up and is occasionally needed for processes like senescence, abscission, and ripening (Schaller 2012). Plant development and cellular defense are regulated by the plant regulator jasmonic acid (JA) and its product, jasmonoyl isoleucine (JA-Ile), which is an amino acid conjugation of JA. Research has shown that jasmonic acid plays a direct role in many biological processes, including as root growth, senescence, and stamen growth. Furthermore, it has been shown to regulate the production of several metabolites, such as terpenoids and phytoalexins (Gomi 2020).

14.2 Global Issues Faced in Establishing Sustainable Agriculture and Exploring the Potential of Phytohormones as Eco-friendly Solutions to Address These Challenges

In the contemporary world, a sustainable farming system is vital as it has the capacity to meet our agricultural needs, unlike traditional agriculture. This agricultural strategy utilises a distinct farming approach that optimises the exploitation of natural resources. This eco-friendly method guarantees the nutritional value and safety of agricultural products. The continued use of chemical fertilisers, pesticides, and herbicides has resulted in decreased soil fertility, inadequate crop yields, soil degradation, biodiversity loss, low economic returns, and sickness from food and major risks to the environment. The quantity and quality of crop products are affected by the use of such compounds in the farming process (Chandra et al. 2018). Due to the impact of worldwide pollution on the environment, agriculture productivity decreases with abiotic stresses such as salts, heavy metal, and drought. Plant growth-promoting rhizobacteria play an essential part in the abiotic stresses in increasing the effectiveness of plant growth promotion and soil. There are some specific plant growth promotion properties of stress-tolerant plant growth-promoting rhizobacteria, for example, hormone synthesis, abscisic acid synthesis, indole-3-acetic acid production, 1-aminocyclopropane-1-carboxylate deaminase, enzyme production. Potassium, phosphorus, nitrogen fixation, and solubilisation, as an outcome, protect a lot of crops throughout such stress condition and increase the sustainability of the crop. Effective plant growth-promoting rhizobacteria are separated from many stress conditions, which is more valuable in contrast to certain stress (Kumar et al. 2019). For good food production, the stresses of the environment are a big challenge because they decrease the production due to the making of reactive oxygen species which give harm to biomolecules (enzyme, protein, DNA), cell organelle, and some others which goes to the process of apoptosis.

Plant growth-promoting rhizobacteria hormone gives an eco-friendly bid to increase growth, maintain a sustainable agriculture system, and give new ways to synthetic agrochemicals. As the plant growth is promoted, rhizobacteria dwell in the rhizosphere and show a positive relation with the roots of the plant. Many-sided benefits give these organisms to plants by various mechanisms, for example, solubilisation of mineral phosphates, phytohormones, siderophore production for iron sequestration, nitrogen fixation, stress, and protection against various pathogens. Plant growth-promoting rhizophore has to control the opposing outcome of the numerous stresses, for example, heavy metals, salinity, floods, drought, and additional pressures on plants by inducing the making of antioxidant enzymes, for example, superoxide dismutase, catalase, and peroxidase (Chandran et al. 2021). For the productivity of agriculture, salinity is the main factor. In the last 20 years, this significant abiotic stress worsened due to the high water demands in semi-arid and arid areas. The use of phytohormones, osmoprotectants, phytoprotectants of microbial origin, antioxidant metabolism-related compounds, and melatonin are mitigators of salt stress in a lot of plant species, and as a result, significant crop quality and crop yield can be obtained which directly have positive impact on the security of food (Acosta-Motos et al. 2020).

14.3 Role of Phytohormones in Abiotic Stress Tolerance (Salt, Heavy Metal, Heat, and Others)

The current time of climate change is characterised by various abiotic stress factors, such as flooding, drought, extreme cold, excessive heat, ultraviolet (UV) radiation, oxidising chemicals, and salt. These conditions pose a threat to plant growth and development, leading to global crop failure. Engineered plants with enhanced resilience to abiotic stressors not only contribute to enhancing food yield and promoting sustainable agricultural growth but also offer opportunities for crop adaptability to future climates. The development and growth of plants entail intricate genetic, hormonal, metabolic, and environmental processes. The regulation of these activities is governed by both external and internal signals, including light radiation, temperature, and phytohormones, which encompass a range of biostimulants and other inorganic and organic elicitors. Phytohormones, such as brassinosteroids, gibberellins, abscisic acid, nitric oxide, salicylic acid, jasmonates, and others, play crucial roles in regulating various developmental and physiological processes in plants. These processes include photosynthesis, mitigation of abiotic stress, pollen formation, leaf withering, seed germination, and plant defensive responses. Phytohormones are now in their first phases of development in terms of enhancing plant responsiveness to dynamic climatic changes. A recent study has provided insights into the mechanisms by which plants adapt to their environment. The study highlights the diverse functions of different phytohormones, including salicylic acid (SA), hydrogen sulphite (H_2S), and nitric oxide (NO), in regulating various developmental processes and enhancing stress tolerance (Sadiq et al. 2020). Plants experience different non-living stimuli that produce reactive oxygen species, which ultimately hinder their regular growth and development. Plants gather specific substances, known as osmolytes, in order to protect their cellular machinery from harm caused by oxidative stress. The essential osmolytes necessary for osmoregulation comprise polyamines, proline, glycine-betaine, and sugars. These substances maintain the osmotic balance between the cytosol and the extracellular environment of the cell. Furthermore, the inhibition of the generation of detrimental reactive oxygen species, such as hydrogen peroxide, hydroxyl ions, superoxide ions, and other free radicals, protects the plant cells from oxidative stress. Phytohormones such as brassinosteroids, jasmonates, cytokinin, abscisic acid, ethylene, and salicylic acid are involved in regulating the accumulation of osmolytes. Therefore, it is essential to comprehend the processes that govern the buildup of osmolytes in plants under non-living factor-induced stress by means of phytohormone mediation (Sharma et al. 2019).

14.4 Plant Growth and Development: The Role of Phytohormones in Germination

Plant hormones exert influence over several plant activities, such as the stimulation of growth and the transient inhibition of growth in seeds (Graeber et al. 2010). Prior to germination, a seed must meet certain requirements, including the presence of food stores within the seed. Seeds contain nutrients such as protein, fat, and carbs. The seed embryos contain specific enzymes and undergo metabolic processes to produce these components (Miransari and Smith 2009). Seed germination is a biological phenomenon that initiates the activation of the embryo through morphological and physiological alterations. Seeds experience enlargement and elongation of their embryos by absorbing water before they start to sprout. The radicle represents the final stage of the seed's developing process, which begins with the breaking of the seed membranes (Hermann et al. 2007). Researchers have extensively studied the germination of seeds and the effects of plant hormones on them in several plant families, such as the Brassicaceae (Müller et al. 2006; Hermann et al. 2007). Conversely, when plants are exposed to non-living stressors such as flooding and living stressors such as the presence of pathogens, abscisic acid has a positive effect on the opening and closing of stomata, the dormancy of seeds, and several plant activities. Seed germination can be enhanced by ethylene, gibberellins, brassinosteroids, cytokinins, and other plant hormones, while being inhibited by their antagonistic interaction with abscisic acid.

The impact of ethylene on seed germination remains uncertain, despite its influence on various plant processes, including tissue development, growth, and seed germination. The precise method by which ethylene affects seed germination is still unknown; however, it has been observed to influence many plant functions such as organ growth, formation, and seed germination (Matilla 2000; Petruzzelli et al. 2000; Rinaldi 2000). Auxin is not necessary as a stand-alone hormone for seed fertilisation. According to the research on the expression of auxin-related genes, auxin is consistently present at the tips of the seed radicle during all stages of seed germination (Liu et al. 2007). Auxin is needed for the growth of new seedlings, while it may not be necessary for seed germination (Hentrich et al. 2013).

14.5 Role of Phytohormones in Biotic Stress Tolerance

Due to their immobility, plants are vulnerable to a wide range of abiotic and biotic stressors. Phytohormones, which are small endogenous chemicals with low molecular weight, trigger strong defense reactions against both living and non-living stressors. These phytohormones not only play a role in defensive signaling but also serve as regulators of growth, development, and physiological activities (Checker et al. 2018). Four hormones, namely, jasmonic acid (JA), ethylene (ET), salicylic acid (SA), and abscisic acid (ABA), are well acknowledged for their significant roles in controlling plant defense responses against various pathogens and pests. Salicylic acid (SA) primarily activates the defense response against hemibiotrophic and biotrophic diseases, whereas ethylene (ET) and jasmonic acid (JA) are important for defending against herbivorous insects and necrotrophic infections. ABA exerts a positive regulatory influence on plant defense during the early stages of infection by facilitating the closing of stomata in response to pathogens and promoting the deposition of callose in case the disease evades the initial defensive mechanisms (Fazal et al. 2021).

Jasmonic acid (JA) plays a significant role in the plant's response to biotic stress, particularly in relation to diseases or insects. Advancements have been made in understanding the roles and mechanisms of JAs in how plants respond to biotic stressors, such as pathogens, insects, and herbivores. Jasmonates interact with other plant hormones to modulate the equilibrium between plant growth and defense. Hence, the interplay between jasmonates (JAs) and other plant hormones such as auxin, gibberellic acid, salicylic acid, brassinosteroid, ethylene, and abscisic acid has been observed (Wang et al. 2021). The receptor-active conjugated complex JA-Ile plays a crucial role in the signaling pathways of jasmonic acid (JA) (Marquis et al. 2020). Environmental stimuli are detected by cell surface receptors, which initiate the synthesis of JA-Ile from plastid lipids. This process leads to interactions between transcription factors and promotes plant growth, development, and specific protective mechanisms (Howe et al. 2018). The regulation of plant defense responses to insects involves chemical signals, with the most crucial signaling molecule being JA (Chini et al. 2016). Plants often suffer from insect damage, with many of these insects being herbivores. This leads to a quick rise in the levels of endogenous JA (Al-Zahrani et al. 2020). Remarkably, the application of JA to plants triggers the production of plant secondary metabolites and systemic resistance as a defense mechanism against insect harm. It is worth noting that JA itself does not possess toxicity towards insects, as stated by Zhang et al. (2020) and Qi et al. (2020). Jasmonates (JAs), which are naturally occurring signaling molecules, play a crucial role in plants' defense mechanism against insects and also exhibit a distinct defensive action against necrotrophic diseases (Zhang et al. 2017).

Prior research has demonstrated that the external administration of JA or MeJA can stimulate the activation of plant defense genes. Similarly, the application of JAs therapy led to a considerable induction of many genes in the JA signaling system, resulting in an enhanced resistance to necrotrophic infections (Wang et al. 2012). Abscisic acid (ABA), a plant hormone, has also emerged as a vital regulator under biotic stressors. Extensive research has demonstrated that ABA can have both beneficial and detrimental effects on the immune response of the host (Chen and Yu 2014). Abscisic acid (ABA) is a stress hormone that builds up in response to various non-living and living stimuli. ABA typically reduces transpirational water loss in leaves by causing stomata to close. Simultaneously, it defends against microorganisms by limiting their access through stomatal pores. The citation is from Bharath et al. (2021).

14.6 Role of Phytohormones in Plant-Microbe Interaction

Plant hormones serve as regulators for nearly all facets of plant development and the plant's reactions to its surroundings. Plant hormones play crucial roles in the interactions between plants and beneficial microorganisms due to their ability to function at low concentrations and their precise spatial regulation of synthesis and response. This special issue, titled *The Role of Plant Hormones in Plant-Microbe Symbioses*, presents novel findings on the involvement of hormones originating from both the plant and microbial partner in communication, establishment, and functioning of symbiotic relationships. This encompasses close symbiotic relationships with arbuscular mycorrhizal (AM) fungi, which are established by most terrestrial plants, as well as the more recently developed nodulation, which involves a specific group of plants in the fabid clade and nitrogen-fixing bacteria. The articles in this special issue also investigate the function of hormones in plant interactions with ectomycorrhizae, endophytic bacteria, and fungi, as well as beneficial microbes that form associations with the root or leaf surfaces. Furthermore, numerous hormones have the ability to not only act alone but also to interact with one another in order to regulate the formation of these mutually beneficial relationships. These intricate networks are currently being uncovered (Hirsch et al. 1997). Auxins and cytokinins, which are plant hormones, play crucial roles in various aspects of plant growth and development, including cell elongation, cell division, tissue differentiation, and apical dominance. The production and mechanism of action of auxins and cytokinins are currently being extensively studied. Auxins and cytokinins can be synthesised by both plants and microbes. The precise understanding of phytohormone biosynthesis by microorganisms remains incomplete. However, it has been observed that in certain instances, pathogenic fungi and bacteria utilise these compounds to cause disease in plants. Additionally, beneficial bacteria and associative symbiosis may stimulate root growth through the production of auxin and cytokinin.

14.6.1 Indole-3-Acetic Acid

The genetic process of auxin biosynthesis and its regulation by *Pseudomonas*, *Agrobacterium*, *Rhizobium*, *Bradyrhizobium*, and *Azospirillum* has been extensively researched. In these bacteria, the production of phytohormones has been found to have various physiological effects. *Pseudomonas* and *Agrobacterium*, which are harmful bacteria, synthesise indole-3-acetic acid through the indole-3-acetamide route. The genes responsible for this pathway are located on plasmids. However, they also possess the chromosomally encoded indole-3-pyruvic acid pathway. Furthermore, they possess genes that are capable of either conjugating unbound auxins or breaking down conjugated forms of auxins and cytokinins by hydrolysis. *Agrobacterium* contains many genes, which are situated in close proximity to the auxin and cytokinin biosynthesis genes and have a role in regulating the sensitivity of the transformed plant tissue to auxins and cytokinins. The symbiotic bacteria *Rhizobium* and *Bradyrhizobium* produce indole-3-acetic acid by synthesising indole-3-pyruvic acid. Additionally, the genetic factors responsible for the indole-3-acetamide pathway have been identified; however, their functionality has not been proven. In *Azospirillum*, a bacterium that promotes plant growth, both the indole-3-pyruvic acid and indole-3-acetamide routes are found, similar to *Agrobacterium* and *Pseudomonas*. However, in *Azospirillum*, the indole-3-pyruvic acid pathway is more important. Furthermore, Costacurta and Vanderleyden (1995) have provided biochemical data supporting the existence of an indole-3-acetic acid pathway in *Azospirillum* that does not rely on tryptophan.

14.6.2 Auxin and Cytokinins

Plant hormones is regulated by endophytic fungi, for example, indole acetic acid (auxin) is identified by the ectomycorrhizal fungi such as *Tuber melanosporum* and *Laccaria bicolor* while *Cis*-Zeatin and isopentane (cytokinins) is also produced by the ectomycorrhizal fungi in a assessable amount which results in the alteration of the roots' morphology of host plant (*Cistus incanus*). Production of auxin in plants is activated by ethylene hormone which is released by *L. bicolor* (Boivin et al. 2016).

14.6.3 Jasmonic Acid

LasA (Lasiojasmonate A) is a plant hormone called jasmonic acid ester which is produced by the pathogenic fungi, that is, *Lasiodiplodia mediterranea.* Lasiojasmonate A is changed to JA lle to activate the signaling of jasmonic acid in plant (Nakano et al. 2022).

14.7 Exogenous Synthesis and Application of Phytohormones

Wood, also known as secondary xylem, is an essential circulatory component mostly found in trees. It has a crucial function in facilitating water flow and offering structural support. This material has wide-ranging applicability in several industries, including construction, pulp and paper production, and the bright prospect of generating biofuels in the future. The development of both primary and secondary xylem entails numerous intricate processes. The processes encompassed in this study involve the organisation of primary vascular tissue into clusters, the growth of cells within these clusters or the secondary vascular cambium, the initiation of xylem differentiation, the control of cell expansion, the formation of a secondary cell wall, and the programmed death of cells (Nieminen et al. 2004). The recent findings on xylem differentiation have revealed its intricate nature, in which hormone signaling cascades play a pivotal role. The molecular mechanisms that govern the impact of chemicals like auxin, cytokinin, gibberellin, and brassinosteroid on vascular development are not completely comprehended, despite the thorough investigation of their signaling pathways (Eriksson et al. 2000; Milhinhos and Miguel 2013). Wood creation takes place during the secondary growth stage of plant development and mostly consists of secondary walls, which are made up of cellulose, hemicelluloses, and lignin. The intricate process of secondary wall development requires the synchronised expression of distinct biosynthetic genes, resulting in the production, release, and organisation of secondary wall constituents. The transcriptional network responsible for regulating the genes involved in secondary wall biosynthesis, including cellulose, xylan, glucomannan, and lignin, is controlled by master switches NAC (NAM–ATAF1,2–CUC2) and MYB, along with downstream transcription factors. Moreover, the proteins produced by these genes have been linked to hormone metabolism, transport, and signalling. Although there has been improvement, our understanding of the development of vascular tissue is still not fully complete (Chen et al. 2015). Comprehending the processes that govern the synchronised activation of genes involved in the production of secondary cell walls is essential for enhancing our understanding of the regulation of wood formation.

Phytohormones have a critical role in regulating essential physiological and developmental processes throughout the whole life cycle of a plant. Previous studies have stressed the role of indole-3-acetic acid (IAA) in cambium and wood formation in woody plants. IAA affects cambial growth, including xylem and phloem development. Gibberellins (GAs) and brassinosteroids (BRs) regulate stem elongation and wood growth. They greatly affect cambial and xylem division, growth, and differentiation (13–16). IAA, GA3, and BR promote wood production, although their regulatory mechanisms are unclear. The study focuses on *Xiaohei poplars* (*Populus simonii* × *P. nigra*), a popular model species for investigating angiosperm trees. These poplars have a genome that has been sequenced, known as Populus trichocarpa. The study examines the impact of IAA, GA3, and BR on the growth of poplar trees, assessing characteristics including plant height, stem diameter, cellulose content, and xylem development. In addition, we analyse the transcriptional responsiveness of 27 genes linked with wood production in poplars to the three phytohormones, using tissue-specific and time-series methods. The objective of the study is to identify the predominant hormone that influences growth regulation and gene expression.

Hormones are signalling molecules that are secreted from certain glands to govern physiology and behaviour in organisms. They have a vital role in coordinating cellular processes. The term 'hormones' has traditionally been linked to animals, but it has now been expanded to encompass specific chemical substances produced by certain cells that have an impact on nearby or distant cells. Within the plant kingdom, hormones are not released by glands. Instead, small signal molecules or compounds are internally generated. These substances function at very low levels to precisely control certain cellular processes in specific plant cells (Vildanova et al. 2014). The main emphasis of our research is on five extensively investigated basic plant hormones: auxins, gibberellins, cytokinins, abscisic acid, and ethylene (Gaspar et al. 1996). Hormones interact with specific receptors on targeted cells, triggering alterations in cell function via signal transduction pathways (Jiang and Asami 2018). While certain chemical compounds, such as the herbicide 2,4-D (2,4-dichlorophenoxyacetic acid), were created for agricultural use, plant hormones like gibberellins, cytokinins, and ethylene are also commercially accessible due to their ability to regulate plant growth.

Plant hormones have been found to have potential effects on human health and the ability to fight against different diseases, as indicated by current research (Trypuc et al. 2016). Plant hormones, such as abscisic acid (ABA), are not limited to plants but are also present in cyanobacteria, fungi, animals, and humans. ABA has been recognised as a controller of animal cell proliferation and differentiation, amplifying immunological responses to varied stimuli. Cytokinins are acknowledged for their role in promoting cell growth and development, as well as their function as anti-stress agents (Voller et al. 2017; Rattan and Sodagam 2005). Furthermore, certain plant hormones, such as auxin and gibberellin, have the potential to be used for enhancing human health. Auxin, renowned for its involvement in plant development, has been recognised as an anti-tumour agent in human cells (Hamayun et al. 2017), whereas gibberellin affects antioxidant characteristics. This review focuses on the biosynthesis and chemical production of plant hormones, examining their functional significance in agriculture, microbial cells, and the human body. We want to provide light on the many functions of plant hormones in different biological systems by studying their potential uses in human health.

14.8 Metabolic Engineering of Plant Phytohormones for Resilience

With the continuous rise in the global human population, there is an imperative requirement for a significant augmentation in agricultural production to guarantee food security. Nevertheless, a multitude of living and non-living factors present considerable obstacles to the efficiency of crop production (Wani and Sah 2014). In order to satisfy the nutritional requirements of an extra 2.3 billion individuals by the year 2050, it is imperative to enhance agricultural output by 70% (Tilman et al. 2011). Plants have complex and unique systems for responding to environmental stress, which is different from how mammals respond. Therefore, it is crucial for plant biotechnologists to have a deep understanding of how plants react to different types of environmental stress. Drought, salinity, and severe temperatures are prominent and influential abiotic stressors. Conventional breeding methods have faced challenges in developing stress tolerance traits, leading to limited achievements. As a result, novel strategies are required to meet the increasing worldwide food requirements. Within this framework, the manipulation of phytohormones in crop engineering presents itself as a promising approach to develop crops that are tolerant to climate conditions and exhibit high productivity.

Phytohormones, functioning as chemical messengers in plants, are produced in little quantities but possess the ability to regulate a broad spectrum of cellular processes (Voß et al. 2014). Phytohormones play a crucial function in coordinating signal transduction pathways that are essential for responding to non-living environmental factors. They effectively control both external and internal stimuli (Kazan 2015). Plants utilise the phytohormone abscisic acid (ABA) as a stress hormone. Plant growth relies on various processes, including seed dormancy maintenance, germination prevention, growth control, stomata closure, fruit detachment, and reaction to living and non-living environmental stimuli (Li et al. 2010). By applying phytohormone principles in engineering, biotechnologists can enhance the nutritional and commercial characteristics of crops. This review provides a comprehensive analysis of phytohormones, elucidating their intricate roles in plant development, growth, and reaction to abiotic stress. Furthermore, the research examines the possibility of using metabolic engineering to alter phytohormones with the aim of enhancing crop plants' ability to withstand abiotic stress. Ultimately, this would enhance food production in terms of both its quantity and quality. In order to enhance the development of robust and productive crops, researchers are striving to understand and manipulate these crucial signaling molecules. This will aid in mitigating the challenges arising from the increasing demands of the global population and environmental considerations.

The worldwide population is experiencing exponential growth, and recent shifts in global climatic patterns require a substantial increase in agricultural production. Projections indicate that agricultural production will need to expand by 70% in order to fulfil the growing demands of the global population by the middle of the century (Francini and Sebastiani 2019). Moreover, it is crucial to acknowledge the influence that climate fluctuations have on agricultural crop production. Both biotic and abiotic stresses exert a substantial influence and serve as crucial factors that restrict overall productivity and agricultural yields (Mangal et al. 2022). Comprehending the intricate mechanisms through which plants respond to different environmental stimuli is a crucial undertaking that necessitates the cooperation of plant biologists and agronomists. Bani et al. found that certain prevalent and noteworthy factors, such as salt, drought, and heat stress, have a substantial influence on the growth and development of plants. The complex characteristics associated with stress tolerance traits hinder the effectiveness of traditional breeding methods, thereby necessitating the creation of novel technologies to bridge the gap between global food production and consumption. In light of the present situation, it is crucial to prioritise the development of innovative and effective methods. Utilising phytohormones is a potent method for cultivating crops that exhibit great productivity and resilience to diverse environmental circumstances. Phytohormones have lately been acknowledged as a viable method for enhancing the resilience of horticulture plants against non-living environmental influences.

Plants produce phytohormones, chemical regulators that play a crucial role in coordinating growth, development, and responses to environmental challenges. Their capacity to coordinate many signal transduction pathways explains their importance under abiotic stress conditions. In addition, phytohormones play a crucial role in regulating various stimuli originating from both the external environment and internal factors within the plant. This regulation has a substantial impact on the growth and development of plants. Given the challenges posed by climate change and population growth, phytohormones present a viable approach to cultivating crops that can withstand extreme weather conditions. This makes them an eco-friendly method to enhance agricultural output.

14.9 The Importance of Phytohormones in Biofertilisers

The optimal growth and development of plants necessitate the presence of vital chemical elements, which can be classified as non-mineral and mineral elements. The non-mineral elements hydrogen (H), oxygen (O), and carbon (C) are present in air and water. Plants require these elements as fundamental components for the process of photosynthesis to take place. This crucial mechanism converts carbon dioxide (CO_2) and water (H_2O) into glucose and complex carbohydrates using solar energy. Starch and complex sugars are the main sources of energy for the plant's growth and biosynthetic processes (Rolland et al. 2002). On the other hand, mineral nutrients are obtained from the soil. Macronutrients are divided into basic nutrients (nitrogen, phosphorus, potassium) and secondary nutrients (calcium, magnesium, sulphur) based on their content in plant tissues. Crucial nutrients, necessary for the growth and survival of plants, are frequently lacking in soil that has been depleted of nutrients. In natural settings, secondary macronutrients are typically abundant and need minimal supplementation. Micronutrients, such as boron, copper, iron, chloride, manganese, molybdenum, and zinc, are necessary in lesser quantities and are often known as trace elements (Hänsch and Mendel 2009). Their accessibility for absorption by plants is controlled not only by the nutrient content of the soil but also by the composition and physicochemical qualities of the soil. The soil, an intricate medium, has a considerable influence on plant life, making it crucial to have optimal soil conditions for favourable plant growth and increased agricultural output (Berendsen et al. 2012). The arrangement, composition, and acidity level of soil are crucial elements that impact the movement of nutrients, the wearing away of soil, the ability of roots to penetrate, and the exchange of gasses. These processes are directly linked to the growth of plants. The nutrient and water retention in soil are influenced by its texture, which is governed by the relative quantities of sand, silt, clay, and organic matter. The presence of aggregated soil particles in an optimal soil structure facilitates the passage of water and air, enhances nutrient availability, and promotes microbial activity. As a result, soil fertility and productivity are improved.

The pH of soil, which measures its level of acidity or alkalinity, has a direct impact on the availability of nutrients. The optimal pH range of 6.0–6.5 facilitates root development, boosts microbial functions, and improves the solubility of nutrients. To achieve sustainable agriculture, which involves maximising crop yields while minimising harm to the environment, it is crucial to comprehend and enhance soil fertility. Conventional agriculture has indeed boosted crop production, but it has also resulted in negative environmental consequences, including the pollution of groundwater and the degradation of soil quality. Organic farming, which is becoming increasingly popular worldwide, strives for sustainability through the reduction of fertiliser and energy inputs. Mineral nutrients, which are essential for both farming systems, are taken up by plant roots. Nevertheless, the quantity of nutrients that may be readily absorbed by the body may be inadequate, therefore requiring the use of external vitamin supplements. On a global scale, the utilisation of fertilisers has consistently increased, highlighting the persistent difficulty of maintaining a balance between agricultural efficiency and ecological sustainability. The study of soil fertility and the management of nutrients continue to be a key focus in the current discussion on sustainable agriculture and its different approaches.

14.10 Conclusion

To summarise, the sophisticated mechanisms involved in wood creation, specifically the growth of secondary xylem, entail intricate hormonal signalling cascades. The control of primary and secondary xylem development, which involves the production of cellulose, hemicelluloses, and lignin, depends on the coordinated activation of particular genes regulated by key transcription factors including NAC and MYB. Although there has been notable advancement, our comprehension of the molecular mechanisms that regulate the formation of vascular tissue, particularly the complex processes involved in secondary wall production, is still lacking.

Phytohormones, including indole-3-acetic acid (IAA), gibberellins (GAs), and brassinosteroids (BRs), play a crucial role in controlling stem growth and the development of wood. Although the significance of these phytohormones in promoting wood production is recognised, the exact regulation mechanisms are still not completely understood. The study analyses how IAA, GA3, and BR affect growth metrics and gene expression in Xiaohei poplars. This study seeks to understand these trees' key hormone that controls growth and gene expression. When we consider their vast influence, plant hormones regulate various physiological and developmental processes throughout a plant's life cycle. The five important plant hormones—auxins, gibberellins, cytokinins, abscisic acid, and ethylene—coordinate cellular processes through complicated signal transduction pathways (Jiang and Asami 2018). These hormones affect plant development and human health, demonstrating their multitudinous functions. The evaluation of changing plant metabolic processes to increase phytohormone production underscores the need for higher agricultural output to feed a fast-growing global population (Tilman et al. 2011). Abiotic stresses including dryness, salt, and high temperatures make traditional breeding methods problematic, requiring phytohormone modification. Stress hormone abscisic acid (ABA) affects plant development and environmental response (Li et al. 2010). Engineering with phytohormones can improve crop resilience to abiotic stressors and food quantity and quality. Soil fertility and nutrient management emphasise the importance of 16 chemical elements, including non-mineral and mineral elements, for plant growth. The availability of nutrients and plant growth depends on soil texture, structure, and pH. Sustainable agriculture aims to maximise crop production while minimising environmental damage. This method uses new organic agricultural methods. Growing global interest in organic farming highlights its ability to reduce nutrient and energy inputs and preserve soil fertility. The intricate interaction of phytohormones, wood development, and soil fertility illustrates plant biology and agriculture. Understanding these complexities requires ongoing research and creative strategies to find sustainable ways to balance agricultural output, environmental conservation, and human well-being.

References

Acosta-Motos JR, Penella C, Hernández JA, Díaz-Vivancos P, Sánchez-Blanco MJ, Navarro JM, Gómez-Bellot MJ, Barba-Espín G (2020) Towards a sustainable agriculture: strategies involving phytoprotectants against salt stress. Agronomy 10:194

Al-Zahrani W, Bafeel SO, El-Zohri M (2020) Jasmonates mediate plant defense responses to *Spodoptera exigua* herbivory in tomato and maize foliage. Plant Signal Behav 15:1746898

Berendsen RL, Pieterse CMJ, Bakker PAHM (2012) The rhizosphere microbiome and plant health. Trends Plant Sci 17:478–486

Bharath P, Gahir S, Raghavendra AS (2021) Abscisic acid-induced stomatal closure: an important component of plant defense against abiotic and biotic stress. Front Plant Sci 12:615114

Binenbaum J, Weinstain R, Shani E (2018) Gibberellin localization and transport in plants. Trends Plant Sci 23(5):410–421

Boivin S, Fonouni-Farde C, Frugier F (2016) How auxin and cytokinin phytohormones modulate root microbe interactions. Front Plant Sci 7:1240

Brackmann K, Qi J, Gebert M, Jouannet V, Schlamp T, Grünwald K, Wallner E, Novikova DD, Levitsky VG, Agusti J, Sanchez P, Lohmann JU, Gre T (2018) Spatial specificity of auxin responses coordinates wood formation. Nat Commun 9:1–15

Chandra D, Barh A, Sharma IP (2018) Plant growth promoting bacteria: a gateway to sustainable agriculture, Microbial biotechnology in environmental monitoring and cleanup. IGI Global, pp 318–338

Chandran H, Meena M, Swapnil P (2021) Plant growth-promoting rhizobacteria as a green alternative for sustainable agriculture. Sustain For 13:10986

Checker VG, Kushwaha HR, Kumari P, Yadav S (2018) Role of phytohormones in plant defense: signaling and cross talk. In Molecular aspects of plant-pathogen interaction, pp 159–184

Chen L, Yu D (2014) ABA regulation of plant response to biotic stresses. In Abscisic acid: metabolism, transport and signaling, pp 409–429

Chen J, Chen B, Zhang D (2015) Transcript profiling of Populus tomentosa genes in normal, tension, and opposite wood by RNA-seq. BMC Genomics 16:1641–1679. https://doi.org/10.1186/s12864-015-1390-y

Chini A, Gimenez-Ibanez S, Goossens A, Solano R (2016) Redundancy and specificity in jasmonate signalling. Curr Opin Plant Biol 33:147–156

Cholodny N (1928) Beiträge zur hormonalen Theorie von Tropismen. Planta 6:118–134

Costacurta A, Vanderleyden J (1995) Synthesis of phytohormones by plant-associated bacteria. Crit Rev Microbiol 21(1):1–18

Darwin C (1880) The power of movement in plants. John Murray, London

Du M, Spalding EP, Gray WM (2020) Rapid auxin-mediated cell expansion. Annu Rev Plant Biol 71:1–24

Eriksson ME, Israelsson M, Olsson O, Moritz T (2000) Increased gibberellin biosynthesis in transgenic trees promotes growth, biomass production and xylem fiber length. Nat Biotechnol 18:7847–7888. https://doi.org/10.1038/77355

Fazal A, Ur RA, Abdul R, Nisar A, Mohammad A, Akhtar R, et al. (2021). The role of phytohormones in combating biotic stress. In Plant growth regulators for climate-smart agriculture, pp. 187–206. CRC Press

Francini A, Sebastiani L (2019) Abiotic stress effects on performance of horticultural crops. Horticulturae 5:67

Gaspar T, Kevers C, Penel C, Greppin H, Reid DM, Thorpe TA (1996) Plant hormones and plant growth regulators in plant tissue culture. In Vitro Cell Dev Biol Plant 32(4):272–289

Gomi K (2020) Jasmonic acid: an essential plant hormone. Int J Mol Sci 21(4):1261

Graeber K, Linkies A, Müller K, Wunchova A, Rott A, Leubner-Metzger G (2010) Cross-species approaches to seed dormancy and germination: conservation and biodiversity of ABA-regulated mechanisms and the Brassicaceae DOG1 genes. Plant Mol Biol 73:67–87

Gray WM (2004) Hormonal regulation of plant growth and development. PLoS Biol 2(9):E311

Hamayun M, Hussain A, Khan SA, Kim HY, Khan AL, Waqas M, Irshad M, Iqbal A, Rehman G, Jan S, Lee IJ (2017) Gibberellins producing endophytic fungus Porostereum spadiceum AGH786 rescues growth of salt affected soybean. Front Microbiol 8:686. https://doi.org/10.3389/fmicb.2017.00686

Hänsch R, Mendel RR (2009) Physiological functions of mineral micronutrients (Cu, Zn, Mn, Fe, Ni, Mo, B, Cl). Curr Opin Plant Biol 12(3):259–266

Hentrich M, Böttcher C, Düchting P, Cheng Y, Zhao Y, Berkowitz O et al (2013) The jasmonic acid signaling pathway is linked to auxin homeostasis through the modulation of YUCCA 8 and YUCCA 9 gene expression. Plant J 74(4):626–637

Hermann K, Meinhard J, Dobrev P, Linkies A, Pesek B, Heß B et al (2007) 1-Aminocyclopropane-1-carboxylic acid and abscisic acid during the germination of sugar beet (Beta vulgaris L.): a comparative study of fruits and seeds. J Exp Bot 58(11):3047–3060

Hirsch AM, Fang Y, Asad S, Kapulnik Y (1997) The role of phytohormones in plant-microbe symbioses. Plant Soil 194:171–184

Howe GA, Major IT, Koo AJ (2018) Modularity in jasmonate signaling for multistress resilience. Annu Rev Plant Biol 69:387–415

Jiang K, Asami T (2018) Chemical regulators of plant hormones and their applications in basic research and agriculture. Biosci Biotechnol Biochem 82(8):1265–1300. https://doi.org/10.1080/09168451.2018.1462693

Kazan K (2015) Diverse roles of jasmonates and ethylene in abiotic stress tolerance. Trends Plant Sci 20(4):219–229. https://doi.org/10.1016/j.tplants.2015.02.001

Kumar A, Patel JS, Meena VS, Ramteke PW (2019) Plant growth-promoting rhizobacteria: strategies to improve abiotic stresses under sustainable agriculture. J Plant Nutr 42:1402–1415

Li X, Yang M, Chen H, Qu L-Q, Chen F, Shen S (2010) Abscisic acid pretreatment enhances salt tolerance of rice seedlings: proteomic evidence. Biochim Biophys Acta (BBA) 1804(4):929–940. https://doi.org/10.1016/j.bbapap.2010.01.004

Liu PP, Montgomery TA, Fahlgren N, Kasschau KD, Nonogaki H, Carrington JC (2007) Repression of AUXIN RESPONSE FACTOR10 by microRNA160 is critical for seed germination and post-germination stages. Plant J 52(1):133–146

Mangal V, Lal MK, Tiwari RK, Altaf MA, Sood S, Kumar D et al (2022) Molecular insights into the role of reactive oxygen, nitrogen and sulphur species in conferring salinity stress tolerance in plants. J Plant Growth Regul:1–21

Marquis V, Smirnova E, Poirier L, Zumsteg J, Schweizer F, Reymond P, Heitz T (2020) Stress- and pathway-specific impacts of impaired jasmonoyl-isoleucine (JA-Ile) catabolism on defense signalling and biotic stress resistance. Plant Cell Environ 43:1558–1570

Matilla AJ (2000) Ethylene in seed formation and germination. Seed Sci Res 10(2):111–126

Milhinhos A, Miguel CM (2013) Hormone interactions in xylem development: A matter of signals. Plant Cell Rep 32:8678–8683. https://doi.org/10.1007/s00299-013-1420-7

Miller CO, Skoog FS, Okomura MH, Von Saltza H, Strong FM (1955) J Am Chem Soc 77:1392–1393

Miransari M, Smith DL (2009) Alleviating salt stress on soybean (Glycine max (L.) Merr.) Bradyrhizobium japonicum symbiosis, using signal molecule genistein. Eur J Soil Biol 45:146–152

Miyazaki S, Hara M, Ito S, Tanaka K, Asami T, Hayashi K et al (2018) An ancestral gibberellin in a moss Physcomitrella patens. Mol Plant 11:1097–1100

Morffy N, Strader L (2020) Old town: routes of auxin biosynthesis across kingdoms. Curr Opin Plant Biol 55:21–27

Müller K, Tintelnot S, Leubner-Metzger G (2006) Endosperm-limited Brassicaceae seed germination: abscisic acid inhibits embryo-induced endosperm weakening of Lepidium sativum (cress) and endosperm rupture of cress and Arabidopsis thaliana. Plant Cell Physiol 47(7):864–877

Nakano M, Omae N, Tsuda K (2022) Inter-organismal phytohormone networks in plant-microbe interactions. Curr Opin Plant Biol 68:102258

Nieminen KM, Kauppinen L, Helariutta Y (2004) A weed for wood? Arabidopsis as a genetic model for xylem development. Plant Physiol 135:6536–6559. https://doi.org/10.1104/pp.104.040212

Parsons KJ, McWhinnie K, Pilakouta N, Walker L (2020) Does phenotypic plasticity initiate developmental bias? Evol Dev 22:56–70

Petruzzelli L, Coraggio I, Leubner-Metzger G (2000) Ethylene promotes ethylene biosynthesis during pea seed germination by positive feedback regulation of 1-aminocyclo-propane-1-carboxylic acid oxidase. Planta 211:144–149

Qi X, Chen M, Liang D, Xu Q, Zhou F, Chen X (2020) Jasmonic acid, ethylene and ROS are involved in the response of cucumber (*Cucumis sativus* L.) to aphid infestation. Sci Hortic 269:109421

Rattan SI, Sodagam L (2005) Gerontomodulatory and youth-preserving effects of zeatin on human skin fibroblasts undergoing aging in vitro. Rejuvenation Res 8(1):46–57. https://doi.org/10.1089/rej.2005.8.46

Rinaldi LMR (2000) Germination of seeds of olive (Olea europaea L.) and ethylene production: effects of harvesting time and thidiazuron treatment. J Hortic Sci Biotechnol 75(6):727–732

Rolland F, Moore B, Sheen J (2002) Sugar sensing and signaling in plants. Plant Cell. 2002;14 Suppl(Suppl):S185-205.

Sabagh AEL, Islam MS, Hossain A, Iqbal MA, Mubeen M, Waleed M et al (2022) Phytohormones as growth regulators during abiotic stress tolerance in plants. Front Agron 4:765068

Sadiq Y, Zaid A, Khan MMA (2020) Adaptive physiological responses of plants under abiotic stresses: role of phytohormones. In Plant ecophysiology and adaptation under climate change: mechanisms and perspectives I: general consequences and plant responses, pp 797–824

Schaller GE (2012) Ethylene and the regulation of plant development. BMC Biol 10(1):1–3

Sharma A, Shahzad B, Kumar V, Kohli SK, Sidhu GPS, Bali AS, Handa N, Kapoor D, Bhardwaj R, Zheng B (2019) Phytohormones regulate accumulation of osmolytes under abiotic stress. Biomol Ther 9:285

Shu W, Zhou H, Jiang C, Zhao S, Wang L, Li Q, Yang Z, Groover A, Lu M (2019) The auxin receptor TIR1 homolog (PagFBL 1) regulates adventitious rooting through interactions with Aux/IAA28 in *Populus*. Plant Biotechnol J 17:338–349

Sponsel VM (2016) Signal achievements in gibberellin research: the second half-century. Annu Plant Rev 49:1–36

Tanaka J, Yano K, Aya K, Hirano K, Takehara S, Koketsu E et al (2014) Antheridiogen determines sex in ferns via a spatiotemporally split gibberellin synthesis pathway. Science 346:469–473

Tilman D, Balzer C, Hill J, Befort BL (2011) Global food demand and the sustainable intensification of agriculture. Proc Natl Acad Sci 108(50):20,260–20,264

Trypuc JA, Matejczyk M, Czerpak R (2016) N6-benzyladenine and kinetin influence antioxidative stress parameters in human skin fibroblasts. Mol Cell Biochem 413(1–2):97–107

Vildanova MS, Savitskaya MA, Onishchenko GE, Smirnova EA (2014) The effect of plant hormones on the components of the secretory pathway in human normal and tumor cells. Cell Tissue Biol 8(5):407–415

Voller J, Maková B, Kadlecová A, Gonzalez G, Strnad M (2017) Hormones in ageing and longevity. In Plant hormone cytokinins for modulating human aging and age-related diseases. Springer, Cham, pp. 311–335

Voß U, Bishopp A, Farcot E, Bennett MJ (2014) Modelling hormonal response and development. Trends Plant Sci 19(5):311–319

Wang Z, Tan X, Zhang Z, Gu S, Li G, Shi H (2012) Defense to *Sclerotinia sclerotiorum* in oilseed rape is associated with the sequential activations of salicylic acid signaling and jasmonic acid signaling. Plant Sci 184:75–82

Wang Y, Mostafa S, Zeng W, Jin B (2021) Function and mechanism of jasmonic acid in plant responses to abiotic and biotic stresses. Int J Mol Sci 22(16):8568

Wani SH, Sah SK (2014) Biotechnology and abiotic stress tolerance in rice. J Rice Res 2(2):e105

Went FW (1928) Wuchsstoff und Wachstum. Receuil des Travaux Botaniques Neerlandais 25:1–116

Werner T, Motyka V, Strnad M, Schmülling T (2001) Regulation of plant growth by cytokinin. Proc Natl Acad Sci 98(18):10,487–10,492

Wybouw B, De Rybel B (2019) Cytokinin–a developing story. Trends Plant Sci 24(2):177–185

Zhang L, Zhang F, Melotto M, Yao J, He SY (2017) Jasmonate signaling and manipulation by pathogens and insects. J Exp Bot 68:1371–1385
Zhang PJ, Zhao C, Ye ZH, Yu XP (2020) Trade-off between defense priming by herbivore-induced plant volatiles and constitutive defense in tomato. Pest Manag Sci 76:1893–1901

Index

© The Author(s), under exclusive license to Springer Nature Singapore Pte Ltd. 2024

M. Faizan, S. Hayat (eds.), *Plant Growth Regulators: Resilience for Sustainable Agriculture*, https://doi.org/10.1007/978-981-97-2918-0

P

R

S

T

U

SPRINGER NATURE

GPSR Compliance

The European Union's (EU) General Product Safety Regulation (GPSR) is a set of rules that requires consumer products to be safe and our obligations to ensure this.

If you have any concerns about our products, you can contact us on ProductSafety@springernature.com

In case Publisher is established outside the EU, the EU authorized representative is:

Springer Nature Customer Service Center GmbH
Europaplatz 3
69115 Heidelberg, Germany

The manufacturer's authorised representative in the EU is Springer Nature Customer Service Centre GmbH, Europaplatz 3, 69115 Heidelberg, Germany. If you have any concerns regarding our products, please contact ProductSafety@springernature.com

Printed and bound by CPI Group (UK) Ltd, Croydon, CR0 4YY
01/10/2026
02225923-0012